FISH FARMING HANDBOOK
FOOD, BAIT, TROPICALS AND GOLDFISH

FISH FARMING HANDBOOK
FOOD, BAIT, TROPICALS AND GOLDFISH

E. Evan Brown

Department of Agricultural Economics
College of Agriculture
University of Georgia
Athens, Georgia

John B. Gratzek

Department of Medical Microbiology
College of Veterinary Medicine
University of Georgia
Athens, Georgia

AVI PUBLISHING COMPANY, INC.
Westport, Connecticut

Library of Congress Cataloging in Publication Data

Brown, E Evan.
    Fish farming handbook.

    Includes index.
    1. Fish-culture. I. Gratzek, John B., joint
author. II. Title.
SH151.B84   639'.31   79-23177
ISBN 0-87055-341-0

Printed in the United States of America
BCDE 321098765

To Dr. Stanislas F. Snieszko for his service to the aquaculture industry and his encouragement to others to engage in fish health research

Contents

Contributors

HOPKINS, MARGARITA L., Department of Agricultural Economics and Fisheries and Allied Aquacultures, Auburn University, Auburn, Alabama

LOVELL, RICHARD T., Department of Fisheries and Allied Aquacultures, Auburn University, Auburn, Alabama

McCOY, EDWARD, Department of Agricultural Economics and Fisheries and Allied Aquacultures, Auburn University, Auburn, Alabama

REINERT, ROBERT E., School of Forestry Resources, University of Georgia, Athens, Georgia

SOCOLOF, ROSS, Socolof Enterprises, Bradenton, Florida 33506

Preface

There exist numerous publications about some specialized areas and species of finfish aquaculture in the United States. For other areas and species relatively little is known or published. The casual reader, the neophyte, the expert, and the serious student all have the same problem. This is one of having a readily available source covering the important technical aspects of culturing fish, be they food fish, bait fish, tropical fish, or goldfish. This book is an attempt to fill this informational need.

The book is organized into 8 distinct chapters. These are: (1) Introduction—which presents and discusses the relative numbers of producers and the farm values of food fish, including catfish, trout, big mouth buffalo, tilapia, eel, salmon, and sea ranching. Then the bait, tropical, and goldfish sectors are presented. The relative importance of each of these sectors may come as a surprise to many readers. (2) Environmental Factors—which includes information on oxygen, temperatures, photoperiod, and toxic materials. (3) Types of Culturing Facilities—covering various aspects of pond, raceway, tank, cage, and pen construction. (4) Maintenance and Improvement of Ponds—covering aquatic weed identification, prevention, and controls, pH and alkalinity, fertilization, oxygen, and temperatures. (5) Methods—covering how to do it, steps in culturing various species from selecting brood stock to harvesting the finished product. Places to order select types of culturing equipment are given. This chapter includes a section on tropical fish farming written by one of the experts in the field. The section presents information which has never previously been made public. (6) Nutrition and Feeding—a chapter written by one of America's foremost authorities. Nutrient requirements, diet formulation, and processing and feeding practices are all presented. (7) Common Fish Diseases and Their Control—presents information which should be extremely valuable to aquaculturalists in recognizing various problems and how to prevent or control these problems. (8) Processing and Marketing—This last chapter not only gives marketing methods used by various food fish and non-food fish farmers, but also includes advice on how a beginning fish farmer should find markets for his product. This section includes advice for fee fish-out operations as well as food and non-food producers.

The authors have tried to present in one book a wide range of subjects for private and public aquaculturalists, as well as for all those who have an interest in fish culture, whether as a large scale enterprise or as a hobby, and for individuals who wish to learn more about an American industry which may well generate as much as $1 billion at all levels. It should be recognized that no one book will answer every individual question or problem; however, it is hoped that the book will be of aid to most individuals with an aquaculture interest.

The authors are pleased to acknowledge the assistance given by five contributing authors. We have also been given assistance by a host of individuals. Special recognition is due to the Fisheries Department of Auburn University, Auburn, Alabama, and to its members who graciously shared their personal libraries, and to Dr. Mayo Martin, Fish Farming Experiment Station, Stuttgart, Arkansas, who read selected parts of the manuscript. We should also like to recognize the aid of Mrs. Mary T. Prather for diligence and persistence in interpreting and typing the manuscript.

E. EVAN BROWN
JOHN B. GRATZEK

Related AVI Books

CHEMISTRY AND BIOCHEMISTRY OF MARINE FOOD PRODUCTS
Edited by *Martin et al.*

ELEMENTARY FOOD SCIENCE, 2nd Edition
Nickerson and Ronsivalli

ENCYCLOPEDIA OF FOOD SCIENCE
Edited by *Peterson and Johnson*

FISHERIES ECOLOGY
Pitcher and Hart

FOOD AND BEVERAGE MYCOLOGY
Edited by *Beuchat*

FOOD AND YOUR WELL BEING
Labuza

FOOD PROCESSING WASTE MANAGEMENT
Green and Kramer

FOOD PRODUCTS FORMULARY, VOL. 2, 2nd Edition
Meats, Poultry, Fish and Shellfish
Long, Komarik, Tressler

FOOD SCIENCE
Potter

FOODBORNE AND WATERBORNE DISEASES: *Their Epidemiologic Characteristics*
Tartakow and Vorperian

FRESHWATER CRAYFISH V
Edited by *Goldman*

FUNDAMENTALS OF FOOD CANNING TECHNOLOGY
Jackson and Shinn

INTRODUCTION TO FRESHWATER VEGETATION
Riemer

METHODS IN FOOD AND DAIRY MICROBIOLOGY
DiLiello

PRINCIPLES OF FOOD PACKAGING, 2nd Edition
Sacharow and Griffin

SOURCE BOOK OF FOOD ENZYMOLOGY
Schwimmer

THE TECHNOLOGY OF FOOD PRESERVATION, 4th Edition
Desrosier and Desrosier

WORLD FISH FARMING: Cultivation and Economics, 2nd Edition
Edited by *Brown*

FISH FARMING HANDBOOK
FOOD, BAIT, TROPICALS AND GOLDFISH

1

Introduction

Mass production of fish creates environmental conditions which have physical, nutritional, physiological, and pathological manifestations. While these are species specific, they can be grouped into categories and discussed with meaning. This book focuses on these requirements for food fish, bait fish, tropical fish, and goldfish.

Food fish culture is widespread throughout the world. The Peoples' Republic of China produces more than one-half of the total. Other significant countries are India, Thailand, Indonesia, Philippines, Taiwan, Japan, Italy, Denmark, France, Germany, and the United States.

In the United States the size of the cultured fish industry comes as a surprise to most Americans. As with ornamental and bait fish culture, the exact size of the food fish segment is not known.

Food fish production includes various species of catfish, buffalo, tilapia, trout, salmon, and eel. A brief introduction to each of these species follows.

FOOD FISH FARMING

Channel Catfish (Ictalurus punctatus)

Channel catfish is the principal warm freshwater food fish cultured in the United States. The principal states of production are the Mississippi Delta states, some of the southeastern states and a few others. Major states, not in order of importance, are Arkansas, Mississippi, Louisiana, Texas, Oklahoma, Kansas, Missouri, Illinois, Indiana, Tennessee, Alabama, Georgia, South Carolina, Florida, California, and Idaho. Catfish may be cultured to some extent in 6−8 other states.

A survey made in 1975 (Martin 1978) indicated that there were 1934 enterprises engaged in the commercial production of catfish. To this number must be added 1000 to 2000 enterprises engaged in fish-out

1

operations and perhaps 150,000 farm operators who have some recreation fishing in private ponds. The survey indicated 18,802 ha (46,441 acres) of commercial ponds, plus numerous cage and tank operations and raceway operations. It is estimated that 96% of production comes from ponds and the remaining 4% from cages, tanks, and raceways.

Probably 60% of the commercial production waters are harvested yearly with an average of about 2.2 MT per ha (2000 lb per acre). Hence, the annual production is approximately 25,330,000 kg (55,730,000 lb). Based on a weighted average value of $1.54 per kg ($0.70 per lb) for food fish and for catch-out prices, the total annual farm value would be approximately $39 million.

A typical foodsize fish producer buys his fingerlings from a specialized producer. The specialized fingerling producer keeps brood fish and spawns and raises the fry to fingerling sizes. The size of fingerlings stocked by foodsize fish producers (grow-out operations) ranges from 6 to 22 cm (2 to 9 in.) depending on when he stocks the fish and when he hopes to harvest at 454 g (1 lb) size. In pond culture, most fingerlings stocked are between 13 and 20 cm (5 and 8 in.). They are usually stocked in March, April, or May. These sizes reach market weights in the fall. The raceway and cage producers buy slightly larger fingerlings. Producers who plan to harvest in late spring or early summer of the following year buy smaller fingerlings.

The primary producing states in 1976 were essentially in the southeastern United States, an area blessed with ample water and a long grow-out season with warm conditions. Exceptions to this were California and Idaho. California produces fish in certain areas because the market price structure is conducive to higher cost grow-out operations. In Idaho, warm water springs and wells are used for year-round growth. The fish are marketed on the west coast.

Double Cropping.—One rather interesting development taking place is the raising of channel catfish and rainbow trout in the same facility at different times of the year. Trout fingerlings are stocked in the fall of the year when water temperatures drop below 21°C (70°F). They are then fed out until the spring of the year when water temperatures increase to above 21° or 22°C (70° or 72°F). By stocking the proper sized fingerlings, the fish can reach market sizes in 120 days or less. After the trout are harvested, channel catfish fingerlings are stocked in the warm water. They are then fed out to the fall of the year when it is time to restock rainbow trout again. By stocking the proper sizes of catfish fingerlings, the fish reach market sizes of 454 g (1 lb) during an eight month season. This work, which was begun in Georgia by the authors and two co-workers in raceways, is slowly being developed commercially.

The major advantage to double-cropping is the sharing of fixed expenses by two crops of fish. This lowers the production costs of both species. Net returns can be increased by 100% or more using this technique.

Rainbow Trout (Salmo gairdneri)

Rainbow trout is the principal cold freshwater fish cultured by the private sector for food usage in the United States. According to the latest information, culturing takes place in every state except Louisiana, Mississippi, and Texas.

A survey made in 1975 (Martin 1978) indicated more than 660 private commercial trout enterprises. Those enterprises contained 774 ha (1913 acres) of ponds and 1,458,959 m^3 (15,698,400 cu ft) of raceways. The leading states in numbers of commercial growers are Idaho, Wisconsin, Colorado, Michigan, Pennsylvania, and North Carolina. Of nearly 4000 fee-fishing enterprises in the U.S., it is believed that there are between 1000 and 2000 trout enterprises. The volume of trout raised domestically probably approaches 12,700 MT (28 million lb), with about 9300 MT (20.5 million lb) processed, and the remainder, 3400 MT (7.5 million lb), sold through fee fish-out ponds.

Idaho contributes about 90% of all the processed trout. The remaining processed volume is contributed by other states in which the small scale producers process small amounts for sales to nearby restaurants or hotels and unprocessed sales go to nearby fish-out ponds.

The primary reason for the overwhelming importance of Idaho as the major trout-producing state is adequate water of proper temperature. This happy event is due to a geological accident referred to as the Southern Idaho Aquifer. The water originates in the mountains and enters the vast lava plain in southern Idaho. This lava plain extends about 240 km (150 mi.) from east to west and about 80 to 120 km (50 to 75 mi.) from north to south. This rock is extremely porous and the mountain runoff is absorbed. Water emerges from this aquifer from the side of a deep fissure cut by the Snake River. In essence, underground rivers emerge to the surface at the canyon wall. The water is 14.4°C (58°F) and temperatures fluctuate only by about 0.5°C (1°F). Maximum water flow through one hatchery is 10,300 liters per sec (147,000 gal. per min).

Producers in other states are not as fortunate in having abundant supplies of cold, good quality waters. Hence, production units are smaller and marketing is generally in the local area through fish-out ponds or to restaurants or hotels desiring fresh, quality trout. Idaho, on the other hand, has several large processors who freeze and distribute their product to stores, restaurants, etc., on a national basis.

The farm value of the processed, fresh fish and fish caught from fish-out operations is about $30 million.

The trout industry in the United States has specialists dealing with one phase of production. These can be classified as:

(1) Egg producers who raise fish to sexual maturity—usually 2−3 years—and spawn them for 3 or 4 years before they are either sold for processing or fee-fishing stock. Fertilized eggs may be sold as eyed-eggs or incubated and hatched. Fry and fingerlings are sometimes sold.

(2) Fingerling producers who either raise their own fry or purchase them in sizes varying from 2.5 to 12.5 cm (1 to 5 in.) and sell them at sizes ranging from 10 to 20 cm (4 to 8 in.). These fish are sold alive to individuals who raise them to marketable sizes for human consumption or to individuals who operate fee fish-out facilities.

(3) Market fish producers who usually buy eyed-eggs to be incubated and hatched. Fry are then raised to market sizes of 25 to 35 cm (10 to 14 in.). At that time they are transferred alive to processing plants, which may or may not be on the premises. Some of these producers may buy fingerlings for growing-out. Depending on water temperatures, a marketable rainbow trout may take from 10 to 18 months after hatching.

(4) Grow-out producers who may raise fish for another producer. The grow-out operator raises 10 to 15 cm (4 to 6 in.) fingerlings to market size without ever taking title to the fish. He is paid on the basis of weight gained. Time required to do this is usually 4 to 6 months. This depends on water temperatures, size of fingerlings stocked, size of marketable fish desired, and management.

(5) Fee fish-out operators who usually buy catchable sized fish which can be caught within a short period of time by fishermen. Sometimes these operators buy fingerlings and raise them for 4 to 6 months and then transfer them to the fish-out facility.

(6) Live haulers who haul live fish from one farm to another, or to a processing plant or fee-fishing facility. Some of these haulers may transport live fish 750 to 3000 km (500 to 2000 mi.).

(7) Processor who may or may not have a fish production facility. He receives live fish for processing into fresh and frozen items.

There are many individuals who perform more than one of these activities and a few who may perform nearly all of them. There is no way of estimating how many individuals are engaged in each type of activity.

Big Mouth Buffalo (Ictiobus cyprinellus)

The big mouth buffalo fish is in some demand in the southern United States. As much as 454 MT (1 million lb) are produced yearly. The fish are usually produced by extensive rather than intensive culturing methods because the price is low. Farm prices are about one-half that of catfish. By using extensive production methods little or no feed is fed and feed costs are minimal. Yields may be only about 100 kg per ha (100 lb per acre); they are usually grown in storage reservoirs. They are also adaptable to polyculture with catfish or bait minnows.

Tilapia (Tilapia aurea)

Tilapia can be cultured with catfish. However, because they are an exotic species there is limited demand and prices are low. Also they are a tropical fish and to survive the winters, except in thermal ponds, the brood fish must be overwintered in heated ponds or hatchery houses. Tests at Auburn University indicate that tilapia quit feeding at approximately 15.5°C (60°F), with 100% mortality at 8°−9°C (47°−49°F).

American Eel (Anguilla rostrata)

The United States is not an eel consuming country. Although several thousand tons of wild eels are captured annually for export, culturing is still in its infancy. There are probably fewer than 50 eel farms in the country. Annual production would certainly not be over 200 MT (220 ST), which translates into a domestic production of less than $1 million yearly. However, the prospects for a large viable industry are bright and for this reason eels have been included in this book under the various chapters.

Marine Culture—Salmon

Salmon—Coho or Silver (*Oncorhynchus kisutch*)
        Chinook or King (*Oncorhynchus tshawytscha*)
        Chum or Dog (*Oncorhynchus keta*)
        Sockeye or Kokanee or Red (*Oncorhynchus nerka*)
        Pink or Lost (*Oncorhynchus gorbuscha*)

Salmon are cultured for food in three states in the USA. These are Washington and Oregon on the West Coast and Maine on the East Coast. In 1975 there were two farms in Washington specializing in fingerling production only, three farms producing fingerlings and pan-sized cultured salmon, and three farms producing only pan-sized salmon. Two of these eight farms were also engaged in sea ranching. Two farms were in Maine and one was under construction.

Production of pan-sized salmon for 1976 for Washington and Oregon was estimated at 681,000 kg (1 1/2 million lb). Production in Maine was estimated at 50,000 kg (110,000 lb). Value of farm production is about $2 million.

Salmon culture in Washington and Oregon has had many obstacles to overcome. The initial culturing was begun by the Bureau of Commercial Fisheries in 1967 in floating net pens. The predecessor of the Bureau of Commercial Fisheries is the National Marine Fisheries Service. In 1970 early attempts at private commercial culturing were initiated. By 1973–74 production was about 350 MT (385 ST). Two years later, in the 1975–76 season, production was estimated at 681 MT (750 ST).

In Washington, private growers are not allowed to spawn wild fish. In Oregon, private spawning of chum salmon was first permitted in 1971. In 1973, coho and chinook salmon were also added to the list of species that could be privately spawned.

Culturing has evolved primarily into culturing only coho salmon in Washington, Oregon, and Maine. Only small numbers of chinook and chum are cultured. Interest has centered on coho because of its resistance to disease and willingness to accept pelleted dry feed. This is fortunate because coho is one of the more desired species and sells for a relatively high price.

Cultured production is less than 1% of the wild catch. However, there is no competition between wild and cultured fish. The cultured fish are in-dividual portions or pan-sized and appeal to the gourmet market. Some 1 kg (2.2 lb) fish are cultured and marketed. These take up to 18 months in the 9°–13°C (50°–56°F) water. Most of the cultured fish are sold be-tween December 1 and April 1 periods when the wild catching season is closed. However, some cultured fish are sold in every month of the year.

Sea Ranching

Sea ranching is another culturing technique. However, instead of feeding the fish throughout the production process, the smolts are released into the sea. The mature fish return after 2 to 5 years to the point of release. Returning fish may be as high as 20% but are usually less than 5%. With this technique, production cost may be only one-third that of fish fed out in pens or ponds.

Sea ranching by individuals or corporations is an extension of public restocking efforts which have been in process for years. The only difference is that the individual harvests all of the returning fish that come back to the privately owned point of release and rearing. Sea ranching of salmon is legal in Oregon, California, and Alaska. It is not authorized in Washington.

BAIT FISH FARMING

Bait fish farming, sometimes known as minnow farming, is nearly a unique American enterprise. The size of the industry is a surprise to many investigators. As with nearly all of the different segments of American aquaculture, the exact size is not known.

Surveys on a state by state basis place the average between 20,250 and 28,340 ha (50,000 and 70,000 acres) with more than 790 producers (Martin 1978). The leading state is Arkansas with more than 8500 ha (21,000 acres), followed by Minnesota with more than 5260 ha (13,000 acres), Louisiana with 1215 ha (3000 acres), Missouri with 810 ha (2000 acres) and Mississippi with 729 ha (1800 acres). Nearly every state east of the Rocky Mountains has some minnow farming as well as Arizona and California west of the continental divide.

Species of fish include golden shiners (*Notemigonus crysoleucas*), fathead minnows (*Pimephales promelas*), goldfish (*Carassius auratus*), carp suckers (*Catostomus commersonii*) in Minnesota, and bluntnose minnows (*Pimephales notatus*). In Arkansas, the most important state, acreage is about 85% shiners, 11% fathead minnows, and 4% goldfish. Yields average 429 kg per ha (383 lb per acre) for shiners, 336 kg per ha (300 lb per acre) for fathead minnows, and 632 kg per ha (564 lb per acre) for goldfish. In total kilograms (poundage) of production, shiners accounted for 85.2%, fathead minnows for 8.8%, and goldfish for 6%. Shiners sold for $3.39 per kg ($1.54 per lb), fathead minnows for $3.56 per kg ($1.62 per lb), and goldfish for $3.48 per kg ($1.58 per lb). In total value, shiners accounted for 84.7%, fathead minnows for 9.2%, and goldfish for 6.1%. Total value of minnow production in 1975 in Arkansas was more than $12.5 million.

Assuming that Arkansas's production rate per acre and price per pound of minnows sold is representative of the USA, the total U.S. farm value would vary between $29.5 million and $41.3 million, and the retail values would then be between $60 and $80 million.

The advent of minnow farming was preceded by many years of harvesting of wild minnow stocks. Minnows were the predominant species captured and used but often small fish of any species were also used. As wild stocks became more difficult to find and harvest, as a result of overharvesting, pollution, and other factors, interest grew in raising bait fish. Initial beginnings were in areas where wild bait fish in sufficient numbers were hard to capture.

Following World War II, with increased disposable income, rapid population growth and increased construction of pay lakes and public reservoirs, the bait fish industry started a rapid expansion.

One of the earliest minnows in production was goldfish. This species had long been grown for hobbyists and some of the fish were not too desirable for such use. Some of the early producers had stock which were unable to transmit good coloration. The result was some poorly colored and brown goldfish. These were sometimes known as "Indiana" or "Baltimore" minnows because of the area of production. Goldfish are still used as bait fish for crappie fishing in the southeast. Their major use is as troutline bait in the area east of the Mississippi River. Most of these fish are oversized for aquarium use, over 3 in., brown, and poorly colored. A few producers raise such fish primarily for bait but most of them are a by-product of the ornamental fish industry. For this reason bait and ornamental fish cannot be discussed as separate industries without recognizing some cross-over effects.

Some problems involved with producing goldfish bait for troutline fishing are that sales are concentrated in the warm water months, and price has to be very low. Coupled with these factors are wide swings in demand from the troutline industry, which is affected unduly by floods, and droughts, which may severely restrict fishing.

The fathead minnow was the earliest native bait fish cultured, probably because of its wide geographic distribution, ready acceptance by fishermen, and easy spawning habits. Moreover, there is still considerable competition for the fathead minnow bait market from wild or semiwild stocks. The semi-wild stocks are obtained from the states north of Missouri from small shallow ponds which freeze over or freeze solidly in the winter, resulting in total loss or near total loss of the minnows. In these lakes or ponds, fathead minnows are stocked in the spring and trapped in the fall or early winter. After trapping they are moved to warmer waters and sold as opportunities present themselves.

The predominant bait fish grown is the golden shiner. These were cultured as early as 1950 in Mississippi, Arkansas, and Missouri.

TROPICAL FISH FARMING

In 1976 an estimated $225 million was spent in the USA in aquarium shops, department and variety stores, discount stores, drugstores, supermarkets, and others for pet fish. If it can be assumed that the marketing margin was 400%, then the farm value would have approximated $56 million. Of this amount about $10 million went for marine fish and $46 million for tropicals and goldfish. This figure would not include goldfish sold for bait fish.

Included in the $46 million figure are an estimated $12–$15 million of imported ornamental fish. These imports are primarily from Hong Kong, Thailand, Singapore, Taiwan, Philippines, and Japan in Asia, and from Colombia, Guyana, Peru, and Brazil in South America.

In the USA there are outdoor production facilities for tropical fish in Florida, California, Hawaii, Louisiana, and Texas. The largest production state is Florida with approximately 250 growers, producing 90% of the domestic supply. Of this production, 80% is in the Tampa area and 20% in the Miami-Vera Beach area. Over 100 species are raised.

One major factor for the industry concentration is the accessibility of air freight terminals. Both Tampa and Miami are major air shipping points for pet fish. From these cities, 750,000 cartons of fish, 200 fish per carton, were shipped out of Florida in 1972 (Kerr 1973). This accounted for 5.06 million kg (over 11 million lb) of freight, making tropical fish shipments the largest class of air freight moved by airlines in the state. Part of these shipments were transshipments of South American fish arriving at the Miami airport.

The classification of firms by type of facilities is difficult because indoor glass aquariums, concrete tanks and outdoor ponds were all utilized. Individual operations ranged from a few ponds with 1 worker to hundreds of ponds with over 100 employees.

It is extremely difficult to determine with any accuracy the farm value of domestically produced tropical fish. The authors estimate that the industry produces a farm value of $15 million per year, but other writers estimate as high as $150 million. Either the latter figure is grossly overestimated or sales data gathered by the pet industry are grossly undervalued.

GOLDFISH FARMING

An estimation of the farm value of goldfish produced for ornamental purposes is nebulous. If it can be assumed that estimates of pet sales for tropicals, marine, and goldfish are correct, and if estimates of the marketing margin, tropicals, and marine value are fairly accurate, then farm production values of goldfish would be between $18 and $21 million. Production takes place in outdoor ponds in approximately 40 states. Important states are Missouri, Indiana, Ohio, Maryland, Arkansas, Pennsylvania, Tennessee, Texas, Oklahoma, Mississippi, Louisiana, Kentucky, and others.

It is estimated that there are several hundred farmers raising goldfish for ornamental sale as well as bait use.

OTHER

Martin (1978) in a survey of all types of fish farming enterprises found 1759 producers in "other" fisheries production. "Other" included largemouth bass, smallmouth bass, redear sunfish, bluegill, crappie, northern pike, carp, *goldfish*, crawfish, frogs, green sunfish, Israeli carp, *eels*, mullet, *suckers, salmon*, longear sunfish, tadpoles, *tropical fish*, and muskellunge.

Of these 1759 producers, perhaps 500—600 are engaged in goldfish, eel, sucker, (bait), salmon, and tropical fish production; to avoid over-counting, this number must be subtracted from the 1759. By even the most liberal discounting there would probably be 1000—1100 producers of "other" fish.

TOTALS

Using the above estimates, we arrive at about 5000 commercial fish farms and about 4000 fee fish-out operations. Coupled with these are more than 150,000 farmers who raise fish for recreational use. The minimum farm value is at least $200,000,000 and may reach $300,000,000.

These are only part of the U.S. aquaculture picture. We must also include processors, haulers, feed processors, suppliers of all kinds, wholesalers, and retailers to arrive at final employment and aqua-business economic values. To try to estimate the total numbers is impossible, but we may be talking in terms of $1,000,000,000 spent at retail and in all other steps involved in production and marketing.

REFERENCES

ALTMAN, R.W. and IRWIN, W.H. (Undated.) Minnow farming in the south-west, Oklahoma State University, Stillwater, Okla. Okla. Dep. Wildl. Conserv., Oklahoma City.

ANON. 1968. Proc. Commer. Bait Fish Conf., Texas A & M Univ., March 19—20, College Station, Texas.

ANON. 1977. The Fifth Annual State of the Pet Industry Report: Pets, Supplies, Marketing. Harcourt Brace Jovanovich, New York.

BAILEY, W.M., GIBSON, M.D., NEWTON, S.H., MARTIN, J.M., and GRAY, D.L. 1975. Status of commercial aquaculture in Arkansas in 1975. (unpublished)

BROWN, E.E., 1977. World Fish Farming: Cultivation and Economics. AVI Publishing Co., Westport, Conn.

BROWN, E.E., HILL, T.K., and CHESNESS, J.L. 1974. Rainbow trout and channel catfish—a double-cropping system. Ga. Agric. Exp. Stn., Res. Rep. *196.*

BROWN, E.E., LaPLANTE, M.G., and COVEY, L.H. 1969. A synopsis of catfish farming. Ga. Agric. Exp. Stn., Res. Bull. *69*.

DONALDSON, J.R. 1976. Salmon aquaculture in Oregon. Proc. Conf. Salmon Aquaculture Alaskan Fish. Community, Cordova, Alaska, Jan. 9—11. Univ. Alaska Sea Grant Program Rep. *76-2*.

FLICKINGER, S.A. 1971. Pond culture of bait fishes. Coop. Ext. Serv., Colo. State Univ., Fort Collins, Bull. *478A*.

KERR, J.R. 1973. Exploratory Survey of the Tropical Fish Industry in Florida. Fla. State Univ., Tallahassee.

KINNAMAN, P.V. 1975. Weyerhauser: the salmon growing company. Aquaculture Fish Farmer *2* (4) 4—18.

KLONTZ, G.W. and KING, J.G. 1975. Aquaculture in Idaho and Nationwide. Idaho Dep. Water Resources, Boise.

MARTIN, M. 1978. Personal correspondence. Fisheries Exp. Stn., Stuttgart, Ark.

MERYMAN, D.C. 1978. Farm production of introduced ornamental fishes in Florida. Symp. Culture Exotic Fishes. Fish Culture Section, Am. Fish. Soc., Atlanta, Jan. 4.

MEYER, F.P., SNEED, K.E., and ESCHMEYER, P.T. 1973. Second Report to the Fish Farmers, Resource Public, 113. U.S. Fish Wildl. Serv., Bur. Sport Fisheries Wildl., Washington, D.C.

NAEF, F.E. 1972. Salmon culture. Am. Fish Farmer World Aquaculture News *3* (4) 12.

NYEGAARD, C.W. 1973. Coho salmon farming in Puget Sound. Wash. State Univ. Coop. Ext. Serv. Ext. Bull. *647*.

SOCOLOF, R. 1977. Florida fish continue to die. Aquarium Ind. Mag. (March) *4*, 8-16.

2

Environmental Factors

Robert Reinert

Generally, success in raising and caring for an animal is directly related to the amount of knowledge one has concerning how the animal functions in its environment. Unfortunately fish live in an aquatic environment that is quite different from the terrestrial one in which we live, and because of this we find it harder to relate to the fish's needs than to ours or those of other terrestrial animals.

In this chapter we will discuss some of the aspects of how fish react to four environmental factors that are very important in regulating their activities. Three of these, oxygen, temperature, and photoperiod, are part of the fish's natural environment. The fourth, toxic substances, although not a natural phenomenon, nevertheless can have a very strong influence on the activities of fish.

It is our hope that knowledge of the effect of these factors on fish will result in a better understanding of how fish function in their environment and this will help the reader to a better understanding of how to raise and care for fish.

OXYGEN

Fish like other animals must have an adequate supply of oxygen in their tissues so oxidation, which results in the release of energy from food, can occur. It is this energy that, after suitable biochemical transformations, becomes available to do the biochemical work required for the myriad of physiological functions that together constitute life.

Because most fish spend their entire life in water they must rely on this medium for their supply of oxygen, and relative to air, water is a difficult medium from which to obtain oxygen. Water is about 800 times denser than air and at saturation contains only 3% as much oxygen.

The principal respiratory organ in most fish is the gill, which is superbly designed to extract the small quantities of oxygen from water.

13

However, for fish to get enough oxygen they must keep a constant supply of water flowing over their gills. This is accomplished by the fish's taking in water through its mouth and by a series of muscular controlled changes in the volume of the buccal and opercular cavities, timed with the opening and closing of the mouth and opercular valve, and pumping the water across the gills and out through the opercular valve.

The respiratory surfaces of the gills are the feather-shaped gill filaments which extend from each of the gill arches. Along both sides of each gill filament are numerous very small, thin, plate-shaped structures called lamellae. These delicate structures are the main site of respiratory exchange, and their total surface area is many times the surface area of the rest of the fish.

As water flows across the gill, oxygen diffuses through the lamellae into the blood. About 99% of the oxygen in blood is carried in the red blood cells, which are very efficient carriers of oxygen. One unit volume of fish blood can carry the oxygen contained in 15 to 25 times that volume of water.

Blood from the heart reaches the gills by way of the ventral aorta from which a single vessel goes to each gill arch where it branches into a fine network of capillaries that bring the blood very close to the gill surface. The bright red color of the fish's gills is good evidence of the richness of the blood supply and the extremely thin nature of the lamellar membrane.

Another feature of the gill that adds to its effectiveness as a respiratory organ is that the blood flowing through each gill lamella flows in a direction that is opposite to that of the flow of water. This creates a "countercurrent" exchange system that is much more efficient than it would be if the blood and water flowed in the same direction. Commercial heat exchangers work on this same principle. The overall efficiency of the fish gill is such that up to about 80% of the oxygen in water that takes about 1 sec to flow over the gill can be taken up by the blood and carried to the tissues.

Another important function of the gills and blood in respiration is in ridding the body of CO_2. In the tissues as oxygen diffuses from the blood, CO_2 generated by metabolic processes in the cells diffuses into the blood where it is carried to the gills through which it diffuses into the water.

Unlike terrestrial animals that are almost always enveloped in a virtual ocean of oxygen, fish often find themselves in water with very low oxygen. In most situations fish will avoid areas of low oxygen; however, this is not always possible. For example, in summer a series of warm, still cloudy days can result in a severe decrease in the concentrations of oxygen in shallow lakes and ponds. Various pond culture practices can also cause a decrease in the concentration of oxygen in shallow bodies of water. Both overfeeding fish and aquatic weed control programs can cause an increase in the amount of decaying organic material on the bottom of a pond, and these decaying materials have a high oxygen demand. Also, lowering the water level of a body of water, which cuts down the surface area and concentrates the organic material in a smaller volume of water, can decrease the concentration of oxygen. Often a combination of weather and one or more of these pond culture practices will cause the oxygen concentration in a pond to decline to below the critical level, and a fish population that has taken the pond owner years to develop will be wiped out in several hours.

In home aquaria the two most common practices that cause oxygen problems are overfeeding and overcrowding fish. A small aquarium pump will generally supply enough air to alleviate such problems. In large outdoor ponds, however, constant aeration can be an expensive process. In such situations the best safeguard against low oxygen problems is a portable oxygen meter. With such an instrument the pond owner can monitor the oxygen concentration and in most instances regulate his culture practices so as to minimize problems caused by low concentrations of oxygen.

The dissolved oxygen requirements of fish vary from species to species, with the coldwater fish having higher requirements than the warmwater species. Most biologists accept 6 ppm as the minimum concentration that will fulfill the oxygen requirements for fish; however, some feel this concentration may be too low to maintain healthy reproducing populations of some species. Many species of fish can survive for varying periods at concentrations well below 6 ppm; however, prolonged exposure to low concentrations of oxygen can indirectly affect fish by decreasing their resistance to disease, decreasing fecundity, inhibiting growth, and increasing the fish's susceptibility to other unfavorable conditions such as temperature fluctuations and the toxicity of various materials.

TEMPERATURE

A fish's body temperature, with a few exceptions, is about 0.5°C (1.0°F) above that of its environment. Consequently, the body temperature of fish is governed by the water temperature and as a result, water temperature plays a very important role in regulating the activities of fish. For the reader to understand how temperature affects the activities of fish it is necessary to briefly discuss some of the factors involved in the regulation of biochemical reactions in fish.

In any organism, life is the result of thousands of integrated rate-regulated biochemical reactions. Protein catalysts called enzymes are extremely important in the regulation of these reactions. At the relatively narrow temperature range at which life can exist, without enzymes, rates of biochemical reactions would occur much too slowly for life. Enzymes, however, lower the free activation energy of the reactants, which simply stated means that in the presence of enzymes, reactions can occur at the rates necessary for life at our low earthly temperatures. It is somewhat of a paradox that enzymes, which are so important in the relationship between temperature and regulation of biochemical reactions, are among the most sensitive of the biochemical materials to temperature change. In man, this normally is not a problem because the body temperature generally remains constant regardless of the environmental temperature. However, if the body temperature of man does change several degrees, it is a matter of concern, and if the body temperature changes much more, it can be fatal, one of the reasons being the deterioration of the enzymes and the resulting breakdown in the integrity of critical biochemical reactions.

If indeed enzymes are so sensitive to changes in temperature, then what biological strategies have fish evolved that allow them to live so successfully in an environment where their body temperature may vary a great deal? Of the several strategies that have been suggested, the one that has been demonstrated is that fish can produce different variants of enzymes such that in cold water a fish produces the variant of an enzyme that is more efficient at colder temperatures and in warm water fish produce a variant that is more efficient in warm water.

In nature, changes in water temperature generally occur gradually over weeks or months, which allow a fish time to produce the necessary enzyme variants and make whatever other physiological adjustments are necessary for it to function within a given temperature range. However, if fish are subjected to sudden changes in water temperature, as often occurs in situations involving home aquaria or where fish are moved from one location to another, the fish may die before they can make the necessary physiological adjustments.

Because of the effects of sudden temperature changes on fish when they are moved from one location to the other, it is a good practice to set the temperature of the water to which the fish are moved to within at least several degrees of the water from which they were taken. Once this is done the fish can slowly be acclimated to the desired temperature by changing the water temperature several degrees per day. In such situations, there is always the temptation to increase the temperature more than several degrees per day, and, depending on the species of fish and its past thermal history, this may be possible. The best practice, however, is to acclimate fish slowly because not only are there the direct effects due to temperature that we have discussed but also when fish are moved from one location to another they may be subjected to stresses due to handling and changes in water quality such as pH, hardness, and dissolved oxygen. These factors either singly or in combination with temperature changes may affect the immune response in fish such that they may become much more susceptible to disease.

Different species of fish have different ranges of temperature tolerance, and both temperature tolerance and temperature preferences are dependent on the past thermal history of the fish. For example, the higher the acclimation temperature, the higher will be the temperature tolerance. The maximum temperatures to which goldfish, bullhead, and brook trout can be acclimated are 41°, 37.5°, and 23.8° to 24.1°C (105°, 100°, and approximately 75°F), respectively. Ordinarily, fish that live in an environment with a diverse range of temperature are much less sensitive to temperature changes than fish that inhabit an environment with a relatively stable temperature.

The fact that fish can be acclimated to withstand very high or very low temperatures for extended periods does not mean that fish can survive indefinitely at these temperatures. For example, although channel catfish have a very wide zone of temperature tolerance [from several degrees above 0°C (32°F) to about 35°C (95°F)], they are not very active below about 10°C (50°F), they grow best at 30°C (86°F), and in nature they spawn at between 20° and 23°C (68° and 74°F). In contrast, brook trout have a narrower zone of thermal tolerance, which extends from several degress above 0°C (32°F) to about 21°C (70°F). Their optimum temperature for growth is 7° to 18°C (45° to 64°F), and in nature they generally spawn at about 5° to 13°C (41° to 55°F).

PHOTOPERIOD

Light and temperature can act through the nervous system and affect many phases of the metabolism of fish. Perhaps the best documented evidence of the importance of light on the activities of fish is the effect that photoperiod has on the reproductive cycles in some fish. This is especially true for fish that breed just once each year.

Although the exact mode of action is not completely understood, it is generally supposed that light affects sexual behavior in the following manner. Changes in the duration of light are sensed by the brain, which in turn stimulates the pituitary gland, a very small organ located at the base of the brain. The pituitary secretes substances called hormones, which are carried by the blood to the various organs they affect. The hormones that affect reproduction act on the gonads and are called gonadotropins. Sex hormones are produced by specialized cells in the gonads at concentrations that are determined by the concentrations of the gonadotropins.

In fish the development and elaboration of eggs and sperm and secondary sex characteristics such as changes in coloration and the maturation of gonopodia depend on the concentration of gonadotropins and sex hormones. Also, sexual behavior in fish appears to be strongly influenced by these hormones.

Although the exact mode of action of hormones has not been established, two possibilities are: (1) that hormones can stimulate the production of specific enzymes or groups of enzymes which in turn may increase the rate at which specific metabolic reactions occur; and (2) that hormones are in some way involved in the transport of a variety of substances across cell membranes and in doing so can influence the rates at which certain reactions occur.

The knowledge that day length does influence reproduction in some fish has been used by fish culturists to alter reproductive cycles. For example, it was found that by first increasing and later decreasing light duration in advance of the natural photoperiod change, the spawning season for brook trout could be advanced by as much as four months.

Another method of altering reproduction cycles in fish has been by the injection of pituitary extracts obtained from other animals. This method of inducing spawning has been used successfully in fish culture operations involving such fish as sturgeon, salmon, trout, carp, and catfish. Although the best results are generally obtained using extracts from glands of the same species, extracts from other species have been used with good success.

TOXIC MATERIALS

There are hundreds of materials that have been demonstrated to be toxic to fish. Among these the groups that have received the most notoriety are the synthetic organic insecticides, the heavy metals, especially mercury, the polychlorinated biphenyls (PCBs), and two materials that often cause problems for the home aquarium owner, ammonia and chlorine.

Because we will be discussing the significance of what concentrations of these materials in water mean in terms of their toxicity, it is important that the reader have some idea of what the concentrations mean in terms of actual amounts of material. Concentrations of materials in water are expressed as parts per million (ppm) or milligrams per liter (mg/liter); parts per billion (ppb) or micrograms per liter (μg/liter); and parts per trillion (pptr) or nanograms per liter (ng/liter). To give the reader a better idea of what these concentrations represent: 1 ppm or 1 mg/liter is equivalent to about 1 oz of material in 1 railroad tank car of water; 1

ppb or 1 μg/liter is equivalent to about 1 oz of material in 1000 railroad tank cars of water, and 1 pptr or 1 ng/liter is equivalent to about 1 oz of material in 1,000,000 railroad tank cars of water.

There are two ways that wild or cultured fish can accumulate toxic materials. One is by direct uptake from the water and the other is through their food. Many toxic materials can be taken up from water by the fish's gills. The large exposed surface area, the thin membrane, the profuse blood supply, and the large volumes of water that pass over the gills, all of which make the gills very effective at removing oxygen, also make the gill a very efficient mechanism for filtering certain toxic materials from water. Accumulation from food in a field situation can result in a buildup or biomagnification of a toxic material as it moves up the food chain from lower to higher levels. In situations where fish are being fed any of the commercial fish foods there is always the possibility that one or more of the ingredients were at one time exposed to a toxic material and that this material can be passed along to the fish. In most instances in field situations, accumulation of a toxic material is the result of accumulation of these materials from both food and water.

Toxic materials that have caused the most problems in aquatic environments are those that are not easily broken down in the environment. Because of their resistance to breakdown these materials tend to accumulate in the environment and can build up to very high concentrations in fish. This is especially true of toxic materials that are very soluble in fat, such as some of the insecticides and PCB compounds. These materials are taken up very rapidly from water and food by fish and are accumulated in the fatty tissues. For example, in the late 1960s certain fish in Lake Michigan accumulated concentrations of the insecticide DDT, which is very soluble in fat, that were over 1,000,000 times higher than the concentrations in the water.

The effects of toxic materials on fish can generally be broken down into acute and chronic effects. Acute effects are generally the result of fish being exposed to a relatively high concentration of a toxic material and the results of the exposure are evident very shortly after the exposure. The most standard test involves exposing fish to concentrations of the material being studied for a set time, generally 4 days (96 hr), and the concentration that kills half the fish in that time is termed the LC_{50} value. Some examples of LC_{50} values for several insecticides are a 94 hr LC_{50} value for DDT of 27 μg/liter for goldfish, a 96 hr LC_{50} value for toxaphene of 50 μg/liter for goldfish, and a 94 value for dieldrin of 7.9 μg/liter for bluegill. If you remember our analogy for concentrations, the preceding LC_{50} value for dieldrin would be equivalent to about 224 g (7.9 oz) in 1000 railroad tank cars of water. Examples of 96 hr LC_{50} values for other materials are 42 μg/liter for mercury for fingerling rainbow trout, and a concentration range of 0.09 to 0.30 mg/liter for chlorine for yellow perch, largemouth bass, and fathead minnows.

The LC_{50} values are an important measure of acute toxicity; however, they do not indicate the long-term effects of lower concentrations of toxic materials. These effects are determined in long-term or chronic toxicity tests which often involve exposing fish for many months to low concentrations of toxic materials. Such tests have demonstrated that concentrations of toxic materials many times lower than those represented by the short-term LC_{50} values can affect such factors as temperature preference, growth, behavior, and reproduction. These often can be very subtle effects that are very hard to measure in natural fish populations, and yet over time these effects can be more damaging to fish populations than the often well publicized fish kills.

Another problem caused by toxic materials is that fish can accumulate such high concentrations that they are unsafe for other animals, including man, to eat. An example is the deleterious effect that high concentrations of insecticides in fish had on the reproductive success of brown pelicans that ate them. Another example is the deleterious effect of PCB compounds on mink reproduction. The reproductive failure was traced to PCB compounds that the mink obtained from coho salmon from Lake Michigan that were used to supplement their diet. The most widely publicized incident of man's being affected by a toxic material obtained through eating fish occurred in Japan where 121 people died and over 700 were affected after eating fish that had concentrated methylmercury which had been discharged into the water as a by-product of the manufacture of a plastics intermediate.

Chlorine is generally only a problem in situations involving the use of city water that has been treated with chlorine. In situations where city tap water is used in an aquarium it should be aerated for several days to expel chlorine before fish are put into the aquarium. In situations where a constant flow of water is needed, and the only source is city tap water, a commercial charcoal filter capable of handling the desired volume of water is necessary.

Ammonia in water originates from two principal sources. One is the breakdown of organic material by bacteria and the other is excretion by fish and other aquatic organisms. Ammonia is the main form of nitrogen excreted by most aquatic animals. In fish most of the ammonia which is a result of the breakdown of nitrogen containing compounds such as proteins is excreted at the gills.

As a general rule ammonia concentrations (total NH_4^+) in water should not exceed 0.1 ppm. Some of the chronic effects resulting from the exposure of fish to ammonia are a reduction of growth and stamina, gill hyperplasia, and an increased susceptibility to disease. As might be expected, problems concerning ammonia generally occur in fish culture operations where fish are crowded and the water turnover rates are not adequate to keep the ammonia concentration at a safe level. The best remedy is to keep fish tanks clean, not to overfeed fish, and not to overcrowd them.

SUMMARY

In this chapter we have tried to show the reader how fish are affected and how they respond to certain factors in their environment. For the purposes of clarity, we have discussed each of these factors separately. In nature, however, these factors and some that we have not discussed are always interacting with the result that the responses of the fish are generally dictated by the combined effects of these factors. For example, temperature and pH can influence the toxicity of various materials. The effects of low oxygen are more serious at high than at low temperatures, and it is generally the combined effects of changes in temperature and photoperiod that regulate the annual reproductive cycle in many fish.

Each species of fish has specific environmental requirements in which it can best grow and reproduce. Because of this, anyone interested. in keeping fish, be it for fun or profit, should make every effort to obtain all available information concerning the environmental requirements of the fish in which he is interested and should attempt to maintain such an environment for the fish.

3

Types of Culturing Facilities[1]

POND CULTURE

Several kinds of water facilities may be needed for growing fish. The kind of facility constructed will depend on the size of the farm and type of fish farming program to be followed. Commonly used structures include ponds, raceways, tanks, and aquariums. Some farmers use only ponds; other farmers use raceways; and still others use a combination of two or more of the water facilities in their fish farming operations.

The most commonly used water impoundments for growing fish are ponds, except for trout and salmon. The term "pond" is frequently used interchangeably with "lake" even though many persons make a distinction between the two. Pond is used to refer to the smaller structures, whereas lake refers to the larger structures, with more hectarage (acreage) in water. In this handbook, pond is used when describing any size of earthen structure for holding a standing body of water, that is, the water does not flow or has limited flow.

There are several kinds of ponds for farming. The kinds are based on the function performed by each in fish production. The kinds of ponds are:

(1) Holding Pond.—A holding pond is usually 0.40 ha (1 acre) or less. A farm may need more than one holding pond, depending on the number of broodfish kept. Many fish farmers prefer to have several small holding ponds of about 0.10 ha (1/4 acre) each so that a disease outbreak among the broodstock can be more easily kept from affecting all broodstock.

(2) Spawning Pond.—A spawning pond is a small pond in which the broodfish are placed for spawning. The broodfish may be placed in pens or in the open pond.

(3) Rearing Pond.—A rearing pond is a small pond in which fry are placed for growing into fingerlings of suitable size for stocking. The

---

[1] Serious readers of this chapter would do well to obtain copies of the references, particularly Grizzell (1971), Lee (1971), and Meyer et al. (1973).

amount of hectarage (acreage) necessary in rearing ponds depends upon the number and size of fingerlings desired.

(4) Growing Pond.—This is the pond in which fish are grown to the size of food fish. The size and arrangement of growing ponds vary considerably.

(5) Catch-out Pond.—A catch-out pond is a pond in which catfish are grown or held until caught by the sport fisherman. Size and arrangement vary considerably.

Three types of ponds may be used for fish culture. These are ravine, excavated, and levee type ponds. The ravine pond is constructed in a deep gorge, gully, or other similar place. An excavated pond is constructed on any fairly level land not subject to overflow from creeks and rivers. A levee pond is usually constructed on flat land with the levee on all sides. Each type has advantages and disadvantages for use in fish farming.

Ravine Ponds

Hilly land is best suited to ravine ponds. The ravine pond is made by damming up a dry ravine or a creek. The dam is usually constructed of earthen materials moved in from nearby higher levels. One disadvantage of the ravine type, since its location may be in wooded areas, is that protective grasses and shoreline vegetation may not grow readily. The dam may need to be reinforced with woodwork, logs, or rocks to prevent erosion during heavy rains. Diversion ditches and levees may need to be constructed to divert hillside drainage around the pond. Another factor that should be considered while in the process of construction is a drainage outlet. This outlet should be made of concrete, tile, or corrosion-resistant iron pipe and should be included in the plans.

Ravine-type ponds must have spillways. Spillways are outlets in dams to release excess water from the pond. The spillways of ravine ponds are usually constructed of wood or concrete, preferably concrete, and are located at the water level which is the maximum fullness that the dam will tolerate. A screen should be provided in the spillway to prevent the loss or escape of fish during the times of overflow.

Excavated Ponds

Excavated means that the earth is removed and used for building embankments or levees for the pond. This type of pond may be constructed on any fairly level land that is not subject to overflow or flooding from the surrounding area. The excavation should be to a depth that will allow a gravity flow of water from some other source; however, the bottom should be at an elevation sufficient for complete gravity drainage. The water to supply this type of pond may be obtained from creeks, surface runoff, or wells. A cut-off valve or gate should be constructed to control the amount of water going into a pond and to be closed during the spawning season because the clay and silt carried by the water will settle on the eggs and prevent hatching. Also, as with the ravine pond, it is necessary to place a screen over the intake valve to prevent undesirable species of fish from entering the pond.

Levee Ponds

Levee ponds, which can be constructed without excavation, have been found to be very successful in areas with flat land. This type of pond can be constructed on any type of agricultural land. The levee pond is similar to an irrigated rice field, except that the pond levee must be considerably higher because of the necessary depth of water. All intakes should be carefully screened with sixteen-mesh wire of copper or other durable metal, or saran filter. Also, all connections between ponds should be carefully screened to prevent fish from traveling from one pond to another.

Great care should be taken in selecting pond sites. Economy of construction, ease of use, and productivity of ponds are dependent upon location.

Several factors should be considered before building a fish pond. One of the first is proper drainage. Each pond should be constructed so that it can be drained individually and completely. Some ponds are constructed so that daily water draw-down can be used to reduce the risk of oxygen problems. The water removed from the pond is replaced with oxygenated water from nearby ponds or wells. Another factor that should be considered is whether the pond will be free from overflow or flooding from surrounding areas. This is especially true when low land near creeks is used. It may be necessary to cut diversion ditches around the pond to prevent flooding.

If the land or watershed area has been used for crops, the soil should be tested for harmful chemical residues of certain pesticides that are particularly undesirable. Most pesticides are toxic to fish. Some are not readily broken down in the soil and residues may have accumulated that will kill fish. A thorough test of the soil should be made if a fish operation is planned on cropland where chlorinated hydrocarbon pesticides have been used.

Extreme care must be taken if aerial pesticide spraying occurs in the vicinity. Pesticides may drift into ponds and kill fish. Fish ponds should never be planned where run-off can enter them from cotton fields or other areas where pesticides are used.

A soil test should be conducted for the purpose of determining whether or not the soil is capable of holding water, especially during dry seasons. A technician of the Soil Conservation Service (SCS) will check for gravel layers, soil structure, depth of water-holding material, rock fissures, indications of sand strata, and other soil characteristics that might interfere with good water-holding qualities. A plentiful supply of good water should be available for long dry periods.

When selecting a pond site, there are other questions that arise, such as, "Should ponds be constructed on good level land, which is more expensive to purchase, or on rolling land, which can be purchased at a lower price?" This will be determined by considering all factors. Although the level land may be more expensive, it may be more profitable in the long run.

After a decision of where to build the pond has been reached, a technician of the SCS will do the survey work and draw up plans and specifications.

The size of a pond for fish production may be a case of necessity, not choice, and will vary according to the slope and size of the site available.

If there is a choice of site, the size of the pond may vary. All sizes have their advantages and disadvantages. The size will depend on whether rotations of fish and agronomic crops are planned. If rotations are anticipated, the ponds should be large enough to economically use machinery in harvesting the agronomic crops. The other crops most frequently grown in rotation with fish are soybeans followed by rice.

Authorities do not agree on the best size for fish ponds. Some fishery biologists recommend that the size of ponds for growing food fish should be 4.0 to 8.1 ha (10 to 20 acres), or larger. Others prefer smaller ponds of 0.40 to 2.0 ha (1 to 5 acres). Ponds of less than 0.40 to 40 ha (1 acre to more than 100 acres) have been used for producing catfish. Large ponds cost less per hectare or acre to construct than small ones; however, large ponds also have several disadvantages. Levee or dam erosion is a serious problem with the larger ponds. Water movement, and hence erosion, is caused by the wind. However, wind action helps to prevent oxygen problems. If parasites or diseases break out in a large pond, it is difficult to treat the fish. Resulting losses can be large. A 20 ha (50 acre) pond is usually stocked with 3700 to 7400 fingerlings per ha (1500 to 3000 per acre), which means a pond contains 75,000 to 150,000 fish to be harvested at one time—a difficult job. Another limiting factor is getting the large quantity of fish to the market at one time. In contrast to large ponds, small ponds of 0.40 to 2.0 ha (1 to 5 acres) provide more flexibility for management, harvesting, overcoming oxygen shortages, and treating for disease and parasites. Also, the small ponds can be drained and refilled quickly and in time will pay for the added cost. Smaller ponds allow for a more gradual harvest with better care for the fish as they are removed. Thus, the commercial operation should govern the size of ponds by harvesting conditions and market demands. Also, the size of ponds for satisfactory fish farming is dependent on the supply of good quality water that is available.

FIG. 3.1. SHAPING FISH FARM PRODUCTION POND

Ponds for fish farming should be arranged for maximum efficiency of production. Some of the factors affecting arrangement, such as natural boundaries, property lines, highways, power lines, and topography, are not easily controlled. The plan of pond layout within these factors is very important.

One of the first steps in planning pond arrangement is to decide on the type of fish farming program to be followed. In the arrangement of growing ponds, consideration should be given to the source of fingerlings—whether produced or purchased. Fingerling production requires three types of ponds: holding, spawning, and rearing. These ponds usually occupy a small percentage of the total land in a food fish farm.

Economics of construction and harvesting should be considered in planning the arrangement of ponds. A square shaped pond requires less levee than a rectangular pond for the same number of hectares or acres of water. The levee requirement increases proportionately as the pond shape changes from square to rectangular. A square 0.40 ha (1 acre) pond requires 257 m (835 ft) of levee; a 0.40 ha (1 acre) pond 31 m wide by 134 m long (100 ft wide by 435 ft long) requires 329 m (1070 ft) of levee. It is obvious that it is more economical to construct square shaped ponds.

Economy of harvesting, however, usually favors rectangular shaped ponds. Less seine is required for harvesting the fish from rectangular ponds as compared with square ponds of the same hectarage (acreage). Feeding from the dam or levee is also facilitated by rectangular ponds. Fish farmers who have several ponds would do well to construct all of them about the same width so that additional seine will not be required for the larger ponds. The length of ponds may vary as long as the width remains the same. Also, by constructing ponds next to each other, both sides of the dam are functioning in holding water, thus reducing cost of construction per unit of water.

In planning arrangement prior to construction, consideration should be given to location of the water well, arrangement of drainage ditches and pipes, and accessibility with motor vehicles. Regardless of arrangement, maximum utilization of water supplies and drainage facilities should be made.

Many factors should be considered in constructing ponds for fish farming. A complete discussion of these is beyond the scope of this book. However, certain basic considerations need to be studied to ensure that ponds will adequately meet the needs of fish culture.

Site Preparation

Pond sites should be cleared of all trees, stumps, roots, and other obstacles before construction is initiated. Such obstacles and trash will interfere with harvesting operations and may give fish an undesirable flavor. Trees and brush should be cut back at least 4.6 m (15 ft) from the water line to allow for the movement of feeding and harvesting equipment. Sometimes a preliminary survey is made to locate the area to be cleared. After clearing, the location and slope of the dam or levees should be surveyed and staked out. An SCS technician will assist the owner in doing this.

Dam Construction

The heart of any pond is the dam! The soil located at the base of the dam or levee should be removed down to the mineral soil. This procedure is to assure a good bond between the mineral soil and the soil to be filled for the dam. The mineral soil area is called the parent material horizon.

The topsoil that is removed from the dam site should be piled to one side and, when the dam is completed, spread over the top of the dam in a layer 5 to 15 cm (2 to 6 in.) thick. It is not necessary to spread topsoil below the water line on the front side of the dam. It is usually best to construct dams during summer and fall. This allows ample time for the soil to settle during rainy seasons in late fall and winter before stocking with fish in the early spring. There is usually about 15% shrinkage in a levee due to settling. All roots, sod, mulch, vegetation, wood, and other decomposable matter should be removed from the soil that is being used in forming the levee.

Filling the above-ground portion of the dam is the most expensive and important operation in building a pond. Soil with a high proportion of clay should go into the core of the dam. The soil with the second highest proportion of clay should go into the section of the dam on the water side. The soil lowest in proportion of clay should be placed on the downstream side of the dam. The soil should be applied in 15 cm (6 in.) layers with each layer being well packed before another layer is added. Packing is accomplished by driving the heavy construction equipment over the dam during construction.

Dams should have a minimum slope of 3-to-1 on both sides. A 3-to-1 slope means that for every 30 cm (1 ft) in height, there should be 120 cm (3 ft) of levee extending toward both the upstream and downstream sides of the dam. On ponds larger than 4.1 ha (10 acres), a 4-to-1 slope on the upstream side and 3-to-1 on the downstream side is advisable. On flat land where the ponds have levees all around them, a minimum top width of 3 to 4 m (10 to 12 ft) is desired. This width will allow the passage of feeding, harvesting, and maintenance equipment. The tops of main levees should be graveled to facilitate travel in all kinds of weather. A freeboard of 0.5 m (2 ft) is recommended. Freeboard is the distance from the normal water level to the top of the levee.

Pond Drainage

All ponds for fish production should have adequate facilities for regulating and draining water. Various kinds of arrangements are used. One of the most popular with fish farmers is a turn-down pipe located at the lowest point of the base of the dam. The turn-down pipe acts as an overflow and drain pipe. Water levels are established by adjusting the pipe.

In constructing a turn-down drainage pipe, a screen should be placed over the end of the pipe inside the pond to prevent the loss of fish and obstruction of the flow of water by turtles. A special anti-seep collar should be placed around the drain pipe inside the dam to prevent water from seeping along the pipe and causing leaks. To prevent unplanned drainage, an upright post with a chain and lock should be used to prevent the pipe from tilting downward, thus draining out the water. The swivel joint should be heavily greased to assure easy operation. A solid footing should support the turn-down pipe assembly, including the drain pipe through the dam. It is recommended that the footing extend 1.25 m (4 or more ft) into the bottom of the pond. The drain pipe through the dam should have a fall of about 25 cm for each 30 m (1 ft for each 100 ft) of distance. Water can be removed from the bottom of a pond by using a doublesleeve device. This device allows water near the bottom that is deficient in oxygen to be removed. At the same time fresh water may be added at the surface if desired.

Various other types of water regulatory devices are used, including gate valves, shear valves, and screw-in plugs. Each of these may be satisfactorily used, but the cost may be greater than the turn-down pipe arrangement.

Size of the drain pipe to use depends on size of pond, speed at which drainage is desired, and the volume of water coming into the pond. A 10 cm (4 in.) pipe will generally take care of small ponds up to 1.2 ha (3 acres) in size. Fifteen to 30 cm (6 to 12 in.) pipes are recommended for ponds of 6.1 to 8.1 ha (15 to 20 acres). As the diameter of a pipe is doubled, water flow increases more than four times. A 10 cm (4 in.) drain will empty a 0.40 ha (1 acre) pond having a maximum depth of 2.8 m (9 ft) and an average depth of 1 m (3 to 4 ft) in about 60 hr. This is assuming that no water is entering the pond during this period. A 15 cm (6 in.) drain will empty the same pond in half the time. The 30 cm (12 in.) drain will require 1/9 as much time as a 10 cm (4 in.) pipe requires to drain a pond.

A spillway may be needed, depending on whether or not a pond has sufficient drainage pipes. Spillways should be large enough to permit the outward flow of surplus water caused by rainfall and filling. A screen will be needed to prevent the loss of fish. For average size ponds with a watershed area of less than 20 ha (50 acres), the spillway size may be found by dividing the total number of hectares or acres in the watershed area by two to obtain an estimated spillway width. It is usually a good procedure to add an additional 3 m (10 ft) to the width as a safety precaution.

The spillway may be located at one or both ends of a dam, or at a convenient point along the sides of a pond. The spillway should be paved with rock or concrete, or completely covered with a good sod. The type of construction to use will depend on location of the spillway, type of soil on which it is to be built, and the anticipated volume of water to flow out of the pond.

Harvest Basin Construction

Ponds should be constructed to facilitate harvesting. If harvesting is to involve water draw-down, a harvest basin or pit is generally constructed. Basins are constructed 45 to 60 cm (18 in. to 2 ft) deeper than the normal level of the pond. The basin should have an extra smooth bottom with sloping sides. A satisfactory harvest basin contains about 10% of the total area in the pond. A harvesting basin is not necessary on land that slopes 60 cm to 1 m per 31 m (2 to 3 ft per 100 ft). A circular basin is most suitable for surrounding fish, but a rectangular basin is satisfactory.

The fish will move to deeper water as the water level is lowered and will be concentrated in the basin for harvesting. Large ponds should be seined while full of water to yield a partial harvest. By partially harvesting the fish before draining, the volume of fish to be handled in the harvest basin will be reduced.

New types of harvest basins outside the levee are being tested. This means the fish would go through the drain pipe into the basin. Outside harvest basins have the advantage of less mud and can be easily reached with equipment.

Dam Maintenance

A vegetative cover to protect the dam from erosion should be established. If the levee is completed in late summer or fall, it will not be possible to establish a sod before winter rains start. In such cases, the levee should be planted with a fast growing temporary winter grass, such as ryegrass, seeded at the rate of 67 kg per ha (60 lb per acre) or wheat, seeded at the rate of 100 to 200 kg per ha (2 to 4 bu per acre). This will prevent any serious erosion from winter rains. During the following spring, the levee should be sowed or sodded with a permanent grass.

To speed the growth of a cover, the topsoil that was removed from the dam site should be evenly spread over the dam before seeding. Occasionally, fertile soil from other sources is hauled in.

Five hundred sixty to 1120 kg (500 to 1000 lb) of 13-13-13 fertilizer should be applied per ha (acre) of levee surface followed by light disking and smoothing. After the sod is established, fertilize annually with 3.4 kg per ha (3 lb per acre) of actual nitrogen. If the seed is sowed on an old sod, streak the levee with a disk and sow on top of the soil. It is best to drill the seed into the soil. Using these fertilizer rates, cattle should not be allowed to graze on the levee. If cattle are permitted to graze, more fertilizer should be used.

Heavy rains may cause severe erosion on newly constructed dams. Where this is a problem, hay or other suitable material may be used to cover the dam to reduce erosion until a vegetative cover can become established.

Diversion Ditch Construction

Not all ponds need diversion ditches. Diversion ditches sometimes resemble terraces and are used to prevent unwanted water from entering ponds. Some ponds are constructed on sites that have excessively large drainage areas. Such ponds are flooded following heavy rains and often stay muddy most of the year. A ditch is laid out around one side of a pond and continues around and below the pond dam where the water may be safely released. The ditch should have 5 cm (0.2 ft) or slightly more fall per 31 m (100 linear ft) of ditch around the side of the pond.

RACEWAY AND TANK CULTURE

Raceway or tank culture is feasible only if large quantities of cheap, high-quality water are available for a "once-through" or "open" system, or if the water can be recirculated and wastes efficiently removed for a "closed" system.

Raceways may be of concrete, block, tile, bricks or other durable material or be earthen. However, because of the usual high volumes and velocity of water flows used, earthen raceways often cannot be used.

Raceways can be used in rearing almost any type of fish but are utilized almost exclusively in trout production and to a lesser degree with catfish. Raceways usually use gravity flow water instead of pumped water because of lower power and pumping costs. Up to a point, fish production can be intensified as flow rates increase. If the velocity of the water is such that the fish have a difficult time in maintaining position and are being swept away, the energy requirements of the fish are increased and feed costs increase. Because flowing water is used, production is greatly intensified to a degree that production can be referred to as super-intensive compared to typical pond culturing when the water flow may only be used to replace seepage and evaporation losses. The water flow replaces the water volume fast enough to remove wastes and replenish oxygen in spite of intensified stocking rates.

It has been widely reported and substantiated by many qualified reports that the most super-intensive production in raceways is on a carp farm in Japan. The farmer uses 16 small raceways. Converted to a hectare or acre basis, production is 2203 MT per ha or 982 ST per acre, which is 1000 times more intensive than many earthen ponds.

Production in any raceway system is largely a function of the rate of water flow and the exchange rate of high quality, well oxygenated water. Brown (1977) reported that researchers in Pennsylvania found that one heavily aerated hatchery had obtained production of 589.6 kg per liter per sec (82 lb per gal. per min). He further reported that another producer in France, using one-third freshwater and two-thirds recirculated water had a production rate of 555 kg per liter per sec (76.5 lb per gal. per min) of water flow. These are in contrast to the more

FIG. 3.2. LARGE SCALE TROUT FARM USING CONCRETE RACEWAYS

FIG. 3.3. CLOSE-UP OF TROUT CONCRETE RACEWAY

normal production rate, which is about one-fourth of these levels.

The shape of raceways may vary. Most are rectangular; however, a few are circular. Some farmers establish a system of raceways making use of the contour of the land. A series of small ponds may be constructed below each other on a hillside. Water is pumped into the pond at the top of the hill and flows out of this pond into the one just below it. The water flows from one pond to another until it reaches the last pond at the bottom of the hill.

Raceway segments may be as short as 10 m (40 ft) or as long as 100 m (400 ft), widths may vary from 1 to 10 m (3 to 30 ft), and depths may vary from 0.5 to 15 m (18 in. to 5 ft).

Semi-raceways are usually much larger than raceways and are of earthen construction. Semi-raceways may be 16 m wide and up to 365 m long (50 by 1200 ft).

The reduced water velocity of semi-raceways does not present the same problems of erosion and excessively muddy water as with earthen raceways. Crushed stone may be used to line the banks of earthen semi-raceways to reduce erosion.

Water should be removed from the bottom of raceways and semi-raceways. Removing water from the bottom aids in carrying away metabolic wastes and in removing water with a low level of dissolved oxygen. The runoff from rainfall should be diverted away from raceways.

The major disadvantage of raceways is the high initial cost of construction. Construction costs vary with size and number of raceways, materials used in construction and availability of good water at an economical rate.

FIG. 3.4. EARTHEN RACEWAY SHOWING ONE TYPE OF AER-ATOR

In a study of commercial trout farmers in Idaho, the maximum water flow for any hatchery was 10,300 liters per sec (325 cfs); the minimum water flow for any hatchery was 190 liters per sec (6 cfs). This shows the wide range in water flow rates between hatcheries. The number of

FIG. 3.5. EARTHEN RACEWAY SHOWING AERATORS AT CONCRETE HEADWALLS

FIG. 3.6. CLOSE-UP OF EARTHEN RACEWAY HEADWALL SHOWING AERATOR AND
DRAINHOLE WHICH CAN ALSO BE USED IN HARVESTING

raceway segments constructed essentially of concrete but including a few
earthen raceways varied from 5 to 101. Water replacement rates varied
from 21 min to 1939 min. This information serves to point out the fact
that no two individual raceway-type facilities are alike from the stand-
point of raceway design, water utilization, feeding practices, fish density
per unit of water space or per unit of water flow or in fish husbandry
methods.

Tanks can be constructed of concrete or fiberglass. Sizes vary from 1 to
10 m in diameter (3 to 30 ft) with varying depths. In general the tanks

are circular instead of square or rectangular.

The advantages of the circular pond are: (1) less water required; (2) nearly a uniform pattern of water circulation throughout the pool, with fish more evenly distributed instead of congregating at the head end; and (3) center outlet, with circular motion of water producing a self-cleaning effect. Even though less water is required, sufficient head or pressure must be available to force and maintain the water in a circular motion. In operating a circular pond, advantage can be taken of an interesting phenomenon. Since the vortex of a whirlpool in the northern hemisphere always rotates in a counterclockwise direction, the water in a circular pond should also rotate in a counterclockwise direction.

To take advantage of the self-cleaning effect of the circular pond, it is necessary that the proper type of screen and outlet pipe be used. In effect, the self-cleaning screen consists of a sleeve larger than the center outlet or standpipe. This sleeve fits over the outlet pipe and projects above the surface of the water. The sleeve may have a series of slots or perforations near the bottom which act as an outlet screen for small fish or may be in the form of a narrow opening between the pool bottom and the lower end of the sleeve for larger fish. This opening must be adjusted according to the size of the fish in the pool. Waste materials are drawn through the opening by the outflowing water.

Circular ponds do not lend themselves well to flush treatment for disease control nor to mechanical fish loading or self-grading devices. The advantages of the circular pond for larger fish are not as great as thought when circular ponds first began to appear. The circular pond, however, is well adapted to rearing fish ranging from newly hatched fry to fish of subcatchable size.

There are many advantages to circular tanks in holding a small number of fish of varying ages and sizes for selective breeding, feed experiments, and many types of research projects. In these cases it is desirable to have a separate water supply for each experiment and a large number of small tanks is very practical.

Both raceways and tanks are suitable for recirculation systems as well as passing the water through and using it only once. Labor requirements for recirculating systems are high due to the need for continuing maintenance and observation. Mechanical failure in any unit of the system can spell disaster. Standby electrical generators and backup pumps are required in the event of power or equipment failures.

Pilot experiments by research organizations and private individuals on closed systems have been made. In these systems, only enough new water may be added to replace losses and water temperatures may be controlled. Some of these experiments have been highly promising—particularly for higher value fish such as tropicals, goldfish and bait fish.

However, the authors know of no economical operation for food fish. High quality water, complete diets, and disease-free fish are almost a must in a closed system.

As is repeated elsewhere in this book, there is the strong recommendation that before constructing raceways or tanks, either with a flow through system, recirculating water or a closed system, a thorough investigation be made of existing operations to avoid pitfalls and problems.

CAGE AND PEN CULTURE

The only difference between cages and pens from the viewpoints of the authors is size. Cages refer to small wire or fiber mesh enclosures over a supporting frame. These cages are then attached to floats and anchored in rivers, lakes, or large ponds. These small cages are in use for production of catfish and/or rainbow trout. Initially, the cages resulted from catfish experiments and the cages were quite small—about 1 m^3 (39 in. deep, wide, and long). Commercial production soon followed, using the same size of cages. In more recent years, larger cages have been used. After initial successes with raising trout in the winter at Tifton, Georgia, in efforts to secure one crop of catfish and one crop of trout from the same production facility during a year, cage use was adopted for trout production in the warmer waters of the southeast.

Pen culture is similar to cage culture except for the size of the units. Pen culture is a common method of growing out salmon in the northwest.

Cages are generally anchored to the bottom of the river, lake or pond in one or two parallel lines although they are sometimes strung along each side of a floating walkway. The common practice is to work with each individual floating cage from a small boat. The salmon pens are much larger and become part of a facility. The individual pens may be 20 m^2 and 10 m deep (approximately 65 ft^2 by 33 ft deep). Whereas the freshwater cages may have only a single layer of net around the framework, the salmon pens have a double layer of netting, with the outside (predator) net made of stronger material. The individual pen segments are arranged in a floating complex, anchored to the bottom as well as at the ends and sides of the complex. Floating walkways extend to each individual pen for ease in feeding, harvesting, and caring for the fish. Because of tides and currents it is essential that each facility be adequately anchored or moored.

The major premise for the use of the cages and/or pens is that water currents or wind action carries away waste products and continually provides fresh oxygenated water. Cages and pens must be anchored so that there is sufficient space between the bottom of the cage or pen and the bottom of the water area. This is to allow solid wastes to settle out of the enclosure.

Cage culture is readily adapted to water areas which cannot be drained or otherwise harvested. Food fish can be cultured in waters already containing populations of wild fish. Fish in cages can be easily observed and farmers can keep accurate checks on their fish. Under ideal conditions, excellent growth has been achieved. Experimental production has been as high as 382 kg per m^3 of fish (24 lb per ft^3) of cage space in the heated effluent water from a powerplant.

Stocking density figures reported for catfish range from 160 to 600 fish per m^3 (5 to 17 fish per ft ^{3}) with marketable fish of 408 to 680 g each (0.9 to 1.5 lb) in 180 days.

Nutritional difficulties were encountered early in cage culture when the standard food ration was used. Diets must be nutritionally complete. Catfish pond diets have been heavily supplemented with trout foods. Because of increased feeding costs, fish survival and feed conversion must be good in order to have a profitable operation.

Oxygen depletions cause catastrophic fish losses whenever they occur, and the cage culture method is not immune from this problem. Fish that are free in ponds can move to areas with higher oxygen concentrations, or disperse when dissolved oxygen levels drop. Caged fish, however, cannot escape from the enclosure; consequently, losses are often complete because the many closely crowded fish deplete the oxygen faster than circulation can supply it.

FIG. 3.7. ANCHORED FLOATING PRODUCTION CAGES

Good water circulation through the cage is essential for aeration and waste removal. This circulation is greatly dependent on wind-induced water currents but is aided by the swimming action of the fish. Cage placement should be such that the longer side is perpendicular to the prevailing wind. Recommended distances between cages will vary with stocking rates, size of cages, water currents and dependability of the winds. A distance equal to at least the length of a cage should be the absolute minimum between cages.

Disease control has been difficult. All diseases commonly found in pond-cultured fish have been observed in caged fish. In addition, other parasitic forms usually found only on wild fish have been noted.

Treatment of parasitized fish is a laborious and expensive task. Effective concentrations of chemicals dissipate rapidly in cages, making disease control difficult or impossible. Each cage of fish must be handled separately, either by removing it or dipping or by enclosing it in some form of container. Labor and time requirements are great and costs are high.

If substandard diets are used and nutritional deficiencies develop, opportunities for disease outbreaks are increased.

REFERENCES

BROWN, E.E. 1977. World Fish Farming: Cultivation and Economics. AVI Publishing Co., Westport, Conn.

COLLINS, R.A. and DELMENDO, M.N., 1976. Comparative economics of aquaculture in cages, raceways and enclosures. FAO Tech. Conf. Aquaculture, Kyoto, Japan, May 26–June 2.

GRIZZELL, R.A., JR. 1971. Pond construction and economic considerations in catfish farming. *In* Producing and Marketing Catfish in the Tennessee Valley. Tenn. Val. Auth. Bull. *Y-38.*

LEE, J.S. 1971. Catfish Farming: A Reference Unit. Miss. State Univ., State College, Miss.

LEITRITZ, E. and LEWIS, R.C. 1976. Trout and salmon culture (hatchery methods). Calif. Dep. Fish Game Fish Bull. *164.*

LINDBERGH, J.M. 1976. The development of a commercial Pacific salmon culture business. FAO Tech. Conf. Aquaculture, Kyoto, Japan, May 26–June 2.

MEYER, F.P., SNEED, K.E., and ESCHMEYER, P.T., 1973. Second report to the fish farmers. U.S. Dep. Interior, Fish Wildl. Serv. Resource Publ. *113.*

MILNE, P.H. 1976. Selection of sites and design of cages, fishpens and net enclosures for aquaculture. FAO Tech. Conf. Aquaculture, Kyoto, Japan, May 26–June 2.

4

Maintenance and Improvement of Ponds[1]

Weed infestations vary in different types of fish ponds. In a new pond, weeds will usually appear during the first year or two if the pond is not fertilized. If water is clear, within a few months weeds will fill major portions of the pond. If water is muddy most of the time, there will be limited weed growth except in shallower, marginal areas.

Ponds with fluctuating water levels, even though fertilized, often become partially or completely filled with weeds. During summer months, when the water level recedes, weeds invade the wet, exposed pond edge. When the water level returns to normal, as a result of winter rains, many species of these weeds continue to grow and begin their invasion of deeper waters.

Hatchery ponds are excellent habitats for all kinds of aquatic weeds since they are drained one or more times during warm months of each year. When refilled with clean, clear water, the ponds rapidly start to fill in with vegetation again.

Food fish derive very little or no substance from vegetation, except for certain carps. Hence, these should be kept out of these ponds. Food plants are often used by ornamental and bait fish for food, shelter and sometimes for spawning.

Unicellular and filamentous algae are plants better adapted for these purposes, but they too may become obnoxious.

Though suitable also for some of these functions, submerged plants should not be encouraged because they are difficult to control and may

---

[1]A classic treatise on the control of aquatic weeds is that by J.M. Lawrence given at the World Symposium on Warm-water Pond Fish Culture in Rome in 1966. The following technology on weed control is based on that report and the contributions of other experts in the field.

take over the whole pond at the expense of fish production. Too many plants will choke the pond to the extent that an oxygen depletion may occur on hot, still nights. Submerged plants often completely cover the shore feeding and nesting areas and make them unsuitable for the fish. A heavy growth of plants uses up a large percentage of the pond fertility in a form that is not available to the fish as food, and a pond choked with weeds is extremely difficult to harvest.

AQUATIC WEED IDENTIFICATION

Aquatic weeds can be defined as those unwanted and undesirable plants growing in an aquatic environment and refer only to those plants which are adapted to grow and reproduce under such aquatic conditions. A knowledge of the identity of several hundred species of aquatic plants which interfere with fish pond operations is necessary if efficient and effective chemical control practices are to be employed. As an aid in both identification and use of control measures, the following simple outline of major plant groups based upon their shape, size, and growth habits has been developed.

Algae
   planktonic, e.g., *(Microcystis, Anabaena)*
   filamentous, e.g., *(Spirogyra, Pithophora, Chara, Nitella)*
Submerged weeds, e.g., *(Potamogeton, Elodea, Ceratophyllum, Utricularia)*
Emersed weeds, e.g., *(Nymphaea, Hydrocotyle)*
Marginal weeds, e.g., *(Juncus, Typha, Carex, Sarpus, Sparganium,* grasses)
Floating weeds, e.g., *(Pistia, Eichhornia, Trapa, Lemna, Salvinia, Azolla)*

PREVENTION OF WEED GROWTHS

The simplest and easiest method for control of aquatic weeds is to prevent their establishment in a pond. Proper construction of the pond is a major step in this type of control. Before construction is started, the proposed pond site should be checked to determine that it meets the requirements for a good pond as described by Lawrence (1949). Construction features include: (1) a dam sufficiently high to produce an impoundment with a minimum of water not less than 0.5 m (19 in.) in depth; (2) deepening of the pond edge to reduce the hazard of marginal and shallow water weed growth; (3) shaping and sodding of the pond edge above water level to reduce the area where marginal weeds could appear; (4) a diversion ditch, if necessary, to carry excessive and/or muddy water around the pond, thus permitting fertilization of the pond from early spring until fall.

After a pond is properly located, constructed, and filled with water, the next step in prevention of aquatic weed growth is proper fertilization of the impounded water. If the pond water clears periodically, because of irregular fertilization or too large an inflow of water, sufficient sunlight often reaches the bottom for submerged weed growth to start. Therefore, it is necessary that regular fertilization of the pond be practiced. A platform method for applying fertilizer, described by Lawrence (1952), is more efficient than the old broadcast method in that less labor is required in applying it, plus the fact that plant nutrients dissolve in the top waters before they come into contact with the soil. Thus, there is a reduced likelihood of phosphorus and potash being bound to clay particles in the bottom muds before phytoplankton can utilize them.

A pond that is properly constructed and fertilized supports a minimum of aquatic weed growth. The few weeds that appear along the pond edge must be removed immediately. Protection from aquatic weed invasion exists only as long as the preventive practices just described are kept in operation.

CONTROL TECHNIQUES

Since the establishment of the Farm Ponds Project on the Alabama Agricultural Experiment Station in the early 1930s, research has been in progress to find means of eliminating unwanted aquatic plant growths in ponds without interfering with fish production. Techniques which have been tried and results that have been obtained are briefly outlined below. For simplicity, these techniques will be separated and classified as biological, mechanical, or chemical methods.

Biological Methods

Inorganic Fertilization.—Application of inorganic fertilizer during winter months promoted growths of filamentous algae over masses of rooted aquatic weeds and resulted in the elimination of many species of submerged weeds when the weather became hot in the spring. Fish production in such treated ponds was of an explosive nature. Elimination of weed cover allowed bass to eat small bluegills, and the released pressure on bluegill food supply plus additional food produced by decomposing plants resulted in tremendous growth of fish. This method cannot be used in ponds receiving large amounts of flood water or those with muddy water during late winter or early spring months.

Applications of inorganic fertilizer to weed-free ponds during warm months has promoted growths of planktonic algae resulting in sufficient shading to prevent establishment of submersed and emersed species of weeds.

Fish

Carp.—Common carp (*Cyprinus carpio*) in sufficient numbers (400 per ha or 162 per acre or more) roiled the bottom muds and the resulting muddy waters prevented submersed aquatic weed growth by shading.

The Israeli strain of common carp in limited numbers (55 per ha or 22 per acre) has controlled the branched alga *Pithophora* and monofilament algae in ponds.

Tilapia.—*Tilapia mossambica* and *T. nilotica* in sufficient numbers have controlled *Pithophora* and other filamentous algae in ponds during the warmer months.

Tilapia melanopleura in sufficient numbers has controlled filamentous algae and a number of submersed weeds in ponds during the warm months.

Grass Carp (Ctenopharyngodon idella).—In limited trials this species has eliminated filamentous algae, submersed and emersed weeds in small pools.

Mechanical Methods

Cutting.—Emersed type weeds and some submersed species have been controlled by periodic cuttings plus regular fertilization. Normally two years of cutting were necessary to completely eliminate the emersed species. If fertilization was not practiced during this two year cutting period, the emersed weeds were replaced by submersed species. Thus, any cutting operation had to include an adequate fertilization program to be successful.

Deepening of Pond Edge.—In old fertilized ponds, repeated removal of marginal weeds including the soil (by shovel, hoe, or mechanical digger) in the shallow water gradually deepened the marginal water areas and prevented weed reinfestation by elimination of suitable habitat. Thus, a deepened pond edge (no water less than 0.5 m or 19 in. in depth) was recommended as a construction feature for all ponds.

Beating.—Several monofilament forms of filamentous algae have been eliminated from fertilized ponds by beating with a cane pole or agitation of the floating algae masses.

Shading.—*Dyes.*—Partial control of filamentous algae and submersed weeds has been obtained by shading with dyes in the early spring.

However, the dyes used (nigrosin and pontamine green) were unstable in pond waters and the color faded rapidly. Thus, the shading effect was temporary, and repeated (weekly) applications of dyes had to be made to maintain the desired shading effects.

Silt.—Submersed and emersed weeds have been eliminated from ponds in which the water periodically was muddy. Such muddy water shaded the pond bottom and deposited silt on leaves and stems of plants which aided in their control.

Chemical Methods

The techniques presented under this section are those developed or tried at Auburn University, and do not imply that the same results would be obtained by the same concentrations of a given chemical on the same or other plant species in different areas of the world, or in the same areas with different soil, water, or climatic conditions.

Copper sulphate

| | Toxicity to Fish | | |
| | Concentration in ppm Safe To | | |
Formulation	Bass[1]	Bluegill[2]	Fathead[3]
Fine crystals	1 to 3	1 to 3	1

Application range: 0.1 to 1.0 ppm
[1] Bass—*Micropterus salmoides*
[2] Bluegill—*Lepomis macrochirus*
[3] Fathead—*Promelas pimephales*

Periodic treatments at 10 to 14 day intervals at rates of 0.7 to 1.0 kg per surface ha (0.6 to 0.9 lb per acre) and applied to the surface layer of water have effectively controlled the abundance of most blue-green algae (primarily *Microcystis* and *Anabaena*). Minimum rates of application were used to prevent too rapid and too extensive kills of these algae and subsequent death of fish from oxygen depletion because of plant decomposition.

Applications of 1 ppm or more have been fairly effective in controlling *Chara*. However, in certain waters 1 ppm has been toxic to fish.

Applications of 1 ppm have been unpredictable in their effectiveness as an algicide for the filamentous algae *Oedogonium* spp., *Zygnema* spp., *Hydrodictyon* spp., and *Rhizoclonium* spp.

Concentrations of copper sulphate that can be tolerated by fish have been ineffective as an algicide for *Pithophora* spp.

Sodium arsenite

	Toxicity to Fish		
Formulation	Concentrations in ppm Safe To		
	Bluegill	Bass	Fathead
Salt	18	12	8

Application range: 2 to 4 ppm As_2O_3

The concentrations indicated have provided excellent control of many branched, net and monofilament algae species as well as most submersed weed species. Reliable results have been obtained over a wide range of pond conditions.

Concentrations in excess of those indicated have been ineffective on *Chara* spp., *Nitella* spp., slender spikerush, needle rush (*Juncus roemericanus*), and southern watergrass.

Herbicidal activity of sodium arsenite on the filamentous algae *Hydrodictyon* and *Pithophora* has been variable depending upon the stage of growth when the chemical was applied.

Concentrations greater than 4 ppm As_2O_3 have reduced warmwater fish production in treated ponds.

Arsenic was found to accumulate in plankton and in bottom muds. Its accumulation was apparently due to replacement of large amounts of phosphorus by arsenic in both the plankton and muds of treated ponds. Arsenic concentrations in bottom muds were reduced by repeated draining and refilling of ponds.

Fish living in arsenic-treated water accumulated arsenic in scales, fins and in liver tissue, but did not accumulate it in muscular and connective tissue.

2,4-D—2,4-Dichlorophenoxyacetic acid

2,4-D	BE	-do-	butyl ester
-do-	ME	-do-	methyl ester
-do-	IPE	-do-	isopropyl ester
-do-	ICE	-do-	isoacetyl ester
-do-	BEE	-do-	butoxyethanol ester
-do-	EE	-do-	ethyl ester
-do-	PGBEE	-do-	propylene glycol butyl ether ester
-do-	ACA	-do-	acetamide
-do-	AA	-do-	alkanolamine
-do-	DMA	-do-	dimethylamine
-do-	Dacamine	-do-	duomeen-o-amine
-do-	Emulsamine	-do-	oil soluble amine

| | Toxicity to Fish | | | |
| Formulation | Concentration in ppm Safe To | | | |
	Bass	Bluegill	Fathead	Trout[1]
Acid	10	10	10	10
Na salt	400	200	—	112
NH$_4$	400	100	—	—
AA	—	4	—	—
IPE	1	1	1	—
BE	2.5	2.5	—	—
PGBEE	2	2	2	—

Application range: 4.5 to 22 kg per ha (4.0 to 19.6 lb per acre)
[1] Rainbow trout—*Salmo gairdneri*

The lower rate has provided effective control of emergent and marginal weeds by repeated spraying with an ester formulation in diesel fuel as carrier.

Selective control of certain broadleafed emergent and marginal weeds has been obtained by spraying with ester or amine formulation in water carrier. With most species much more effective control has been obtained when 0.25% of a good emulsifying agent was added to the 2,4-D solutions.

Many submerged weeds have been controlled by applications of impregnated granules of either 2,4-D ester or amine at the maximum rate indicated.

Growths of slender spikerush (needle rush) and southern watergrass have been successfully controlled in ponds by draining, removing the rank growth and allowing the weed to start regrowth on the empty pond bottom. This regrowth was then sprayed with an ester formulation in diesel fuel as carrier, and resprayed within one week to control plants missed by the first application. The pond bottom was then flooded.

All formulations have been ineffective as control agents for *Pithophora* spp. as well as for most other forms of algae.

2,4-D acid was found to be rather non-toxic to fish (no kills of warmwater species at 10 ppm). The butyl ester was non-toxic to these same species at 2.5 ppm whereas the propylene glycol butyl ether ester was non-toxic at 2.0 ppm. Solvents and emulsifiers used with the various ester formulations of 2,4-D varied in toxicity to warmwater species of fish, but minimum toxic concentrations were in the range of 5 ppm or less.

Silvex—2-(2,4,5-Trichlorophenoxy) propionic acid

-do-	BEE	butoxyethanol ester
-do-	PGBEE	propylene glycol butyl ether ester
-do-	K	potassium salt

| | Toxicity to Fish | | |
| Formulation | Concentration in ppm Safe To | | |
	Bass	Bluegill	Fathead
Acid	10	10	10
BEE	2	2	2
PGBEE	2	2	2
K	10	10	10

Application range: 4.5 to 22 kg per ha; 1 to 5 ppm (4.0 to 19.6 lb per acre)

Lower rates have provided control of several hard-to-kill emersed and marginal weeds by spray application of an ester formulation in diesel fuel or water solution. The addition of 0.25% of a good emulsifying agent has increased the effectiveness of both spray solution combinations.

At concentrations of 2 to 5 ppm, ester formulation in water has provided excellent control of most submersed and emersed weeds in ponds, and has provided sufficient soil residual to prevent reinfestation for periods up to 3 years. The same results have been obtained using maximum rates of an ester formulation impregnated onto clay granules.

All formulations, even at maximum rates, have been ineffective as control agents for *Pithophora* spp., *Chara* spp., and most other forms of filamentous algae.

2,4,5-T—2,4,5-Trichlorophenoxyacetic acid

| -do- | BEE | -do- | butoxyethanol ester |
| -do- | PGBEE | -do- | propylene glycol butyl ether ester |

| | Toxicity to Fish | | |
| Formulation | Concentration in ppm Safe To | | |
	Bass	Bluegill	Fathead
Acid	10	10	10
BEE	2	2	2
PG	3	3	3

Application range: 4.5 to 9 kg per ha (4.0 to 8.0 lb per acre)

Primarily effective as a control agent for woody marginal species of plants, best results have been obtained by spraying an ester formulation in a diesel fuel carrier with 0.25% of an emulsifying agent added.

It is ineffective as a control agent for most emersed and submersed weeds and algae in ponds, even at the maximum rate.

MCP—2-Methyl-4-chlorophenoxyacetic acid

-do-   AA   -do-   alkylamine salt

	Toxicity to Fish		
	Concentration in ppm Safe To		
Formulation	Bass	Bluegill	Fathead
Acid	10	10	10
AA	710	710	710

Application range: 4.5 to 22 kg per ha (4.0 to 19.6 lb per acre)

Specified rates provided adequate control of marginal and floating weeds by spray application of ester formulation in diesel fuel as carrier. All chlorophenoxyacetic compounds (2,4-D; silvex; 2,4,5-T; MCP) are volatile, the ester formulation being more volatile than the amines. So, due care has to be exercised in spraying these chemicals to avoid damage to desirable surrounding vegetation.

Dichlone—2,3-Dichloro-4-naphthoquinone

	Toxicity to Fish		
	Concentration in ppm Safe To		
Formulation	Bass	Bluegill	Fathead
Wettable powder	0.1	0.1	0.1

Application range: 0.15 ppm

At rates non-toxic to warmwater fish (less than 0.1 ppm), this chemical has been ineffective as a control agent for blue-green algae.

Delrad (rosin amine D-acetate)—Dihydroabietylamine acetate

	Toxicity to Fish		
	Concentration in ppm Safe To		
Formulation	Bass	Bluegill	Fathead
Acetate	0.6	0.6	0.6

Application range: 0.25 to 0.50 ppm

This chemical has given varied results as an algicide in ponds. Under certain conditions a single application has given fair control of the branched algae *Pithophora* spp. and the net algae *Hydrodictyon* spp., and in other situations the chemical was practically inactive. The chemical was fairly rapidly deactivated in pond waters, thus adding to its inability to give reliable algae control. This deactivation was believed to have been due to exposure to ultraviolet light.

This chemical was fairly toxic to warmwater species of fish (maximum safe concentration was 0.6 ppm). When making marginal applications to ponds, young bluegills and bass caught in fairly high concentrations of the chemical have been killed. However, when the chemical was applied by the float technique, three applications of 0.3 ppm in ponds during the period bass spawned had no harmful effect upon the eggs or young fish.

In other pond tests, no difference in fish production in delrad-treated and untreated ponds was detected. This chemical was not toxic to the green or blue-green plankton algae present in treated ponds.

Diquat—1,1′-Ethylene-2,2′-dipyridylium dibromide
-do-   dichloride

| Formulation | Toxicity to Fish | | |
| | Concentration in ppm Safe To | | |
	Bass	Bluegill	Fathead
Dibromide salt	20	20	20
Dichloride salt	20	20	20

Application range: 1 to 4.5 kg per ha; 0.2 to 1.0 ppm (cation) (0.9 to 4.0 lb per acre)

This chemical at rates of 1 to 2 kg (cation)/surface ha (0.9 to 1.8 lb per surface acre) has provided excellent control of many submersed and floating species of weeds. Herbicidal activity has usually been noted within a few hours. This herbicide was adsorbed onto the mud and organic matter on the pond bottom within 14 days.

Diquat cannot be used in muddy or silty waters because of rapid adsorption onto suspended clay and organic particles. It is relatively ineffective on submersed weeds whose leaves and stems are silt- or algae-laden.

This chemical has produced a kill of the blue-green alga, *Anabaena*, within 24 hr at a concentration of 0.5 ppm. It appeared to be very toxic to monofilamentous algae and produced 80% control of *Pithophora* at concentrations of 0.7 ppm.

Small amounts, e.g., 0.2 to 0.5 kg per ha (0.2 to 0.5 lb per acre) of this chemical in combination with 2,4-D, fenac, or similar compounds have produced more rapid and complete herbicidal activity on many submersed weeds than could be obtained with either chemical alone.

Mixed with water and 0.25% wetting agent, diquat at lower rate has provided control of marginal weeds, including grasses, for periods of six to eight weeks.

Paraquat—1,1'-Dimethyl-4,4'-bipyridylium di-(methyl sulphate) dichloride

| | Toxicity to Fish | | |
| | Concentration in ppm Safe To | | |
Formulation	Bass	Bluegill	Fathead
Di-(methyl sulphate) salt	5	10	10
Dichloride salt	5	10	10

Application range: 1 to 4.5 kg per ha (0.9 to 4.0 lb per acre); 0.1 to 1.0 ppm (cation)

This chemical at rates of 1 to 2 kg (cation)/surface ha (0.9 to 1.8 lb per surface acre) has provided more complete control of submersed weeds than diquat, but it is less effective as a control agent for floating weeds.

Herbicidal injury from paraquat is slower in appearing than with diquat, but under adverse conditions of an algae covering on plants, etc., is more certain to be achieved. Likewise the persistence of paraquat in water may be twice as long as for diquat.

As with diquat, this chemical is rapidly adsorbed onto clay and organic particles; thus it cannot be used in muddy or silty waters.

Paraquat has controlled the blue-green algae *Anabaena* and *Pithophora* at concentrations ranging from 0.5 to 1.0 ppm.

Mixtures containing equal parts of diquat and paraquat have provided excellent control of submersed weeds and filamentous algae.

Combinations of paraquat and 2,4-D, fenac, etc., were equally as effective as those using diquat.

Simazine—3-Chloro-4,6-bis-(ethylamino)-S-triazine

| | Toxicity to Fish | | |
| | Concentration in ppm Safe To | | |
Formulation	Bass	Bluegill	Fathead
Wettable powder	40	10	10

Application range: 2 to 11 kg per ha (1.8 to 9.8 lb per acre)

At the higher rate, this chemical controlled all submersed and emersed weeds plus filamentous and plankton algae. Residual simazine was detected in water and soil for 18 months following treatment. Growth of

filamentous and planktonic algae was severely inhibited for two sum-
mers, but for a much shorter period at the lower rate.

This chemical is relatively non-toxic to warmwater species of fish, but
because of its algicidal properties it has interfered with fish production.

Diuron—3-(3,4-Dichlorophenyl)-1,1-dimethylurea

	Toxicity to Fish		
	Concentration in ppm Safe To		
Formulation	Bass	Bluegill	Fathead
Wettable powder	5	10 (laboratory)	15 (laboratory)
		1 (ponds)	1 (ponds)

Application range: 1 to 11 kg per ha (0.9 to 9.8 lb per acre)

Treatments in ponds located on piedmont and coastal plains soils were
effective only at maximum rates on *Pithophora* and many submersed
and emersed weeds. No regrowth of aquatic weeds or filamentous algae
occurred in these ponds for six months.

Diuron at the higher rate seriously interfered with fish production in
experimental ponds. Toxicity to fish became evident 14 to 21 days after
the chemical was applied.

Borascu—Anhydrous polyborate

	Toxicity to Fish		
	Concentration in ppm Safe To		
Formulation	Bass	Bluegill	Fathead
Salt	20	20	20

Application range: 110 to 225 kg per ha (98 to 200 lb per acre)

Pond treatments at rates of 225 kg per ha (200 lb per acre) gave a very
limited degree of control of *Pithophora* and no control of either submersed or emersed weeds.

TCA—Trichloroacetic acid

	Toxicity to Fish		
	Concentration in ppm Safe To		
Formulation	Bass	Bluegill	Fathead
Acid	10 plus	10 plus	10 plus

Application range: 2 to 18 kg per ha (1.8 to 16.0 lb per acre)

The maximum rate provided no control of submersed aquatic weeds. A
combination of minimum rate of TCA with 2,4-D and a good wetting
agent provided effective control of cattail (*Typha* spp.)

Roccal—Alkyldimethylbenzylammonium chloride

	Toxicity to Fish		
	Concentration in ppm Safe To		
Formulation	Bass	Bluegill	Fathead
Salt	1	1.5	1

Application range: 0.25 to 0.5 ppm

This chemical at maximum concentration has given rapid control of monofilamentous algae for short periods (10 to 14 days) when the water temperature was below 21°C (70°F). At higher temperatures the chemical was ineffective as an algicide.

In addition to its algicidal properties, the chemical was an effective bactericide; thus it has been useful in combating certain infections of fish.

Amitrol—3-Amino-1,2,4-triazole

	Toxicity to Fish		
	Concentration in ppm Safe To		
Formulation	Bass	Bluegill	Fathead
Wettable powder	710	710	710

Application range: 1 to 5.5 kg per ha (0.9 to 4.9 lb per acre)

This chemical showed certain herbicidal properties on emersed aquatic weeds, but when used alone provided poor control. In combination with some chlorophenoxyacetic compounds, amitrol has shown certain synergistic properties.

This chemical was relatively non-toxic to fish at fairly high concentrations and has not exhibited any pathological effects on test fish.

Dalapon—2,2-Dichloropropionic acid

	Toxicity to Fish		
	Concentration in ppm Safe To		
Formulation	Bass	Bluegill	Fathead
Wettable powder	1000	80	710

Application range: 11 to 33 kg per ha (9.8 to 29.0 lb per acre)

Maximum rates controlled marginal grasses in empty ponds when applied as a water spray with a watering agent.

AMS (ammata)—Ammonium sulphate

	Toxicity to Fish		
	Concentration in ppm Safe To		
Formulation	Bass	Bluegill	Fathead
Salt	10	10	10

Application range: 11 to 22 kg per ha (9.8 to 19.6 lb per acre)

This chemical at maximum rate controlled certain submersed and marginal weeds in shallow water areas of ponds.

Diesel Fuel.—This material is fairly effective for temporary control of floating weeds when applied as a spray. It also increased the herbicidal activity of chlorophenoxyacetic compounds on emergent growths of aquatic weeds.

Rates of application in excess of 75 liters per ha (20 gal. per acre) have imparted flavors to fish for 4 to 6 weeks.

Diesel fuel application also controlled the air-breathing immature insects inhabiting treated ponds.

Fenac—2,3,6-Trichlorophenylacetic acid

	Toxicity to Fish		
	Concentration in ppm Safe To		
Formulation	Bass	Bluegill	Fathead
Disodium salt	10	10	10

Application range: 5.5 to 11 kg per ha (4.9 to 9.8 lb per acre)

This chemical, which is most effective herbicidally through the root system of plants, has provided partial to complete control of all submersed and emersed weeds in a pond for a period of 18 months or longer. It was most effective at the maximum rate indicated. It was ineffective in the control of *Chara, Nitella,* and filamentous algae.

Due to its lack of algicidal properties, treatments at the highest rate indicated have had no harmful effects on fish production in ponds.

Dichlobenil (casoron)—2,6-Dichlorobenzonitrile

	Toxicity to Fish		
	Concentration in ppm Safe To		
Formulation	Bass	Bluegill	Fathead
Wettable powder	2	2	2

Application range: 2 to 9 kg per ha (1.8 to 7.9 lb per acre)

This is one of the most volatile compounds that has been tested as an aquatic herbicide. Herbicidal effectiveness upon emersed weed growth has been drastically different under laboratory and field conditions. A rate of 2 kg per ha (1.8 lb per acre) applied as a spray completely killed alligator weed within 10 days in the laboratory, while a rate of 9 kg per ha (7.9 lb per acre) only acted as a defoliant in field spray applications.

Limited research to date shows some promise that underwater treatments at the rate of 9 kg per ha (7.9 lb per acre) may control some submerged weeds.

The chemical was relatively non-toxic to fish and apparently decomposed rapidly and produced no effects upon fish production.

Endothal—3,6-Endoxohexohydrophthalic acid
-do-   (TD47, Al-4)   -do-   (dimethylalkylamine) salt

	Toxicity to Fish		
	Concentration in ppm Safe To		
Formulation	Bass	Bluegill	Fathead
Disodium salt	10	10	10
TD47, Al-4	1	1	1

Application range: 5.5 to 11 kg per ha; 1 to 2 ppm (4.9 to 9.8 lb per acre)

Treatments of endothal at maximum rates have provided control of many submersed weeds for periods of 4 to 8 weeks in pools and ponds. This chemical is primarily a contact type; thus, its effects were of a temporary nature if the species present were capable of regrowth from rootstocks. If the species was not capable of rootstock regrowth, then the species could be eliminated by this treatment, provided no seeds were present to repopulate the treated area.

No algicidal properties were noted for this chemical at any rate of application.

The di-(N,N-dimethylalkylamine) salt of endothal enhanced its activity against numerous hard-to-kill water weeds, but this chemical eliminated the fish population wherever it was tested.

Rates of application vary with water quality (hardness, silt turbidity, and temperature), with the species of plant to be controlled, and with the state of the plant's development. If unfamiliar with the action of a chemical, proceed with caution and try lower application rates first. Also, rate of application in ppm is for active ingredients, not for commercial formulation. For aid in computing amounts of chemicals for various treatment purposes, write for a copy of "Treatment Tips" by Dr. Fred. P. Meyer, Fish Farming Experiment Station, Stuttgart, Arkansas 72160.

Chemicals should be mixed with water before applying. For small ponds chemicals are easily applied by slinging them over the water with a dipper, but do not sling directly into the wind. For larger ponds, a venturi boat bailer attached to an outboard motor is excellent for applying chemicals. A valve can be attached to the intake line to regulate the flow of the chemical. Turbulent currents from the boat and motor thoroughly mix the chemicals in the pond.

Ice and snow cover create an additional problem with vegetation by reducing the amount of light that penetrates the water. Plants need light to produce oxygen. Thus, with ice and especially with snow cover the input of oxygen is reduced while the demand is essentially the same. Winterkills due to suffocation of fish can be avoided by having deeper ponds, removing snow, or maintaining an ice-free area by adding water that is above freezing temperature.

It should be pointed out here that only a few chemical compounds have been approved by the U.S. Food and Drug Administration for use on fish whether the treatment is for vegetation control, parasite and disease treatment, or other management purposes. However, at the present time some unapproved chemicals may be used on bait and ornamental fish without objection. If the user is in doubt, prior approval should be received.

pH AND ALKALINITY

Some measurements of the pH of water which you make can be helpful in pond management. But many measurements of pH will mean little unless you can interpret them in relation to factors in the water causing the readings.

If the pH is below 4.0, the water will kill fish. If the pH is above 11.0, the water will kill fish. If the pH is between 4.0 and 5.0, fish may not spawn, their growth rate will be slow, and it will be impossible to get a "bloom" by fertilizing the pond. A pH of 5.0 to 6.0 will affect fish growth and the possibility of getting a bloom to a lesser degree. Waters with a pH of 6.5 to 9.0 are likely to be suited for good fish production. A pH of 10.0 will be unfavorable to fish growth. These are the only positive facts you can determine by measuring pH alone. Any further information will have to come from measuring or interpreting other things which affect the water chemistry of the pond.

For example, you find the water of a pond has a pH of 7.2. Other than in the broad scope given in the preceding paragraph, this measurement alone is meaningless. If the readings were made in early morning, you might be surprised to find the same pond water had a pH of 8.5 by midafternoon if it had been a sunshiny day. Or if a measurement were made again in the same pond before daybreak, the pH might be 6.5. If the pond is less than 4 ha (10 acres) in size, during the summer the pH of the water at a 2 m (6 ft) depth will be lower than at the surface.

On the other hand, if the pH of a pond is 5.0 or lower, you will find that it remains the same regardless of the time of day and over a period of years. This is often also true of pond waters with a pH in the range of 5.0 to 6.0. Obviously, there is something quite different in this pond water from the pond previously described.

Let's look at some of the causes of pH in pond waters—and more particularly—their relationship to fish production. One of the major influences will be the presence or absence of calcium compounds in the water.

Calcium carbonate is one of the more abundant natural minerals. In a nearly pure state it is known as limestone, but it also occurs as particles in many rocks and soils. Calcium carbonate is relatively insoluble in pure water (only 13 parts per million). A saturated solution in pure water will have a pH of 9.3 and is therefore alkaline.

Calcium from calcium carbonate principally gets into solution in ponds and streams through a naturally occurring chemical reaction. First, carbon dioxide from the air (or given off by aquatic plants and animals) goes into solution in the water. Actually this solution of carbon dioxide in water forms a weak acid called carbonic acid:

$$\text{carbon dioxide} \quad + \quad \text{water} \quad \rightleftharpoons \quad \text{carbonic acid}$$
$$CO_2 \qquad\qquad\qquad H_2O \qquad\qquad\qquad H_2CO_3$$

Carbonic acid is very unstable and the double arrows in the preceding equation show that the reaction is easily reversible. This means that in natural waters, a part of the carbon dioxide will always be in the free state. When all of the carbon dioxide that it is possible to get into solution in pure water at normal temperature and air pressure has solubilized, the water will have a pH of about 5.0.

This sets the stage for one of the more important reactions which can happen in pond and stream waters. If a supply of calcium carbonate is available, the carbonic acid reacts with this alkaline material, tending to neutralize both the acidity and alkalinity by forming a new compound, calcium bicarbonate:

$$\text{calcium carbonate} \quad + \quad \text{carbonic acid} \quad \rightleftharpoons \quad \text{calcium bicarbonate}$$
$$CaCO_3 \qquad\qquad\qquad H_2CO_3 \qquad\qquad\qquad Ca(HCO_3)_2$$

Calcium bicarbonate is about 30 times more soluble in water than calcium carbonate so there will be an erosion of limestone or other source of the carbonate as bicarbonate is formed and goes into solution. Since the bones of fish require calcium for growth, this is one way it is made available to the food chain. Most important of all, the calcium bicarbonate serves as a "cupboard" where carbon dioxide can be stored for ready use by aquatic plants.

As shown by the double arrows in the last equation, calcium bicarbonate is unstable and easily breaks up into calcium carbonate and carbonic acid. We have already seen that unstable carbonic acid even more easily breaks up into carbon dioxide and water. Therefore, we have a constant see-saw of reactions taking place in a pond depending upon whether carbon dioxide is being removed by plants through photosynthesis in sunlight or whether there is an excess of carbon dioxide as given off by plants and animals through respiration. In the briefest form we might diagram as follows.

In daytime:
  calcium bicarbonate → calcium carbonate + carbon dioxide ↑

At night or on dark days, the reverse takes place:
  carbon dioxide + calcium carbonate → calcium bicarbonate

We can now see why there is a change in the pH of most pond waters during a 24 hr period. Calcium bicarbonate will cause the water to have a pH of slightly below the neutral point of 7.0. Then, during a day with sunshine, the aquatic plants removing carbon dioxide cause the bicarbonate to break up, leaving calcium carbonate both in solution and precipitated in minute particles in the water. Thus the pH will gradually rise. By midafternoon perhaps most of the bicarbonate has been exhausted and the effect of the calcium carbonate will cause the pH to be around 8.0 to 9.0. During the night the reverse takes place with the formation again of calcium bicarbonate so that before daybreak the pH may be about 6.5.

The dissolved oxygen content of the water varies in a similar manner. During the day, aquatic plants (including the microscopic phytoplankton) are releasing oxygen into the water as a by-product of their manufacture of food.

| chlorophyll of plants | + | carbon dioxide | + | energy from sunlight | → | starches + oxygen ↑ |

By midafternoon on a sunshiny day, the dissolved oxygen content in a pond is usually highest.

At night and on dark cloudy days, there is no oxygen being produced by plants. Both plants and animals use oxygen in their life processes and therefore tend to deplete the supply of dissolved oxygen in a pond if none is being produced.

$$\text{metabolism of plants and animals} \ + \ \text{oxygen} \ \rightarrow \ \text{carbon dioxide}$$

Thus we find that the dissolved oxygen in a pond will be lowest shortly before daybreak (and on cloudy days). This is the reason fish farmers watch their ponds at this time for signs of fish distress indicating an oxygen shortage.

You notice that carbon dioxide is produced in the above. This could be harmful by making the water acid or by being toxic to fish, but we have seen before if the pond has calcium carbonate, much of this carbon dioxide will be tied up in the formation of calcium bicarbonate. We say that such waters are buffered—and materials such as calcium carbonate, calcium bicarbonate, and some others are buffers. They help reduce abrupt or radical changes in the water chemistry which could be harmful to fish. In this age of television commercials lauding the virtues of buffered aspirin, the term "buffer" should be familiar.

So far we have talked only of calcium carbonate and calcium bicarbonate. Depending upon the section of the country, magnesium carbonate may be present in appreciable amounts. For example, dolomitic limestone is rich in magnesium carbonate. Everything we have said about calcium carbonate works the same way with magnesium carbonate. It forms a bicarbonate the same way. In some areas iron carbonate is found, and here again it can form a bicarbonate similar to the way calcium behaved. There are others. For this reason, most water chemists prefer to use the terms "carbonate" and "bicarbonate" without specifying the kind. However, the results of water tests are usually expressed in calcium carbonate equivalent, even though other materials are causing part or all of the effect.

The compounds of calcium, magnesium, and iron are the principal cause of "hard" water. When we measure the total hardness of pond water, we measure the combined amount of all of these compounds in solution. If these compounds are less than 100 parts per million (measured as $CaCO_3$), pond water is regarded as "soft" water and is poorly buffered. Above this there are varying degrees of hardness and in some localities hardness may exceed 200 parts per million.

If you study all of the discussion up to this point, you will have a general knowledge of many of the important factors relative to pond waters throughout the major portion of our country. But, as you might suspect, there are conditions in some localities where not much of the foregoing will be applicable.

In coastal plains ponds, in excavated ponds, and in strip-mined areas, one may encounter waters that are extremely acid, with a pH below 4.0. Extremes as low as a pH of 1.5 occur! Fish, most plants, and phytoplankton cannot live in water that has a pH below 4.0. Characteristically, these waters are very clear and often have a bluish or blue-green hue. It looks pretty, but it is barren.

The cause of the low pH is the presence of strong mineral acids such as sulfuric acid. The acid may be produced through oxidation of sulfides present in the soil and exposed by pond construction. Also, where there is elemental sulfur or sulfur compounds, some types of bacteria use these materials as a source of energy and sulfuric acid is formed as one of the waste products of their metabolism. Pond waters with such influences will continue to have a pH of below 4.0 for an undetermined period of years.

Calcium carbonate and calcium bicarbonate cannot exist at a pH of 4.0. If any calcium is present in these waters it will be in the form of calcium sulfate, which is inert as far as changing the pH. Since there are no plants and no bicarbonates, you can see that there will be no fluctuation in pH during either the day or the night.

These acid ponds have been managed for fish production but this should not be undertaken unless it is realized that the ponds will require an unending series of treatments and close vigilance.

The treatment consists of applying calcium hydroxide (commonly known as builders' lime or slaked lime) at the rate of 36 to 180 kg/ha/m (10 to 50 lb/acre/ft) of water. This neutralizes the sulfuric acid by combining with it to form stable calcium sulfate (gypsum), most of which precipitates to the pond bottom.

calcium hydroxide + sulfuric acid → calcium sulfate↓ + water
$Ca(OH)_2$ H_2SO_4 $CaSO_4$ H_2O

Before fish are stocked in the pond, it does not matter if an excess amount of builder's lime is applied. After all of the acid has reacted, the excess builder's lime will start combining with carbon dioxide going into solution in the water from the air and will change to calcium carbonate, which is precipitated.

calcium hydroxide + carbonic acid → calcium carbonate↓ + water
$Ca(OH)_2$ H_2CO_3 $CaCO_3$ H_2O

After a pH suitable for fish has been established and fish stocked, then applications of builders' lime must be made with care and in conjunction with water tests until the correct amount needed can be derived by experience. If too much is applied, the excess hydroxide in the water will cause the pH to rise above 11.0 and kill the fish before the complete reaction of the excess with carbon dioxide can take place.

In practice, it will be found that the pond will inevitably tend to become acid again. After treatment with the lime, in one month or perhaps six months, the pH slowly decreases into the acid range. At least by the time the pH reaches 5.5, another application of lime should be made. If there is failure to methodically follow through with lime treatments, sooner or later the pH will dip below 4.0 and a fish kill will result.

In coastal plains sections, ponds with a pH of 4.5 to 6.5 are more numerous than those of a neutral or alkaline pH. Here again, the pH remains constant both day and night. These ponds support fish populations but fish growth is slow, spawning of bass and bluegills is unreliable, and there is little response from fertilization.

The acidity of such ponds is usually caused by weak organic acids resulting from the slow drainage of water through humic soils. These ponds are rather easily corrected for better fish production and management by neutralizing the acidity and having an available source of carbonates and bicarbonates.

This can be accomplished by treatment with builders' lime, agricultural lime (calcium carbonate), or basic slag, or combinations of these. The initial correction of ponds with a pH of 4.0 to 5.5 is usually best done with cautious applications of builders' lime. (This material is nearly 2000 times more soluble than agricultural lime). A suitable pH can then probably be maintained by having an abundant supply of calcium carbonate such as from agricultural lime. In these, and the weakly acid ponds with a pH of 5.5 to 7.0, agricultural lime is scattered in shallow water at the rate of 2.5 to 5.0 MT per ha (1 to 2 ST per acre) of pond water. The water currents and wave action in shallow water assist in making the carbonate available. It is not necessary to apply agricultural lime all around the pond—along a part of one side may be sufficient. There is no danger of damaging the pond or harming fish if an excessive amount of agricultural lime is applied.

Basic slag at a rate up to about 560 kg per ha (500 lb per acre) can be used instead of agricultural lime, but it is more expensive. The use of basic slag is better suited to those ponds where access and scattering by a lime truck are not possible because of soft soils, trees, or lack of roads.

Organic stains cause many coastal plains pond waters to have a dark color, but there are some ponds which are murky because of fine particles of organic matter suspended in the water. Some of this matter settles and accumulates as an undesirable sludge on the pond bottom.

A few field trials indicate that it may be possible to control both the murkiness and sludge by using nitrate of soda in the fertilization program for the pond. Bacteria which digest the sludge are favored by the nitrate form of nitrogen. Experience in these field trials has shown that the organic matter has been reduced and also that there is a rise in pH of the water in these normally acid ponds after applications of nitrate of soda. An explanation is that phytoplankton and bacteria use the nitrate portion of the compound, leaving the sodium ions which are basic (alkaline) in effect.

Now we will go completely to the other extreme—there are ponds in which the water has a pH of 10.0 or above. Both fish and fish-food organisms are adversely affected and we have already seen that a pH of 11.0 is the death point for fish.

Strong alkalies as hydroxides are one cause of high pH readings.

These might be introduced by pollution but this is rare in small farm ponds. Sodium carbonate—much more alkaline in effect than calcium carbonate—is present in some well waters. (Pure sodium carbonate is known as "washing soda.") There may be other alkaline materials in well water. When such a well is the primary source of water for a pond, over a period of time evaporation will tend to concentrate these materials and thus increase the pH still further. It may reach the point where fish will be killed.

Almost all of the high pH ponds have wells as their sole source of water. Although it has been mentioned that acid waters are characteristic in the coastal plain, many of the alkaline ponds will also be found there, when artesian water is the source of supply.

Practical experience with ponds of high alkalinity has shown they can be converted into very productive fish ponds. To do this, they are fertilized regularly with sulfate of ammonia instead of conventional fertilizers. Most fertilizer dealers either stock sulfate of ammonia or can obtain it. The cost will be about the same, or less, than mixed fertilizers.

Make applications of sulfate of ammonia at the rate of about 112 kg per surface ha (100 lb per surface acre) of pond water. The frequency of application will usually be about that required for a normal pond fertilization program but will be on a year round basis. There is no substitute for periodic tests of the water's pH. The need for this regular testing of pH will be particularly apparent during the winter months.

Sulfate of ammonia lowers the pH of an alkaline pond through the action of phytoplankton in using the ammonium portion of the compound in their life processes. The sulfate ions are left and, being acid in effect, tend to neutralize part of the alkalinity.

If fish production is increased through the use of sulfate of ammonia and the pond owner becomes negligent in making regular applications when needed, a fish die-off may occur. The pH will rise and the crowded fish crop will be under such stress that they may die at a lower pH than the usual alkaline death point.

Of more importance than pH when considering chemical quality of the water supply is total alkalinity. If total alkalinity is below 25 parts per million (ppm), the water is too poorly buffered to resist great changes in pH values. High carbon dioxide levels could produce acid waters, resulting in fish mortality. The addition of hydrated or agricultural lime will correct this situation, but do not add more than 56 kg per surface ha (50 lb per surface acre) per day if fish are in the pond. At the other extreme, good results cannot be obtained at total alkalinities above 300 ppm.

Liming can be carried out in three different ways: (1) liming water coming into the pond; (2) liming the pond directly; or (3) liming the bottom of a drained pond. When liming, the application should be spread as uniformly as possible either over the water's surface or over the bottom.

Liming water flowing into the pond can be done either manually or by using a lime mill which feeds limestone into the stream.

Liming the pond directly can be done using a boat or a platform drawn by two boats.

Liming the bottom is done in the fall or winter when the pond is dried out. After filling, the pond should not be restocked until the pH has stabilized.

It has been reported that application of liming agents on soils of low calcium content results in greater fish production as well as clearing the water of humic substances. This clearing of the water results in deeper light penetration and greater photosynthesis, resulting in increased zooplankton and phytoplankton production. This in turn results in increasing the benefits of natural foods.

Efforts should be made to obtain pH values ranging from 6.5 to 9.0 before daybreak.

FERTILIZATION

Basically, water and soil fertilization programs are done for the same reasons—to increase the availability of proper foods for animals or plants. However, use of fertilizers, particularly organic fertilizers, is not usually or is seldom done by commercial scale catfish producers where the end product is a food item. This is because of the reputed off-flavors included in the meat. Because fertilization is not essential to successful catfish culture, the producer who is just starting in business is advised to leave his ponds unfertilized.

When monoculture catfish ponds are fertilized, it is usually done to produce a bloom to shade out plants. As a general rule if a Secchi disk is not visible below about 50 cm (18 in.), fertilizer is not necessary.

Fertilization is encouraged with ornamental and bait fish since the animals can obtain a considerable quantity of their nutrients from natural foods in the water and the need for supplemental feeds is decreased. The same consideration applies to nonintensive pond cultures such as exist in many polycultured farm ponds. In these ponds, predominantly bass, bluegills or sunfish, and catfish, usually no supplemental feeds are fed, and if by chance a few catfish at certain times of the year have an off-flavor, there is no significant economic loss. Fertilization of nonintensive, polycultured, recreational-type ponds usually results in greater production. Proper use in some ponds will increase fish production by four or five times. However, in some farm ponds little or no additional fertilization may be needed. Ponds located in used, well managed pastures may require no fertilizers while farm ponds with wooded watersheds may need both lime and fertilizer. The plankton bloom is used as a guide for fertilization.

Fertilizers added to water contribute to the development of microscopic plant organisms (phytoplankton) which serve as food for fish, either directly or indirectly. The small plants may be eaten by microscopic animals (zooplankton) and the zooplanton eaten by the minnows.

Bloom refers to a visible crop of microscopic organisms generally consisting of one-celled algae. Blooms, when formed, increase the turbidity of the water and usually give the water an apparent brownish or greenish coloration. Turbidities thus produced can be used as indicators for the fertility of water.

Just as soil fertilization needs differ for various localities, so will pond fertilization needs differ. Formulas successful in one area will not necessarily produce the same results in another area; therefore, the proper fertilizers for each body of water must be determined.

Because of the variation in pond waters, only general procedures are suggested. Each producer by experimentation can find practices which will give improved results. By keeping records of the fertilizers used, their amounts, and the resulting production, progress can be made.

Fertilization of ponds is done primarily for the purpose of providing nutrients in sufficient quantities for good plant growth. The water may be deficient in certain necessary compounds or later may become deficient from plant use.

Nitrogen, phosphorus, potassium, and calcium are among the chemical elements generally considered as important in plant growth. Little is known of the importance of some of the other elements. A formula of 6-8-4 of an inorganic or so-called complete fertilizer simply refers to the percentage relationship of the nitrogen, phosphorus, and potassium in that order.

Different specific recommendations for the application of fertilizers are given by different authors, but the solution to proper fertilization will rest with the water concerned. Some fisherymen prefer the use of organic fertilizers, some prefer the use of inorganic fertilizers, while most prefer a combination of the two types.

At present it is impossible to make definite recommendations regarding which fertilizer to use. Only by experimentation can you determine which will consistently give best results under given conditions.

Organic Fertilization

Recommendations for the use of organic materials for fertilization include such items as fish meal, cottonseed meal, grasses, leaves, hay and the various kinds of manure. Recommendations for the use of manure range from 35 kg per ha to over 1 MT per ha (30 lb to over 1 ton per acre), depending on the initial fertility of the water and the kind of manure.

Manure containing small quantities of plant fibers will be stronger than equal quantities of grasses, leaves or hay. Also, manures derived from animals such as chickens or hogs which were fed largely on grains will be stronger than manure derived from animals fed upon hay.

Davis (1953) states that organic fertilizer should be applied at the rate of 56 kg per ha (50 lb per acre) at weekly intervals. Should the water become turbid from decay of organic matter, the period between applications should be increased to prevent oxygen depletion.

In Israel greater success has been found using liquid cow manure without any supplemental feed with ponds stocked with common carp, tilapia, silver carp and grass carp. Yields as high as 8 MT per ha (3.6 ST per acre), using daily applications of liquid cow manure, are obtained.

Inorganic Fertilization

Pond fertilization experiments have been conducted quite extensively in Alabama by E.V. Smith and H.S. Swingle but recommendations for that part of the United States may not necessarily produce the same results in other areas.

Fertilized ponds produced 560 to 675 kg of fish per ha (500 to 600 lb per acre). Applications of 112 kg (100 lb) of 6-8-4 fertilizer plus 11 kg (10 lb) of nitrate of soda per ha (per acre) are suggested until the bloom is established. Then only enough fertilizer is added to maintain the bloom.

Combination Fertilization

Barnyard manure should be applied at the rate of 450 to 1125 kg per ha (400 to 1000 lb per acre), depending on pond fertility. Applications should start in the spring 2 or 3 weeks before the pond is stocked with fry so that a heavy bloom is available for the adults or young of plant-eating fish or minnows. Then maintain the bloom by using light applications of commercial fertilizers as often as needed.

A rough index of light penetration determined by the use of a Secchi disk (a metal disk divided into quarters and painted alternately black and white) is considered an index to the fertility of water. This index of fertility is useful only when the turbidity of the water is produced by the presence of plankton and not by the presence of suspended soil particles.

In general, if the Secchi disk becomes invisible in the water at a depth of 8 to 10 in., the bloom in the pond is sufficient for good fish growth.

Experience can permit operators to increase pond fertility and therefore production beyond the standards stated above, but the danger of fish kill is also increased. Only experience can reveal the danger symptoms from overfertilization.

If a green bloom turns brown with dark brown streaks, the algae are dying and oxygen depletion is imminent. Sometimes the addition of 56 to 112 kg per ha (50 to 100 lb per acre) of triple phosphate will stimulate algae growth. Some writers advise using potassium permanganate at the rate of 18.4 to 29.5 kg per ha-m (5 to 8 lb per acre-ft) of water to temporarily relieve oxygen stress. Other individuals feel that potassium permanganate is not beneficial. The best and only sure solution is mechanical aeration or rapidly replacing a portion of the water.

OXYGEN AND TEMPERATURES

In warm weather attention needs to be directed toward meeting oxygen requirements of fish. The warmer the water, the less the amount of dissolved oxygen it is capable of holding. When there are heavy rates of feeding, with some feeds uneaten, decomposition of feces, and dying or dead plankton organisms in water, available oxygen declines rapidly.

Oxygen deficiencies sometimes occur during long periods of cloudy weather when little oxygen is produced by the microscopic plants. It is particularly noticeable immediately before sunrise. Goldfish, although apparently requiring as much oxygen as other minnows, can stand a low oxygen concentration for short periods because of their ability to gulp air at the surface of the water. This characteristic makes possible feeding of higher supplemental feeds to obtain higher production.

When filling ponds from springs, wells or storage ponds, the water should be oxygenated to add oxygen (practically nil in spring and well waters) and to eliminate undesirable gases. The water can be directed into a splash box and then allowed to splash one or more times on boards or other objects before going into the pond. Excess nitrogen can be driven off during this process. The excess nitrogen causes a disease called "pop-eyes." A high iron content can be controlled by pumping the water into a settling pond and allowing the iron oxide to settle out. It can also be precipitated out by running the water over a gravel or rocky bed into a collecting or storage pond.

One problem in maintaining proper oxygen levels in ponds is the building of phytoplankton and then a die off. With the die off oxygen is absorbed and levels for the fish oftentimes are insufficient. Copper sulfate and simazine are widely reported to be good control agents for phytoplankton.

Tucker and Boyd (1978) at the Auburn Department of Fisheries used bi-weekly applications of 0.84 kg per ha (¾ lb per acre) of copper sulfate for control of phytoplankton density and found it to be ineffective. Three periodic applications of simazine reduced density but low dissolved O_2 concentrations followed which resulted in decreased fish yields or poor conversion ratios.

Crance (1963) of the Alabama Department of Conservation used copper sulfate on *Microcystis* and zooplankton in ponds and found no indication that reduction of *Microcystis* after application of 0.8 to 1.0 kg of copper sulfate per surface ha (0.7 to 0.9 lb per surface acre) caused fish to suffer or die from oxygen deficiency.

These findings indicate that extreme care must be exercised in using chemicals to control bloom.

The oxygen solubility of fresh water in equilibrium with air changes with altitude and temperatures (Table 4.1). Temperatures from 4.4° to 23.9°C (40° to 75°F) cover the range usually encountered in streams and

TABLE 4.1. DISSOLVED OXYGEN IN PARTS PER MILLION FOR FRESH WATER IN EQUILIBRIUM WITH AIR

					Elevation in Meters and Feet				
Temperature °C	°F	0 / 0	305 / 1000	610 / 2000	915 / 3000	1220 / 4000	1525 / 5000	1830 / 6000	2135 / 7000	2440 / 8000
4.4	4z13.0	12.5	12.1	11.6	11.2	10.8	10.4	10.0	9.6	
7.2	45	12.1	11.7	11.2	10.8	10.5	10.1	9.7	9.3	9.0
10.0	50	11.3	10.9	10.5	10.1	9.8	9.4	9.1	8.7	8.4
12.8	55	10.6	10.3	9.9	9.5	9.2	8.9	8.5	8.2	7.9
15.6	60	10.0	9.6	9.3	8.9	8.6	8.3	8.0	7.7	7.4
18.3	65	9.4	9.1	8.8	8.4	8.1	7.8	7.5	7.2	7.0
21.1	70	9.0	8.7	8.4	8.0	7.8	7.4	7.2	6.9	6.7
23.9	75	8.6	8.3	8.0	7.7	7.4	7.1	6.8	6.5	6.3
26.7	80	8.2	7.9	7.6	7.3	7.0	6.7	6.4	6.1	6.0

hatcheries. The marked decrease in content of dissolved oxygen with increasing temperature and altitude should be particularly observed.

The saturation level for dissolved oxygen is less in sea water, being approximately 75% of that for the same temperature in fresh water. Intermediate salinities have oxygen saturations that lie almost on a straight-line curve between the limits of fresh water and sea water.

REFERENCES

ALTMAN, R.W. and IRWIN, W.H. (Undated.) Minnow farming in the southwest. Okla. Dep. Wildl. Conserv.

ARCE, R.G. and BOYD, C.E. 1975. Effects of agricultural limestone on water chemistry, phytoplankton productivity, and fish production in soft water ponds. Trans. Am. Fish. Soc. *104* (2) 308-312.

AVAULT, J.W., JR. 1965. Biological weed control with herbivorous fish. Proc. Weed Control Conf. *18*, 590-591.

BARDACH, J.E., RYTHER, J.H., and McLARNEY, W.O. 1972. Aquaculture—The Farming and Husbandry of Freshwater and Marine Organisms. John Wiley & Sons, New York.

BLACKBURN, R.D., LAWRENCE, J.M., and DAVIS, D.E. 1961. Effects of light intensity and quality on the growth of *Elodea densa* and *Heteroanthera dubia*. Weeds *9*, 251-257.

BOYD, C.E. 1976. Fertilizing farm fish ponds. Highlights Agric. Res. *23* (2) 8.

BOYD, C.E. 1976. Water chemistry and plankton in unfertilized ponds in pastures and in woods. Trans. Am. Fish. Soc. *105* (5) 634-636.

BROWN, E.E. 1977. World Fish Farming: Cultivation and Economics. AVI Publishing Co., Westport, Conn.

CRANCE, J.H. 1963. The effects of copper sulfate on *Microcystis* and zooplankton in ponds. Prog. Fish Cult. *25* (4) 198-202.

DAVIS, H.S. 1953. Culture and Diseases of Game Fishes. Univ. of California Press, Berkeley.

DOBIE, J.R., MEECHAN, O.L., and WASHBURN, G.N. 1948. Propagation of minnows and other bait species. U.S. Fish Wildl. Serv. Circ. *12*.

DUPREE, H.K. 1960. The arsenic content of water, plankton, soil and fish from ponds treated with sodium arsenite for weed control. Proc. Conf. Southeast Assoc. Game Comm. *14*, 132-137.

FLICKINGER, S.A. 1971. Pond culture of bait fishes. Colo. State Univ. Coop. Ext. Serv. Bull. *478A*.

HASLER, A.D., BRYNILDSON, O.M., and HELM, W.T. 1951. Improving conditions for fish in brown-water bog lakes by alkalization. J. Wildl. Manage. *15*, 347-352.

HICKLING, C.F. 1962. Fish Culture. Faber and Faber, London.

HUET, M. (Undated.) Textbook of Fish Culture—Breeding and Cultivation of

Fish. Fishing News (Books), Surrey, England.

LAWRENCE, J.M. 1949. Construction of farm fish ponds. Ala. Agric. Exp. Stn. Circ. *95.*

LAWRENCE, J.M. 1952. A new method of applying inorganic fertilizer to farm fish ponds. Prog. Fish Cult. *16* (4) 176-177.

LAWRENCE, J.M. 1958. Recent investigations on the use of sodium arsenite as an algacide and its effects on fish production in ponds. Proc. Conf. Southeast. Assoc. Game Commrs. *11*, 132-137.

LAWRENCE, J.M. 1962. Aquatic herbicide data. U.S. Dep. Agric., Agric. Res. Serv. *231*, 133.

LAWRENCE, J.M. 1964. Plastic pool technique for evaluation of aquatic herbicides. Proc. Southeast Weed Control Conf. *17*, 329-330. (Abstract)

LAWRENCE, J.M. 1966. Aquatic weed control in fish ponds. Proc. World Symp. Warm-water Pond Fish Culture, Rome, May 25–26, F.A.O. Fish. Rep. *44.*

LEITRITZ, E. and LEWIS, R.C. 1976. Trout and salmon culture. Calif. Dep. Fish Game Bull. *164.*

MOORE, G.T. and KELLERMAN, K.F. 1905. Copper as an algacide and disinfectant in water. Bull. Bur. Plant Ind., U.S. Dep. Agric. *76*, 55.

NEELY, W.W. 1968. Some things to know about pond water. U.S. Dep. Agric. Soil Conserv. Serv., Columbia, S.C. Tech. Ref. Paper *14.*

NEESS, J.C. 1948. Development and status of pond fertilization in central Europe. Trans. Am. Fish Soc. *76*, 355-358.

PRATHER, E.E., FIELDING, J.R., JOHNSON, M.C., and SWINGLE, H.S. 1953. Production of bait minnows in the southeast. Ala. Polytech Inst. Agric. Exp. Stn. Circ. *112.*

SMITH, E.R. 1968. Minnow pond construction and water quality. Proc. Commer. Bait Fish Conf., Texas A & M Univ., College Station, Texas, March 19–20.

SMITH, E.V. and SWINGLE, H.S. 1941. Use of fertilizers for controlling several submersed aquatic plants in ponds. Trans. Am. Fish Soc. *71*, 4-101.

SNOW, J.R. 1949. Control of pondweeds with 2,4-D. Prog. Fish Cult. *11* (2) 105-108.

SURBER, E.W. 1931. Sodium arsenite for controlling submersed vegetation in fish ponds. Trans. Am. Fish. Soc. *61*, 142-149.

SURBER, E.W. and MEEHAN, L.L. 1931. Lethal concentrations of arsenic for certain aquatic insects. Trans. Am. Fish. Soc. *61*, 225-239.

SURBER, E.W., MINARIK, C.E., and ENNIS, W.B., JR. 1947. The control of aquatic plants with phenoxyacetic compounds. Prog. Fish Cult. *9* (3) 143-150.

SWINGLE, H.S. 1942. Management of farm fish ponds. Ala. Agric. Exp. Stn. Bull. *254.* (revised)

SWINGLE, H.S. 1945. Improvement of fishing in old ponds. Trans. North

Am. Wildl. Conf., 299-308.

SWINGLE, H.S. 1947. Management of farm fish ponds. Ala. Agric. Exp. Stn. Bull. *254.* (revised)

SWINGLE, H.S. 1947. Experiments on pond fertilization. Ala. Polytech Inst. Agric. Exp. Stn. Bull. *264.*

SWINGLE, H.S. and SMITH, E.V. 1939. Fertilizer for increasing the natural food for fish in ponds. Trans. Am. Fish Soc. *68,* 126-135.

TUCKER, C.S. and BOYD, C.E. 1978. Consequences of periodic applications of copper sulfate and simazine for phytoplankton control in catfish ponds. Trans. Am. Fish Soc. *107*(2) 316-320.

5

Methods

In its entirety this section is rather long—out of necessity. While culturing methods and techniques are presented in some detail, the authors may have omitted some item in which the individual reader is interested. Again, as in several other places in the book, the beginning culturalist is urged to seek practical experience before going into fish raising. No one book or any number of books are a substitute for practical knowledge.

This chapter includes a section on methods of raising catfish, a combined section on trout and salmon because of many overlapping techniques, one for eel, a combined section on bait and goldfish because of overlapping techniques, a section on tropical fish farming, and a brief discussion of polyculture. Where specialized equipment is used, one or more manufacturers are listed to aid in obtaining such equipment. Such listings may not be complete but do serve as a reference point.

All references listed were valuable aids. Among the most important for catfish were: Brown *et al.* (1969), Brown (1977), Lee (1971), and Meyer *et al.* (1973); for trout and salmon: Leitritz and Lewis (1976); for bait and goldfish: Altman and Irwin (undated), Dobie *et al.* (1948, 1956), Flickinger (1971), Guidice (1968), Martin (1955), Prather *et al.* (1953), and Wascho and Clark (1951).

The *Commercial Fish Farmer and Aquaculture News* magazine, 620 E. Sixth, Little Rock, Arkansas 72202, publishes one issue annually which lists different types of equipment used in the catfish industry. Individuals raising catfish should use these issues for securing equipment. Specialized equipment used in trout and salmon rearing is listed under that section of this chapter.

CATFISH (ICTALURUS PUNCTATUS)

Brood Stock

Large catfish usually spawn earlier than smaller ones and produce more eggs. Culturists prefer 1 to 5 kg (2 to 11 lb) broodfish, although stunted catfish can become sexually mature when they weigh only 340 g (0.75 lb). Fish larger than 5 kg (11 lb) are hard to handle. Channel catfish brood stock can be raised and reliably spawned in three years.

Brood stock should be selected prior to feeding. The female ready to spawn should have a well rounded abdomen with the fullness extending past the pelvis to the genital orifice. The ovaries should be palpable and soft, and the genitals swollen and reddish. Less care needs to be used in selecting males. Males with prominent secondary sexual characteristics, such as a heavily muscled head wider than the body, dark pigmentation under the jaw, and large protruding genital papilla, usually have well developed testes. Such males may be used successfully for as many as three times, whereas males with poor secondary sexual characteristics may be capable of fertilizing only one spawn.

Pairing

Successful pairing of fish, which is an essential part of pen and aquarium spawning, depends on the skill of the culturist in sexing and selecting the fish. Sex determination is slightly less important in pond spawning.

Channel catfish fight during the spawning season. Their bites are often deep enough to break the skin, and the resulting wounds may become infected. Fish sometimes die from severe bites. For this reason, special care must be taken to pair fish properly in pens and aquariums. If the female in particular is not ready to spawn the male will fight with her, and in the confinement of an aquarium may inflict enough injuries in 15 to 20 min to kill her. In spawning pens, the condition of the female is not quite as important, since the pen is usually large enough for her to escape from the attacking male. Even so, it is not uncommon to see bite marks on the female in a pen. When fish reproduce in ponds they pair when they are ready to spawn, and fighting is not a serious problem.

Although most culturists prefer or require that the male be slightly larger than the female, biologists who have observed hundreds of pairs spawned in aquariums under different experimental conditions have concluded that males and females of similar size should be used. If the male is considerably larger than the female, spawning is usually successful. If the female is much the larger, however, she usually attacks the male and may not mate with him.

Care and Handling

Immediately after spawning season, brood stock may be placed in a pond at the rate of 150 per 0.40 ha (150 per acre). Although the fish are easily frightened, 90 to 95% of the fish held in small ponds will learn to come for feed within 1 week. If this same number of fish is held in a 0.40 ha (1 acre) or larger pond, 25 or 35% may never come for feed, and consequently will be in poor condition and undesirable as brood stock during the next spawning season. One recommended practice is that of dividing the fish among several ponds to prevent destruction of the entire brood stock by a possible epizootic. Some hatcheries distribute the adult catfish throughout all available ponds during the summer. The scattered fish then need not be fed because sufficient natural food is available. When the ponds are drained in fall and winter, the catfish from several ponds are returned to a single pond and fed until time for spawning. The feeding of brood stock is important because diet quality and quantity largely govern the number and size of eggs, spawning time and general health.

Brood fish that weigh 1 to 1.5 kg (2 to 3 lb) should be stocked at the rate of 136 to 182 kg per 0.40 ha (300 to 400 lb per acre) when additional growth is desired. Larger fish should be stocked at the rate of 364 kg per 0.40 ha (800 lb per acre) if further growth is not wanted. The feeding rate and diet depend partly upon water temperature. Brood fish should be fed 2 to 3% of their body weight on each of 3 or 4 days a week when water temperature is above 13°C (55°F). Since brood stock channel catfish feed sparingly, even when the water temperature is as low as 7°C (45°F), they should be fed only on the warmest days of cold weather periods. When it is very cold, catfish feed better on meat or diets high in animal protein than they do on cereal feeds. Meat diets can be readily utilized by the fish. It is generally accepted that fresh or frozen meat or fish should be included in the brood stock diet.

Number of Eggs Laid

Females weighing 0.5 to 1.8 kg (1 to 4 lb) and in good condition produce about 4000 eggs per 0.45 kg (1 lb) of body weight; larger fish usually yield about 3000 per 0.45 kg (1 lb). Fish in poor condition produce fewer eggs.

Egg Development and Incubation

The number of days required for eggs to reach various developmental stages varies according to temperature. Channel catfish spawn at 21°C (70° to 85°F). The optimum temperature is about 27°C (80°F). The incubation period ranges from 10 days at 21°C (70°F) to 5 days at 29°C (85°F). At incubation temperatures above 29°C (85°F) many deformed fry are produced.

The male channel catfish assumes a position over the eggs after spawning is finished and cares for the eggs during the incubation period. Although the female aerates the eggs during spawning, she is driven away along with other intruders after the male takes possession. The male generally faces in the same direction with his pelvic fins working alternately in a continuous beat. Occasionally he circles away from the eggs and returns. The most striking activity of the male is the vigorous shaking of his body as he presses and packs the eggs with the sides of his pelvic fins in a manner that moves the entire egg mass. Apparently this act helps aerate the developing eggs, especially those deep within the mass, but it may also serve to move the embryos within the eggs.

Many fish farmers who produce fingerlings for sale prefer to collect spawns and hatch them in artificial hatching systems. Good incubation and hatching are obtained at some hatcheries by the paddle wheel method (described later), which simulates the male's agitation of the eggs.

After the eggs hatch, the fry accumulate on the bottom and remain there for about two days before coming to the surface. At this time, the yolk is greatly reduced and the skin pigment is visible. By the third day the fry start to feed and swim actively.

Pond Method of Spawning

In early attempts to induce spawning, tiles, beer kegs, nail kegs, or boxes were partly embedded in the bank of a pond about 0.6 to 1 m (2 or 3 ft) below the water surface. After the brood fish placed in the pond had spawned, the newly hatched fish were removed from the containers and transferred to a clean pond. In later years the egg masses were removed from the pond for incubation and continuous-motion paddles were used to agitate the water and the eggs. This system for hatching catfish eggs is still in use.

The pond method remains essentially unchanged. Brooders are placed in small, usually shallow ponds, ranging up to about 2 m (6 to 7 ft) deep. Equal numbers of males and females are placed in the pond at a stocking rate of 24 to 150 fish per 0.4 ha (1 acre).

Forty liter (10 gal.) milk cans and small drums are popular spawning containers. Ordinarily it is not necessary to provide a spawning receptacle for each pair of fish, since not all fish spawn at the same time. Most culturists allow 2 to 3 receptacles for every 4 pairs of fish. The cans or drums are usually placed with the open end toward the center of the pond. Fish have spawned in containers in water as shallow as 15 cm (6 in.) and as deep as 1.5 m (5 ft). The receptacles can be most easily checked in water no deeper than arm's length.

Frequency of examination of spawning containers depends on the number of brood fish in the pond and the rate at which spawning is progressing. Caution should be used because an attacking male can bite severely. In checking a container the culturist gently raises it to the surface. If this is done quietly and carefully the male is not disturbed. If the water is not clear the container may be slowly tilted and partly emptied until the bottom can be checked for eggs or fry.

Spawns may be handled in different ways by the fish farmer. In the pond method, he may either remove eggs or fry or leave them in the spawning receptacle. Removal of the eggs has several advantages: it minimizes the spread of diseases and parasites from adults to young; provides protection from predation; and may increase the percentage hatched. The main reason for removing fry is to improve control of stocking rates, although it also protects them from predation. Fry or eggs are often removed when spawns are produced for stocking other ponds. Large-scale producers commonly use special brood ponds.

If eggs and fry are left in the pond, the brood stock should be removed with a large-mesh seine. Periodic seining with a small-mesh seine provides information about numbers and growth of fingerlings.

Advantages of the Pond Method.—The pond method is inexpensive because it requires minimal facilities of spawning containers and a pond and does not place demands on the farmer for critically selecting, sexing and pairing his brood stock. The fish in the pond continue to feed and develop until they pair and spawn. If the brood fish are of marginal quality, the pond method is more likely to produce spawn than are the other methods.

Pen Method of Spawning

Pens about 3 m long and 1.5 m wide (10 ft long and 5 ft wide) are commonly used by federal and state hatcheries and by a few private hatcheries. The pens are constructed of wood, wire fencing, or concrete blocks. They may be enclosed on four sides, or the bank of the pond may be used as one side. The sides should be embedded in the pond bottom and should extend at least 30 cm (12 in.) above the water surface to prevent the escape of the fish. Water in the pen should be 0.6 to 1 m (2 to 3 ft) deep.

Location of the spawning receptacle in the pen is not critical, but the opening generally faces the center of the pond and the receptacle should be staked down. Forty liter (10 gal.) milk cans, 45 kg (100 lb) grease drums, and earthenware crocks are popular spawning containers. After spawning, eggs or fry and parent fish may be removed and a new pair placed in the pen. Alternatively, the female may be removed as soon as

FIG. 5.1. PEN SYSTEM OF SPAWNING CATFISH, EGGS TAKEN TO HATCHERY BUILDING

FIG. 5.2. PEN SYSTEM OF SPAWNING CATFISH, FRY RELEASED DIRECTLY INTO REARING POND

an egg mass is found, and the male then allowed to hatch the eggs.

Advantages of the Pen Method.—The pen method has several advantages: (1) it provides close control over the time of spawning, which can be delayed by separating females from males; (2) it offers the advantage of pairing selected individuals; (3) it facilitates removal of spent fish to a separate pond where they can be given special care; (4) it protects the spawning pair from intruding fish; and (5) it allows the use of hormones.

To succeed with the pen method, the culturist must know his fish well enough to be able to pair the right fish at the right time.

Aquarium Method of Spawning

The aquarium method provides still greater control than the pen. A pair of fish is placed in an aquarium with running water and induced to spawn by the injection of hormones. The method capitalizes on limited facilities, use of hormones, and expert brood fish selection. It is an intensive type of culture in which many pairs can be successively spawned in a single aquarium during the breeding season, since eggs are immediately removed to a mechanical hatching trough. The technique is used in federal, state and a few private hatcheries.

In this method only well-developed females nearly ready to spawn should be used. Males need not be injected with hormones, but should be about the same size as the females with which they are paired. If the male attacks the female, he should be removed until after the female has been given 1 to 3 hormone injections. He may then be placed with the female again. Males may be left to attend the eggs in the aquarium; or, preferably, the eggs will be removed to a mechanical hatching trough.

Spawning may be induced in catfish by injecting the female with pituitary material from carp, buffalo fish, flathead catfish or channel catfish. Potency of the pituitaries differs little among these species and is not affected by the date of collection. The total amount of acetone-dried pituitary material required varies widely. However, most females require about 6 mg per 454 g (1 lb)—that is, 3 injections of 2 mg per 454 g of body weight at 24 to 48 hr intervals. Most fish begin spawning within 16 to 24 hr after the injection.

Human chorionic gonadotropin has been used successfully at a dosage of about 800 international units (IU) per 454 g. A single injection is usually sufficient.

Fish spawned by the hormone method are not particularly disturbed by people moving around the area.

Advantages of the Aquarium Method.—The aquarium method has several advantages. (1) Spawn can be obtained at a convenient time. The hormone injections eliminate such environmental variables as spawning areas, light, temperature, and other climatic conditions. (2) The spawning period can be altered within reasonable limits and total spawn-taking time can be reduced. (3) Fish that will not spawn naturally can sometimes be induced to spawn. (4) Culture ponds can be stocked with fry of uniform age and size. (5) Disease transmission from brood stock to offspring as well as predation by adults is minimized.

Controlling Spawning Time

The date and length of the spawning season for channel catfish vary from year to year and among localities. In various natural waters the season may begin as early as April and end as late as August.

In June and early July fish in pens occasionally spawn for a few days and then completely stop. Raising the water level rapidly 5 to 8 cm (2 to 3 in.) will sometimes cause spawning to resume immediately.

Some farmers inject brood females with human chorionic gonadotropin before transferring them to the spawning pond. Others have advanced the spawning time about two weeks by taking advantage of the warmer water of small, shallow brood ponds.

Spawning can be delayed 20 to 30 days by keeping the sexes separated. It may also be delayed by holding the fish at water temperatures of 17° to 18°C (65° to 66°F) during May, June and July.

Hatching Eggs

For egg incubation, temperatures below 18°C and above 29°C (65° and 85°F) should be avoided. Temperatures from 26° to 28°C (78° to 82°F) are considered optimal. In this range the eggs will hatch in about six days.

Color is an important index of the condition and stage of catfish eggs. Under proper conditions the yellow eggs turn pink as the embryo develops and establishes its blood supply. Unfertilized or dead eggs turn white and enlarge.

All hatching devices must provide sufficient agitation to supply the entire egg mass with oxygenated water of a suitable temperature. When eggs are hatched in troughs, they are agitated with paddles driven by an electric motor or a water wheel. The agitation should be sufficient to move the whole spawn, but not enough to throw eggs out of the holding baskets. If well water is used it must be aerated and of suitable temperature and quality. For example, water with a high iron content is not considered desirable. Gravity-flow water should be used if available because this system is not likely to fail.

Trough hatching systems may be constructed from a variety of materials. Aluminum is commonly used, but wood or steel serves equally well. The shaft is fitted with a pulley at one end and it is belt driven by an electric motor (frequently 1/2 hp) at a preferred speed of 30 rpm. Combinations of pulley sizes or a variable-speed gear box may be used to deliver the desired speed of rotation.

Spawn baskets are constructed of 0.25 in. hardware cloth. Each basket is divided into four equal sections and fitted with wire hooks so that it can be hung on the sides of the trough with the top edge 2.5 cm (1 in.)

above the water. A flow of about 10 liters (2.5 gal.) per min of well-aerated water should be provided.

It is important that spawns in each section of the trough be of the same age, because the mixing of spawns of different ages may prevent the use of prophylactic treatments.

A flush treatment of malachite green at about 2 ppm may be introduced at the head of the trough once or twice a day if needed to control fungus. Fungi grow on dead eggs and spread to living ones, eventually destroying the whole spawn. This chemical is not to be applied within 24 hr of hatching. If fry are present it will kill them.

Rearing Fry

To remove fry from the hatching trough the culturist simply siphons them from the trough with a hose into a washtub or pail. To remove fry from a pond spawning receptacle he first removes the male, usually by frightening him away, then lifts the spawning container to the surface and carefully pours out part of the water. The remaining water with the fry can then be emptied into a floating tub that contains 2.5 to 5 cm (1 to 2 in.) of water. Fry should be counted and then moved to either a rearing trough or a pond.

Channel catfish fingerlings are reared in either ponds or troughs. In ponds, low-cost pelleted fish feed may be used, but in troughs a more expensive, balanced feed is required. In the two environments different methods and techniques are used to stock, feed, and harvest the fish. Some farmers prefer to start the fry in rearing troughs and transfer the fingerlings to ponds after they are actively feeding. The choice of method depends on facilities and labor available and on the number and size of fingerlings to be produced.

Regardless of the rearing method selected, attention should be given to the water supply. In the trough rearing method the incoming water

FIG. 5.3. REARING PONDS FOR CATFISH FRY

should range between 24° and 29°C (75° and 85°F) and contain not less than 6 ppm of dissolved oxygen. Factors such as pH, hardness, and dissolved iron content influence production of fish in troughs that are supplied with well water. When water from ponds is used for trough culture, these conditions may be disregarded. Then, however, a saran screen or sand filter between pond and trough is desirable. In pond rearing a major problem is fry-eating insects and fish. Predatory fish can be controlled by filling the pond with fish-free well water or by using a fine-mesh screen to filter water from other sources. Fish-eating insects can be controlled by treating weekly with oil or kerosene until the fish are about 3.8 cm (1.5 in.) long.

Trough culture of fingerling catfish begins with yolk-sac fry from the hatching trough or from spawning containers in the ponds. It is a good practice if time and facilities permit because it gives the culturist complete control of the small fish. When fry about 2 cm (0.75 in.) long are stocked in rearing ponds, 60 to 90% can be harvested the following fall.

Techniques for feeding catfish fry are extremely important, particularly when the fry are learning to feed. For about the first 3 or 4 days after hatching, fry subsist on yolk and remain on the bottom of the pond or trough. After they have absorbed the yolk sac they become known as "swim-up" fry. When they are seen swimming along the sides and surface of the trough in search of food they must be fed at once. Fry that do not learn to feed during the first few days after absorption of the yolk sac will die.

Trough Method.—Rearing troughs may be made of wood, metal, fiberglass, or plastic. Typical troughs are 2.5 to 3.0 m long, 30 cm deep and 20 to 50 cm wide (8 to 10 ft long, 1 ft deep, and i to 20 in. wide). Each trough must be supplied with running water and equipped with a drain and a standpipe. The fry from one or two spawns are put into each trough. A flow of about 20 liters (5 gal.) of fresh oxygenated water per min is sufficient. Standpipes are screened so the fry are not washed over the standpipe and down the drain.

Fry begin to feed shortly after the yolk sac is absorbed and the fish begin to develop a grayish color. This usually occurs at three days of age. It is mandatory that suitable feed be available at this time. Channel catfish fry eat a variety of feeds. Feeding frequencies and particle size are important considerations. Young fry should be fed every 2 to 4 hr around the clock for the first week. Thereafter the fry should be fed about four times a day. Diets for channel catfish are now available commercially in the USA. Fry may be raised to the fingerling stage in troughs or may be moved to a rearing pond at any time.

Pond Method.—Although the area of rearing ponds for channel catfish varies from 1000 m² to 2 ha (0.1 to 5 acres) and larger, it is usually about 0.4 ha (1 acre).

Predatory insects are often a problem in the pond culture of fry. If a pond is not filled until immediately before it is stocked with fry, establishment of the insects is prevented. If water has been standing in the pond for several days, or if surface water is used, it should be treated with a nonresidual insecticide 2 to 3 days before the pond is stocked. The operator should use extreme care because insecticides are dangerous to man.

Fry can either be released directly into the open pond or held for the first few days in floating cages made of screen or in a wooden frame with a screen bottom. They are then protected and can be fed during this vulnerable period. If a pond has a basin, fry may be placed in it and the rest of the pond kept dry. As the fry get larger the pond is gradually filled.

In pond culture either sac fry, swim-up fry or feeding fingerlings may be stocked. Stock rates vary, depending on the fingerling size desired. If the fish are to be harvested at 5 to 10 cm (2 to 4 in.) lengths after about 120 days, they are stocked at the rate of 100,000 to 150,000 per 0.4 ha (1 acre). If 20 cm (8 in.) fish are desired, the stocking rate should be reduced to 14,000 to 20,000 per 0.4 ha. A combination of these methods may be used. The fish can be stocked at the maximum rate initially and then partially harvested for sale or transfer to other ponds when they reach the 5 to 10 cm (2 to 4 in.) size. Young fish are fed daily along most of the shoreline at a rate of about 4 to 5% of their body weight at each feeding.

Table 5.1 presents a general guide that can be used to determine the number of fingerlings that may be produced per ha or per acre. Table 5.2 gives the weights of fingerlings by length.

Fingerlings

A food fish producer must decide whether to grow fingerlings in conjunction with the food fish or to purchase them from growers specializing

TABLE 5.1. RATE OF STOCKING REARING PONDS WITH FRY BASED ON THE SIZES OF FINGERLINGS

Sizes of Fingerlings (Length in 120 Days)		Rate of Stocking Fry	
cm	in.	per ha	per Acre
5.0	2	341,000	138,000
7.5	3	286,500	116,000
10.0	4	235,000	95,000
12.5	5	180,000	73,000
15.0	6	131,000	53,000
17.5	7	74,000	30,000
20.0	8	24,700	10,000

Source: Mitchell and Usay (1969).

TABLE 5.2. THE AVERAGE WEIGHT OF FINGERLINGS BY LENGTH AND NUMBER PER KG AND LB

| Length | | Avg. Weight per 1000 Fish | | No. of Fish per | |
cm	in.	kg	lb	kg	lb
2.5	1	0.6	1.3	1688.9	767.7
5.0	2	1.6	3.5	628.5	285.7
7.5	3	4.5	10.0	220.0	100.0
10.0	4	9.0	20.0	110.0	50.0
12.5	5	14.5	32.0	68.4	31.1
15.0	6	27.3	60.0	37.4	17.0
17.5	7	42.3	93.0	23.8	10.8
20.0	8	50.9	112.0	19.8	9.0
22.5	9	81.8	180.0	12.1	5.5
25.0	10	149.1	328.0	6.8	3.1

in fingerling production. It is impossible to suggest which source of fingerlings is better, but there are several criteria which should be considered before a decision is made. Regardless of the choice, good fingerlings must be available. Factors to consider include (1) size of fish farm, (2) knowledge of fingerling culture, (3) availability of fingerlings, (4) comparative cost, (5) availability of labor, (6) availability of facilities and equipment, and (7) personal preference.

The larger fish farmers usually find it more convenient and economical to produce their own fingerlings rather than to purchase them from other farmers. Smaller farmers ordinarily purchase fingerlings from hatcherymen or specialized fingerling producers. Some small farms may be more profitable if converted entirely to fingerling production rather than attempting to compete with larger food fish producers.

Fingerling producers must have considerable knowledge of the biology and culture of catfish, including selection and care of broodstock, spawning, hatching, feeding, and disease and parasite control. A knowledge of all these areas requires considerable study and experience with fish. It is usually preferable for inexperienced producers who are just beginning food fish production to purchase fingerlings rather than attempt to produce them.

Food fish farmers must have good quality fingerlings available at a reasonable price. Farmers raising their own fingerlings are aware of the health problems that have been encountered. Treatment for diseases and parasites can be administered as needed. Farmers purchasing fingerlings should deal with reputable fingerling producers who will truthfully state the conditions under which the fingerlings have been grown.

Fingerlings are grown for profit. Farmers need to analyze the local situation and decide accordingly. A short supply of fingerlings may result in reduced availability and increased prices. An abundance of fingerlings means that prices may be low, possibly even below the cost of production. If it is cheaper to grow fingerlings than to purchase them, food fish farmers should consider producing the quantity needed. The cost of ponds for holding, spawning, and rearing must be included when making

FIG. 5.4. CATFISH FINGERLINGS HARVESTED FROM POND FOR TRANSFER TO HOLDING TANKS OR DIRECT TO FOOD FISH PRODUCER

FIG. 5.5. CATFISH FINGERLINGS BEING TRANSFERRED FROM HOLDING TANK FOR SHIPMENT TO FOOD FISH PRODUCER

a decision. The larger farmers usually find it more economical to produce the necessary fingerlings and to perhaps produce a few extra for sale to other farmers.

Marketable fingerlings can usually be produced in 4 to 8 months. This is about the amount of time between spawning in late spring and early summer and stocking food fish production ponds in winter and early spring. The amount of time required to produce marketable, or stocking-size, fingerlings will depend upon the environmental conditions in which the fish are grown. The number of days required primarily depends upon four interrelated factors:

(1) Stocking rate of fry.
(2) Length of fingerling desired.
(3) Feeding and management.
(4) Length of growing season.

Rearing Fingerlings to Market Size

A large healthy fingerling, a good environment and conscientious feeding program are necessary for a profitable food fish production program. If market-size fish are to be produced in one growing season, fingerlings 15 to 20 cm (6 to 8 in.) long or longer must be stocked. Such fingerlings will weigh at least 454 g (1 lb) after about 210 days if properly cared for.

Time of stocking is not as critical as some believe. A pond should be stocked whenever it is ready to receive fish. A 5 cm (2 in.) fish stocked in July will be only 24 to 25 cm (9 to 10 in.) long by November and must be reared to market size the next summer.

The poorest stocking months are December and January because of the low water temperature. Fish feed least at this time of the year and sometimes do not resume feeding readily after they are moved. The fish that die after stocking at this time of year may never be seen.

Fish should be fed during winter, but the feeding rate should be reduced as the water cools. Self-feeders are useful under such conditions. At low water temperatures fish move slowly and do not seek out feed as they do when the water is warm. They also consume less food at each feeding and digest it much more slowly.

It is very important that fingerlings start feeding immediately after they are stocked in a pond. Well-fed healthy fish are more resistant than others to parasites, disease, and predators and reach marketable size sooner. Fish soon learn to feed as food is provided along the entire edge of the pond on the day after they are stocked.

Once the fish start eating a good feeding program should be initiated and followed. Food allowances are 3 to 6% per day of the estimated weight of fish in the pond; rates are lower (1.5 to 2%) during unusually hot or cold periods. Feed is offered in the early morning and late afternoon in summer but only in late afternoon during the cooler seasons. If sinking pellets are used it is desirable to scatter them along the shallow areas where feeding activity can be observed. Floating feeds may be scattered over the entire surface of the pond. Feeding activity is a good index of the well-being of the fish; rapid and vigorous consumption of the food suggests good environmental conditions and good health.

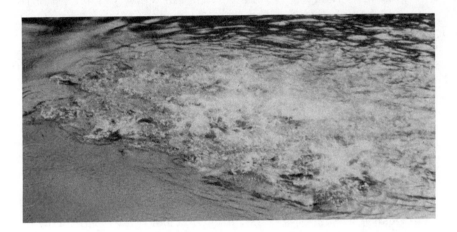

FIG. 5.6. CATFISH IN FEEDING FRENZY

Catfish are being raised successfully in water from many sources. Well water is best, but other uncontaminated supplies such as clean streams or springs are acceptable if they are free of fish and disease organisms.

Oxygen depletion is the greatest fish farming problem. Most oxygen depletion kills are preceded by a phytoplankton die-off and decay. This situation is aggravated by the decomposition of uneaten fish feeds and fecal waste. When excessively thick blooms of phytoplankton (algae) occur, it is desirable to add fresh water to the pond. Feed should be reduced in amount or withheld entirely until the condition has improved.

Catfish culture is also influenced by aquatic vegetation. Although rooted aquatic plants and filamentous algae are not as troublesome in pond rearing of catfish as they are in some other forms of fish culture, they should be removed if they appear. Manual removal and some chemical controls are feasible.

When treating a pond with chemicals, the culturist should be aware that the chemical may be toxic to the fish, and that killing too much vegetation at one time can result in an oxygen-depletion mortality. Ponds should be carefully checked for low oxygen each day for 7 to 10 days after applications of herbicides.

Harvesting Fingerlings

The problem of harvesting fingerlings is not nearly as great as harvesting food fish because the size of rearing ponds is usually considerably less than growing ponds. Rearing ponds are often 0.4 ha (1 acre) or less and rarely more than 2 ha (5 acres) in size. The problem of harvesting is virtually eliminated with trough reared fingerlings.

Growing ponds should be stocked with healthy fingerlings. Fingerling harvesting procedures should be used that will ensure good quality fingerlings. Fingerlings should not be held out of water for more than a few seconds. Fingerlings which "slip their slime," or become dry, are very susceptible to diseases and parasites even if returned to water before death. The seines and equipment used in harvesting should be constructed to minimize bruises and abrasions. Harvesting procedures should include a treatment to aid in disease prevention. It is also preferable to harvest fingerlings in the early morning while the water and air are relatively cool. Persons involved in harvesting should avoid excessively muddying the water.

Most pond reared fingerlings are harvested by seining. A seine is a large rectangular net with floats at the top and weights at the bottom. Fingerlings are concentrated in a small area of water by pulling a seine through the water across a pond. Rearing ponds are usually relatively shallow, thus making it possible for a man at each end of the seine to wade out into the water to pull the seine. The size of the mesh in the seine should prevent the escape of fingerlings through the mesh. Common mesh sizes for harvesting fingerlings are 6 to 12 mm (¼ to ½ in.). The width of the seine should be sufficient to prevent the escape of fingerlings below the seine. Seines for harvesting fingerlings may be constructed of lighter materials than seines for harvesting food fish. It is frequently a good idea to seine around the feeding areas to partially harvest the fingerlings. The fingerlings are dipped out with dip nets or other devices. The harvested fingerlings should be immediately placed in storage or hauling facilities which have adequate aeration.

The fingerlings in a rearing pond may be completely harvested by draining. It is suggested that partial harvests be made before lowering the water level. Seining around the feeding areas may yield 75% of a fingerling crop.

Harvesting usually includes grading fingerlings. Grading is done to ensure uniformity of size when growing ponds are stocked. Grading into lots of uniform sizes simplifies the estimation of numbers. Various kinds of grading devices are used. Spacing between bars may be used to retain fish larger than a certain size and permit the smaller fish to pass through. The grader sizes for channel catfish are shown in Table 5.3.

TABLE 5.3. GRADER SIZES FOR CHANNEL CATFISH

Distance Between Grader Bars	Size of Fingerling Retained
10 ½ mm (27/64 in.)	7.5 cm (3 in.) length or larger
12 ½ mm (32/64 in.)	10.0 cm (4 in.) length or larger
15 ½ mm (40/64 in.)	12.5 cm (5 in.) length or larger
18 ¾ mm (48/64 in.)	15.0 cm (6 in.) length or larger

Source: Meyer *et al.* (1970).

Harvesting Food Fish

Harvesting catfish raised in tanks or cages is relatively easy; they are lifted out with a dip net, loaded into transfer containers, weighed and placed in the truck for transportation.

Raceways can be harvested using a seine. Since the raceway segments are usually not over 6 m wide (20 ft), a seine about twice this length is used. The seine is dragged the length of the raceway segment, crowding the fish against the concrete headwall where they are dipped out. A second alternative is to drain the segment. With this system a fish trap is placed below the headwall and the fish are trapped as they leave the raceway segment.

Pond harvesting is much more complex. The USA has some of the largest, if not the largest, fish culturing ponds in the world. A few ponds are as large as 65 ha (160 acres). Ponds of 16 ha (40 acres) are commonplace.

In current practice extensive preparations are usually required to ready ponds for harvesting the fish. Draining a pond and preparing for final harvesting operations by means of pumps or ditches require several days. During draining, some of the fish may die from being concentrated in a small volume of water with inadequate oxygen. Also, valuable amounts of water are wasted. Coordinating production with market demands is difficult. For example, to harvest and transport to market over 20 MT per day from one pond may not be possible. Hence, for ponds larger than 8 ha (20 acres) the volume of fish in the ponds exceeds the capacity of the market. Since each truckload is about 5 MT (5.5 ST), an 8 ha (20 acre) pond requires the movement of 4 truckloads daily. This is not always possible. Harvesting the fish often must be limited to periods of the year when it does not conflict with other activities on the farm.

The smaller ponds below 16 ha (40 acres) are usually harvested using a haul seine. Ponds of about 16 ha (40 acres) and larger have necessitated innovative harvesting techniques. A mechanical haul seine is commonly used.

Because larger ponds require such a long and heavy seine, a boat or barge is used to set the net around the pond. The seine is retrieved by using a mechanized wire-cable puller or a rope-line puller. The mechanized puller winds the cable or rope onto a double drum winch. As the fish are crowded into a harvesting area, they are dipped out by dip nets or crowded into a fish loading conveyor. When more than 8 ha (20 acres) are to be harvested, even these mechanized aids are not the sole answer due to the inability to sell over 20 MT (22 ST) per day. Thus, the fisheries experiment stations at Stuttgart and Kelso, Arkansas, have developed a new technique.

This new technique consists of having a "fish holding bag" or live car in the middle of the chute formed by the seine net. As the fish are crowded they swim into the live car. Each live car holds about 5 MT (5.5 ST) of fish. When a live car is filled it is detached and an empty live car attached (Fig. 5.7). The filled live car is then floated to deeper water to give the fish more water room and to get them out of the muddy

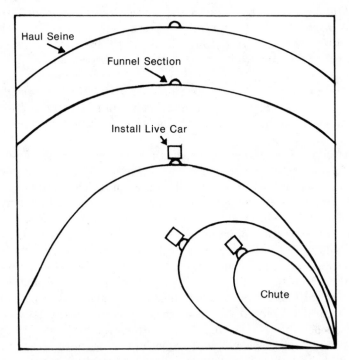

FIG. 5.7. SCHEMATIC DRAWING OF HAUL SEINING PROCEDURE USED WITH LIVE CAR FOR HARVESTING CATFISH

harvesting water. The colder the water, the longer the fish can be held in the live car. If the fish become distressed, aeration can be started, or the fish can be released back into the pond in case of an emergency.

When the pond is harvested and the fish are in the live cars, each live car can be positioned near the bank and a mechanized brailer used to dip the fish out onto a sorting table positioned in the water. The small fish of

Courtesy of Fish Farming Exp. Stn., Stuttgart, Ark.

FIG. 5.8. MECHANICAL REELING IN OF SEINE IN LARGE CATFISH POND

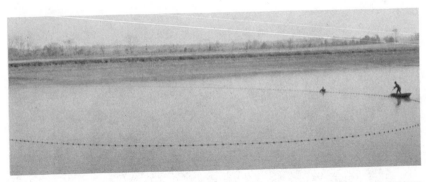

FIG. 5.9. SEINE BEING REELED BY MECHANICAL REEL IN LARGE CATFISH POND

less than 340 g (0.75 lb) can be released back into the pond or transported to another finishing pond. The fish between 340 and 1135 g (0.75 to 2.5 lb) are loaded into the waiting trucks for transport to the processing plant or other market. The larger fish over 1135 g (2.5 lb) are sorted out for use as brood fish or are sold separately for fish fillets.

Courtesy of Fish Farming Exp. Stn., Stuttgart, Ark.

FIG. 5.10. THREE LIVE CARS FILLED WITH CATFISH DURING SEINING OF LARGE POND

Courtesy of Fish Farming Exp. Stn., Kelso, Ark.

FIG. 5.11. BAILING CATFISH FROM LIVE CAR

As many as 90% of the fish in a pond can be captured in a single haul. The remaining fish are permitted to stay in the pond, which is then restocked with fingerlings.

FIG. 5.12. MECHANICAL LOADING OF HAULING TRUCK FROM CATFISH LIVE CARS

Courtesy of Fish Farming Exp. Stn., Kelso, Ark.

Some farmers use trapping devices which may harvest several tons per day. Over a period of several weeks as many as 80 to 90% of the fish may be harvested, and trapping ceases for that year. Use of this system permits a large pond to be harvested slowly so that the market is not depressed by too large a quantity of fish being harvested at one time. Several commercial trapping devices have been invented which, when baited with food, serve as efficient harvesters.

TROUT AND SALMON (SALMONOIDS)

Because many of the culturing techniques are similar for both trout and salmon, they have been combined in this section. This section relies heavily on Leitritz and Lewis (1976). Their booklet is an excellent discourse on trout and salmon cultural methods.

Courtesy of Fish Farming Exp. Stn., Kelso, Ark.

FIG. 5.13. BAILING CATFISH DIRECTLY INTO HAULING TRUCK

Culturing of trout in the USA generally embraces rainbow trout *(Salmo gairdneri)*, brown trout *(Salmo trutta)*, and brook trout *(Salvelinus fontinalis)*. Nearly all the trout cultured for food are rainbows. Salmon culturing centers on coho or silver salmon *(Oncorhynchus kisutch)* with some culturing of chum *(Oncorhynchus keta)* and chinook or king *(Oncorhynchus tshawytscha)*.

Brood Stock

Trout brood stock are commonly raised. It is essential that these fish receive an adequate diet and be handled gently. A female trout or salmon heavily laden with eggs cannot withstand rough treatment. Salmon brood stock are usually from wild fish. In any case, to protect the fish or the egg quality the fish should be dipped out only one or two at a time. Some brood stock are ready for spawning at an earlier age than others. This saves the cost of maintaining brood stock for an extra year. Select-

Courtesy of Fish Farming Exp. Stn., Kelso, Ark.

FIG. 5.14. FOOD SIZED CATFISH READY FOR PROCESSING

ing for large eggs gives the fry a better start and higher survival. Selective breeding has developed strains of rainbow trout which spawn in every month but May and June. Unlike trout, the salmon eggs are obtained from October through December. In general, egg size depends on the size and age of the parent fish, the larger specimens producing more and larger eggs. In California, spring spawning trout brood stock at 2 years of age averaged 15 eggs per g with 102 g of eggs (427 eggs per oz with 3.6 oz of eggs) or a total of 1553 eggs. At 3 years of age the same female averaged 8.9 eggs per g with 256 g of eggs (254 eggs per oz with 9 oz of eggs) or a total of 2210 eggs. The size of eggs increased 40% while the number increased 42%.

Fertilization

Ripe female trout should always be held tail down with the head high. No forceable attempt should be made to expel more eggs than those that flow under gentle pressure. Several manipulations may be required to obtain the eggs. The eggs should be permitted to flow or roll along the oviduct toward the vent. Since the Pacific salmon all die after spawning, the incision method is used. With this method nearly all the eggs can be obtained.

FIG. 5.15. POTENTIAL TROUT WATER FLOWING FROM AQUIFER, SNAKE RIVER CANYON, IDAHO

Usually the female is stripped first and then the milt is added and the eggs gently stirred. If there are no broken eggs, another female is stripped and then another male into the same container. By using a different male each time there is more insurance of fertilization. It is stated that one drop of sperm is enough to fertilize 10,000 eggs. Since there is a limit on the time that the eggs and sperm remain viable, correct timing in spawning is important. Under no conditions should the eggs or sperm be exposed to water for more than 2 min prior to fertilization. Less than 1 min is recommended.

FIG. 5.16.   LARGE SCALE TROUT OPERATION, SNAKE RIVER CANYON, IDAHO

FIG. 5.17.   CLOSE-UP OF LARGE SCALE TROUT OPERATION, SHOWING RACE-WAYS AND FEEDING AND HARVESTING ROADS, SNAKE RIVER CANYON, IDAHO

When eggs are broken during spawntaking, the process of fertilization is hampered if not completely stopped. Broken eggs in the spawning pan will appear as a white creamy substance resembling sperm. This is the albumen from the broken eggs and it seals off the micropyles in the eggs and prevents the sperm from entering unless washed off immediately. When albumen appears in the spawning pan it should be washed off immediately, the sperm added and the pan emptied of eggs before any more are added.

FIG. 5.18. SMALL SCALE TROUT FARM IN EASTERN UNITED STATES.
Adapted to supplying local fee fish–out operations and restaurants.

When eggs are taken by an inexperienced spawntaker or the eggs are extremely soft shelled with many broken eggs, fertilization can be improved with a salt solution. This can be done by dissolving 28 g of common salt in 4.2 liters of water (1 oz in 1 gal.) and adding the solution to the empty pan to fully cover all eggs to be taken.

During spawning the eggs and milt should not be exposed to light, especially direct rays of the sun. The eggs are killed by direct exposure to sunlight for even a few minutes. Incandescent lights are recommended.

Newly taken or green eggs may be measured and handled as soon as they become water hardened and for approximately 48 hr, depending on water temperature.

After the eye spots appear, the eggs can be handled without any great danger of injury as long as they are not subjected to unusual shock, freezing or dehydration. All eggs at all times are subject to mechanical shock.

After the eye spots are clearly visible, the eggs can be shocked so that the infertile eggs turn white and can be separated from the fertile ones. Actually, this amounts to nothing more than agitating the eggs enough to rupture the yolk membrane in the infertile eggs, which causes them to turn white. Eggs can be shocked by stirring them in the basket with the bare hand. However, this may injure some of the good eggs by striking them against the sides and bottom of the basket. By far the safest and best way to shock eggs is to siphon them through a 1.4 m (4 ½ ft) length of common garden hose from an egg basket in a hatchery trough to a pail on the floor. The pail should be perforated near the top edge, to allow the water to run off without carrying away the eggs. The distance from the end of the hose to the water surface in the pail will determine the degree of shock being given and can be varied until the desired results are obtained.

Separating Eggs

Pipette Method.—Unless a fungicide is used to prevent dead eggs from fungusing and the fungus spreading to the fertile ones, it is necessary to remove the dead eggs. This is called egg picking.

Various methods have been developed to separate the white or dead eggs from the fertile ones. At one time, nearly all egg picking was done with a large pair of metal tweezers, a tedious process. A great improvement in technique was made when the pipette became universally adopted for picking eggs. It consists of a length of glass tube about 20 cm (8 in.) long inserted into the bulb of an infant syringe. The inside diameter of the glass tube should be just large enough to pass the eggs. An experienced operator can pick eggs with a pipette quite rapidly. The operation, however, is quite tiring and can cause serious eyestrain.

Siphon Method.—Various types of continuous flow, siphon-type egg pickers have been developed, and great claims have been made for them. The continuous flow, siphon egg picker consists of the pipette with bulb, and a length of flexible hose reaching from the bulb at working height to a pail on the floor. The siphon is started and the egg picker moves the glass tube in the basket of eggs in much the same way as when using the pipette alone. The flow of water through the siphon hose is controlled by applying pressure to the syringe bulb, located between the end of the glass tube and the upper end of the siphon hose and held in the hand. Eggs can be picked faster with a siphon than with a pipette, because it is not necessary to empty out the bulb as is done with the pipette, since the eggs picked up are siphoned into the pail on the floor.

Salt Flotation Method.—The flotation method is one of the older methods still used to some extent. A salt box in which a standard egg basket will fit is used. The salt box is filled with water to nearly overflowing and common stock or table salt is added and stirred into the water until a solution of the proper strength is reached. The amount of salt to be added can best be determined by dipping up a sample of the solution in a glass container, preferably a quart (0.95 liter) fruit jar, and then dropping a few eggs into the jar, using both live and dead eggs. If the solution is right, the dead or bad eggs will float and the live or good eggs will slowly settle to the bottom. If both good and bad eggs float, the solution is too strong and should be diluted by adding water. If both good and bad eggs settle to the bottom, the solution is too weak and more salt should be stirred in. The margin at which the salt solution will separate the good eggs from the bad is quite narrow, so great care must be taken in the preparation of the solution. When a solution of the proper strength has been attained, the basket containing the eggs floating on top is then skimmed off with a small scaph net. Care must be taken in skimming off the dead eggs and any turbulence created by the scaph net must be kept down, since any movement of the water will cause the good eggs to rise to the surface and mix with the bad. The eggs in a basket should not be over 2.5 cm (1 in.) deep for salting, regardless of the size of the eggs. Separation becomes more difficult with more than that amount.

For satisfactory salting, trout eggs should be well eyed. The further the embryo has developed, the more rapidly the eggs will settle in the salt solution. Eggs should be shocked at least 36 hr before salting for good results. As the solution becomes diluted, more salt should be added. A salinometer may be used to determine the strength of the solution.

Once the strength of the solution has been determined, it is easy to maintain it by periodic testing. The optimum strength may vary with different lots, depending on the stage of development and the elapsed time between shocking and salting.

Mechanical Egg Sorter.—Machines have been developed that can sort from 24,000 to 100,000 eyed eggs per hr. The dimensions are about 40 cm × 50 cm × 30 cm (16 in. × 20 in. × 13 in.) deep. The machines can be set up in an area where water and electricity are available. The eggs are placed in the magazine and water washes them into the lower holes of a vertical rotating disc that has a row of holes around its circumference. The machines operate electronically on light sensitivity. As the disc rotates, the opaque, or dead, eggs are blown out of the disc by an air jet into a container. As the disc continues to rotate the clear, or good, eggs are blown out of the holes in the disc by a steady air current. These good eggs go into a container that is supplied with running water by a hose. Discs with different size holes are used for eggs of various sizes. The machines require no attention other than reloading about every 2 hr or less and putting the good eggs into baskets or incubator trays. One man can operate several of these machines or carry on other duties if only one machine is used.

Information about the mechanical egg pickers can be obtained from (1) ROE, Inc., Rte. 1, P.O. Box 149, Palmyra, Wisconsin 53156 or (2) American Distributors and Service Center, Jensorter, Inc., Greg K. Jensen, Rte. 4, Box 480, Bend, Oregon 97701.

Flush Treatment to Eliminate Egg Picking.—Egg picking, once a tedious and time consuming process and still practiced at most hatcheries, has been found entirely unnecessary at others so long as fungus can be controlled. To eliminate egg picking, trout and salmon eggs are flushed once each day with a malachite green solution from the day they are taken until hatching commences. The solution is made by dissolving 43 g (1 ½ oz), dry weight, of malachite green in 4.2 liters (1 gal.) of water. The flow in the hatchery trough is regulated to about 22 liters (6 gal.) per min and then 84 ml (3 liquid oz) of the stock solution are added to the upper end of the trough. As soon as the water in the trough has cleared, the flow is increased to normal. The strength of solution, water flow, and time required to flush are not extremely critical.

After the basket or tray of eggs has finished hatching, the dead eggs are disposed of. By previously having determined the percentage of fertilization, egg losses can be computed quite accurately. At locations where hatchery water flows through outside ponds, the flush treatment should not be done at feeding time.

Hatching

In a few cases an egg basket is still used when hatchery troughs are available. However, the baskets have been basically replaced by incubators.

Basically, fish egg incubators consist of a number of shallow trays stacked one on top of another and spaced apart by guiding strips, much as drawers in a dresser.

Some of the advantages of the egg incubator are the following: less space is required than for troughs; eggs are easily treated for fungus control; egg picking is eliminated; and, due to the small amount of water required, temperature may be controlled.

One of the first incubators was described by Nordqvist (1893). Since then several kinds have been described and used with varied success. They may be divided into two types: drip and the continuous or vertical flow. In the drip incubator, eggs are placed in the trays as soon as they are water-hardened and kept there until ready to hatch, at which time

FIG. 5.19. JAR TYPE HATCHING OF TROUT EGGS

FIG. 5.20. OLDER TYPE TROUT HATCHING BOX

they are transferred to troughs. In the vertical flow incubator they are allowed to hatch there. The alevins remain in the trays until ready to feed. In general, the two types differ mainly in the way in which the water comes in contact with the eggs. The vertical flow incubator is widely used because the alevins can be placed directly into tanks or concrete ponds to start feeding and it is not necessary to have hatchery troughs at the installations.

Vertical Flow Incubator.—Incubators have several advantages over the conventional trough and basket method for incubating and hatching

FIG. 5.21. TRAY HATCH-
ING OF TROUT EGGS

eggs. The flow of water through a 16 tray incubator varies from 11 to 36 liters (3 to 10 gal.) of water per min. The flows are sometimes increased for eyed eggs and alevins as they require more oxygen than the green eggs. Larger flows are used when a tray is loaded with an extra large number of eggs. The average flow generally used in a 16 tray incubator is 18 liters (5 gal.) per min. This small amount of water permits recirculating and either heating or cooling the water to speed up or delay development, whichever may be desirable.

To control the temperature, a portion of the recirculated water passes through a chiller for cooling water or a gas fired water heater and a filter. Another method of heating the recirculated water is to drain the water from the incubator into a well in the floor in a heated room. The room heat usually keeps the water at a very desirable temperature as it is recirculated.

In the vertical flow incubator, the trays actually consist of two compartments: first, the basket with cover in which the eggs are held; and second, the tray in which the basket rests. The trays are stacked one on top of another. Water flows from a tray to the tray immediately below, until it is drawn off at the bottom. The water is introduced to the bottom of each tray in such a manner that it wells up through the basket containing the eggs. This provides sufficient aeration for their development. In the vertical flow incubator, the eggs are allowed to hatch and the alevins remain in the trays until ready to feed. For this reason, the trays must be covered to prevent the newly hatched fish from escaping.

Fungus is controlled with malachite green flush treatments. A stock solution of malachite green is made by dissolving 43 g of dry malachite green in 4.2 liters (1 ½ oz in 1 gal.) of water. The flush treatment consists of pouring 85 g (3 oz) of the stock solution into the top tray of the incubator stack when the water flow is 21 to 25 liters (5 or 6 gal.) per min. This flush treatment is generally continued on a daily basis. Most hatcheries discontinue the treatment when the eggs start to hatch, while others continue the treatment until the alevins are ready to feed and are removed from the incubator.

When the water is recirculated through the incubator, the malachite green flush treatment is done only two or three times per week. The treated water must be wasted and the system filled with fresh water which will probably change the temperature in the incubator for a few hours.

When the fish have absorbed their yolk sac and are ready to feed, the tray units can be moved to troughs, tanks, or ponds, placed on the bottom, cover screens removed, and the fish allowed to swim out. Dead eggs and deformed fish will remain in the egg tray and can be counted and then discarded. This accounts for the entire egg and fry loss to the feeding stage. There is no need to remove the dead eggs earlier as the malachite green controls the fungus. When eyed eggs are to be shipped, the dead eggs should be removed before packing them so fungus will not develop while the eggs are in transit.

The number of eggs put into each incubator tray varies with conditions and facilities available. Salmon eggs have been hatched successfully with 4.5 kg (160 oz) or 12,000 fry per tray, although the California average is about 3.1 kg (110 oz) or 8000 fry per tray. Rainbow trout eggs have been hatched successfully with 2.8 kg (100 oz) or 45,000 fry per tray, although the California average is about 2.0 kg (70 oz) or 25,000 fry.

Information about incubators can be obtained from Health Tecna Corporation, 19819 84th Ave. South, Kent, Washington 98031.

Packing Eggs for Shipment

Green Eggs.—Green eggs may be shipped for a period of up to 48 hr after being taken. Normally, when green eggs are transported only a short distance they are placed in 42 liter (10 gal.) milk cans filled with water. This method is quite satisfactory as long as severe jolting and rise in temperature are prevented. Care must be taken not to place too many eggs in a can. It must be remembered that trout and salmon eggs increase in size by about 20% during the water-hardening process, and if too many eggs are placed in a can the eggs near the bottom may be killed by the resulting pressure from the eggs above. A simple way to prevent excessive pressure on the eggs in the bottom portion of the can is to measure out eggs in amounts of 1.4 kg (50 oz) each and place them on pieces of mosquito netting about 45 cm (18 in.) square. The four corners of the netting are then picked up and tied with a string. The eggs can then be suspended at different depths in the can by varying the length of the suspending string.

Eyed Eggs.—Eyed salmonoid eggs have been shipped to all corners of the world in a variety of shipping containers. In selecting shipping containers, one should bear in mind that eyed eggs must be kept cold to prevent their hatching en route.

Good judgment is necessary on the part of the hatcheryman when making egg shipments. Eggs must be kept moist to prevent dehydration and containers should be light to keep shipping charges at a minimum and should protect the eggs as much as possible against sudden shock. Eggs should not be so far advanced that they will hatch en route with the resulting alevins lost. The temperature in a shipping case is usually higher than that of most hatchery waters and fungus will grow very rapidly on dead eggs if they are left in the shipment. For shipping purposes, eggs should be sufficiently eyed to permit shocking, so that the infertile eggs can be removed.

The term eyed eggs is used rather loosely in hatchery practice and is usually applied to the eggs any time after the eye spots are visible to the unaided eye.

Egg shipping containers can be obtained from Health Techa Corporation, 19819 84th Ave. South, Kent, Washington 98031.

Use of Wescodyne.—As a further disease prevention measure, all eggs received at a hatchery are treated with Wescodyne as they are removed from the egg case. Wescodyne rapidly destroys bacteria, viruses, molds, and fungi. Toxicity to man is exceptionally low.

Each lot of eggs, upon arrival, should be treated with a 1:300 solution of Wescodyne. Either green or eyed eggs should be completely submersed in the disinfecting solution for 10 min and then rinsed in fresh water. A 1:300 solution is formulated by using 12.6 ml of Wescodyne in 4.2 liters (in 1 gal.) of water (4 ¼ fl oz in 10 gal.). The volume of eggs to disinfecting solution should be about 1:5 (25 oz of eggs per gal. of disinfectant). The resulting solution may be used until the amber color fades to a light yellow. It should then be discarded and replaced with a fresh one. This formulation should also contain sufficient oxygen at all times.

In extremely soft water hatcheries, i.e., those where the alkalinity is less than 35 ppm, a buffer should be added with the Wescodyne to prevent a drastic drop in the pH to under 6. If in doubt, check the pH prior to treatment. Sodium bicarbonate is recommended as a buffer at the rate of

0.5 g per 4.2 liters (per 1 gal.) of disinfecting solution.

Wescodyne at a concentration of 1:20,000 is harmful to fish; therefore, care must be taken during disposal. Do not introduce it back into the hatchery system or into any septic systems dependent on bacterial decomposition.

Wescodyne may be obtained from: West Chemical Products, Inc., 8125 36th Ave., Sacramento, California 95824.

Temperature Units for Hatching

For many years temperature units were used by hatcherymen to estimate the length of time required for trout eggs to develop, and it was believed that the total number of units was constant for any species regardless of the average temperature throughout the incubation period.

One temperature unit (TU) equals 1°F above freezing (32°F) for a period of 24 hr. For example, if the average water temperature for the first day of incubation was 55.9°F it would constitute 23.9 TU (55.9°F minus 32°F). If, on the second day, the temperature averaged 52°F, it would add 20 more TU, or a total of 43.9 TU to the end of the second day. In the Celsius scale, each 1°C of temperature is calculated as 1 TU. Hence, if hatching water is a constant 10°C (50°F), each day counts as 10 TU's and it would take approximately 31 days for rainbow trout to hatch.

Temperature units vary considerably at different water temperatures (Table 5.4). It clearly demonstrates that they are not a safe measure for the trout culturist. They are included here only for the purpose of illustration.

In any species of trout, the incubation period or average hatching time of the eggs is not immutably fixed for a given temperature but may vary as much as 6 days between eggs lots taken from different parent fish.

TABLE 5.4. NUMBER OF DAYS AND TEMPERATURE UNITS REQUIRED FOR TROUT EGGS TO HATCH[1]

Water Temperature		Rainbow			Brown			Brook		
		Number Days to	Temperature Units		Number Days to	Temperature Units		Number Days to	Temperature Units	
°C	°F	Hatch	°C	°F	Hatch	°C	°F	Hatch	°C	°F
1.7	35	—	—	—	156	260	468	144	240	432
4.4	40	80	356	640	100	444	800	103	458	824
7.2	45	48	347	624	64	462	832	68	491	884
10.0	50	31	310	558	41	410	738	44	444	799
12.8	55	24	307	552	—	—	—	35	447	805
15.6	60	19	296	532	—	—	—	—	—	—

[1]Spaces without figures indicate incomplete data rather than a proved incapability of eggs to hatch at those temperatures.

Hatching Troughs

Different sizes of hatching troughs are used, depending on need. These troughs are used for placing the young fry when they come out of the incubatory tray, if they are not placed directly into rearing ponds. In general, the industry has shifted to aluminum alloy troughs. Each trough alloy is identified by number and the manufacturer can identify the various alloys.

It is very important for anyone contemplating the installation of aluminum troughs to first obtain a complete chemical analysis of the water at the location where the troughs are to be installed. This analysis should be made available to a reliable aluminum manufacturing company, which will usually furnish competent engineering assistance in selecting the proper alloy.

Trough Carrying Capacity.—It is generally agreed that the oxygen requirements for salmon and trout are nearly the same for all species, with rainbow possibly requiring the highest amount, brown trout the next higher, and eastern brook the lowest. Oxygen requirements are increased by activity, feeding, and rise in temperature. While it is desirable to maintain an oxygen level of 7 ppm for holding salmon and trout, the absolute minimum for this purpose is 5 ppm. It has been shown that at 7.2°C (45°F) temperature, a yearling rainbow will consume 3 cc of oxygen per hr, while at 20.0°C (68°F) consumption will increase to 12 cc per hr, an increase of 300%. The increase in oxygen with rising temperature, crowding, feeding, weighing, and activity must be considered when determining the number of fish to place in hatchery troughs and ponds (Tables 5.5 and 5.5A).

There are minimum water requirements in a hatchery trough (Table 5.6). Data for troughs are not applicable to rectangular or circular ponds, in which reaeration from the surface is large in comparison to oxygen directly available from the water supply.

Even though it is not general practice to install hatchery trough aerators, the oxygen level in troughs can be considerably increased by their use.

Trout and Salmon Ponds

Pools or ponds for rearing trout and salmon have over the years been as varied in design as French fashions. One can understand that in large, deep, still pools fish rest most of the time and food conversion is normally better than in narrow, swift ponds, where a good portion of a fish's energy is used in maintaining its position in the pond. The ideal pond is

TABLE 5.5. TROUGH CARRYING CAPACITY IN G OF TROUT PER LITER PER MIN OF WATER FLOW; INLET WATER SATURATED; OUTLET WATER OXYGEN CONTENT 5 PPM[1]

Temperature °C	Elevation (m)								
	0	305	610	915	1220	1525	1830	2135	2440
	Grams of Trout								
7.2	2704	2570	2366	2232	2096	1928	1792	1622	1522
10.0	1420	1318	1218	1150	1082	980	912	812	744
12.8	892	846	784	724	676	628	562	514	466
15.6	628	573	540	486	454	420	378	338	304
18.3	446	412	386	344	312	284	250	224	202
21.1	352	324	298	264	244	210	190	162	148

TABLE 5.5A. TROUGH CARRYING CAPACITY IN OZ OF TROUT PER GAL. PER MIN OF WATER FLOW; INLET WATER SATURATED; OUTLET WATER OXYGEN CONTENT 5 PPM[1]

Temperature °F	Elevation (ft)								
	0	1000	2000	3000	4000	5000	6000	7000	8000
	Ounces of Trout								
45	400	380	350	330	310	285	265	240	225
50	210	195	180	170	160	145	135	120	110
55	132	125	116	107	100	93	83	76	69
60	93	85	80	72	67	62	56	50	45
65	66	61	57	51	46	42	37	33	30
70	52	48	44	39	36	31	28	24	22

[1]At lower temperatures and altitudes, actual crowding of fish would become a limiting factor, rather than available oxygen. Thus, at 7.2°C (45°F) and at sea level, 20 liters (20 gal.) per min would supply 54,080 g (8000 oz) of trout with sufficient oxygen, but that weight of fish would require more than a standard trough to prevent overcrowding.

TABLE 5.6. MINIMUM WATER REQUIREMENTS IN HATCHERY TROUGHS IN LITERS OR GAL. PER MIN FOR EACH 1000 OZ OR 6760 G OF TROUT[1]

Temperature		Elevation in m or ft								
°C	°F	m 0	305	610	915	1220	1525	1830	2135	2440
		ft 0	1000	2000	3000	4000	5000	6000	7000	8000
		Liters or gal. per min								
7.2	45	2.5	2.7	2.9	3.0	3.2	3.5	3.8	4.2	4.5
10.0	50	4.8	5.4	5.6	5.9	6.2	6.9	7.4	8.3	9.1
12.8	55	7.6	8.0	8.6	9.3	10	11	12	13	15
15.6	60	11	12	13	14	15	16	18	20	22
18.3	65	15	16	18	20	22	24	27	30	33
21.1	70	19	21	23	26	28	32	36	41	45

[1]Values above 10 are shown only to the nearest liter or gal. For tandem troughs with good aerators between upper and lower, the water requirements can be lowered about 10%.

one that can be operated rather deep and with little current most of the time, but can be readily converted to a shallow, swift pond when necessary. Deep, still pools have the disadvantage of not lending themselves to flush or prophylactic treatments, whereas long, shallow, raceway type ponds are ideal for this purpose. Very often the type of pool selected depends on several factors, such as the amount of water available, surrounding terrain, and accessibility. In fish culture as it is practiced

FIG. 5.22. AUTOMATIC FEEDERS FOR TROUT FRY

FIG. 5.23. OUTSIDE TROUT FRY REARING TROUGHS

today, ponds must lend themselves to ease in flush treatment for prevention of disease, automatic grading, accessibility to mechanical loading and feeding equipment, and simplicity in overall management. In general, concrete raceway ponds have proven to be the most practical.

Rearing Pond Capacities.—The carrying capacities of rearing ponds depend, to a large extent, on conditions existing at the various hatcheries. Capacities must be established at the hatchery itself to be of benefit in planning the station program. Some of the factors which influence or determine the carrying capacities of rearing ponds are: (1) water quality, (2) water temperature, (3) water volume, (4) rate of flow, (5) rate of change, (6) reuse of water, (7) degree of pollution of water supply, (8) kind of fish held, (9) size of fish held, (10) frequency of grading and thinning, and (11) diseases encountered.

The best way to determine the proper holding capacities of rearing ponds at a particular installation is to examine the results of several seasons of production for which accurate records are available regarding numbers and weights of fish held, growth, food conversion, types of feed used, incidence of disease, and mortality—all plotted against the basic factors, such as water volume and temperature, which might influence holding capacities.

Circular Ponds.—The advantages of the circular pond are: less water required; nearly a uniform pattern of water circulation throughout the pool, with fish more evenly distributed instead of congregating at the head end; and a center outlet, with circular motion of water producing a self-cleaning effect. Even though less water is required, sufficient head or pressure must be available to force and maintain the water in a circular motion. In operating a circular pond, advantage can be taken of an interesting phenomenon. Since the vortex of a whirlpool in the northern hemisphere always rotates in a counterclockwise direction, the water in a circular pond should also rotate in a counterclockwise direction.

To take advantage of the self-cleaning effect of the circular pond, it is necessary that the proper type screen and outlet pipe be used. In effect, the self-cleaning screen consists of a sleeve larger than the center outlet or standpipe. This sleeve fits over the outlet pipe and projects above the surface of the water. The sleeve may have a series of slots or perforations near the bottom which act as an outlet screen for small fish or may be in the form of a narrow opening between the pool bottom and the lower end of the sleeve for larger fish. This opening must be adjusted according to the size of the fish in the pool. Waste materials are drawn through the opening by the outflowing water.

While carrying capacities will vary among installations, circular tanks 4.3 m (14 ft) in diameter, with a water depth of 75 cm (30 in.) and an inflow of 210 liters (50 gal.) per min, will safely carry 182 kg (400 lb) of trout at water temperatures up to 15.5°C (60°F). With the comparatively small amount of water used in a circular pool, the rate of water exchange is quite slow and extreme care must be taken in feeding.

Circular ponds do not lend themselves well to flush treatment for disease control nor to mechanical fish loading or self-grading devices. The advantages of the circular pond for yearling and larger trout are not as great as thought when circular ponds first began to appear. The circular pond, however, is well adapted to rearing trout ranging from newly hatched fry to fish of subcatchable size.

There are many advantages to circular tanks in holding a small number of fish of varying ages and sizes for selective breeding, feed experiments, and many types of research projects. In these cases it is desirable to have a separate water supply for each experiment. Generally there are only a small number of fish involved in each experiment and a large number of small tanks is very practical.

Raceway Ponds.—The earthfill raceway type rearing pond with concrete cross dams was at one time the most widely used pond. They were constructed 30.5 m long, 1.2 m deep, 3 m wide at the bottom, and 9.1 m wide at the top (100 ft long, 4 ft deep, 10 ft wide at the bottom, and 30 ft wide at the top) with sides sloped 2 ½ to 1. The gradient was 15 cm in 30.5 m (6 in. in 100 ft). They were usually built in a series of 4 to 8 ponds long with a roadway along one side of each series.

These earthfill ponds require considerable maintenance as weeds and plants grow on the banks and protrude into the pond. The pond banks erode making the banks between ponds and roadways quite narrow unless they are reshaped annually. Flush treatment for disease control in ponds with irregular widths is not satisfactory as they have poor circulation.

Concrete raceway type rearing ponds have been built in newer hatcheries and have become the standard type rearing ponds replacing earth type ponds in most areas.

The standard concrete raceway type rearing pond is 30.5 m long, 1.1 m deep, and 3.0 m wide inside measurements, with a 15 cm gradient per 30.5 m (100 ft long, 42 in. deep, and 10 ft wide inside measurements, with a 6 in. gradient per 100 ft). The top of the pond wall is generally 45 cm (18 in.) above ground level. The depth of water in the pond at the lower end varies from 60 to 75 cm (24 to 30 in.) depending on various conditions at a particular hatchery, i.e., size and number of fish per pond, water flow, water temperature, diseases present, and fish's increased desire to jump out of the ponds at certain times of the year. The flow of water in a series of six ponds is usually about 84 liters per sec (3 cfs). Pond series are constructed from 4 to 10 ponds in length but general practice is to build only 6 ponds in a series. A water flow of 140 liters per sec (5 cfs) should be available for a series 10 ponds in length. In a series of 6, and particularly 10, ponds the water should be pumped through an aerator midway in the series so the lower ponds can operate at full capacity.

About 2 ppm oxygen can be added to a series of ponds by operating a

Fresh-flow aerator in the upper end of pond 5 of a 6 pond series, thereby increasing the production in ponds 5 and 6. This type of aerator is operated by a 1 horsepower motor and sprays about 28 liters per sec (1 cfs) of water into the pond.

Two series of ponds are normally built side by side with a common wall between them. There is a roadway between each pair of ponds. The roadways give access to all ponds for the mechanical pellet blower to feed the fish, the fish pump, and tank trucks. A battery of ponds as described above can be constructed in a much smaller area than earthen ponds.

FIG. 5.24. FEED BIN FOR LOADING TROUT FEEDING EQUIPMENT

Keyways in the concrete walls at 30.5 m (100 ft) intervals provide for screens and checkboards. At these points metal posts with keyways are placed in recessed holes in the middle of the concrete floor. Thus two screens about 1.5 m (5 ft) long are used instead of one 3 m (10 ft) screen. The water drops 15 cm (6 in.) over the checkboards at the end of each pond which provides some aeration. It is advantageous to have keyways in the concrete walls at 7.6 m (25 ft) intervals in the upper ponds of a series so small fish can be crowded together for proper feeding.

Fry are taken from the incubator trays when they reach swim-up stage to start feeding. The number of fry to stock an entire series of ponds can be placed in the upper end of the series and kept crowded until they are feeding well. They are then released into additional ponds in the series or moved to other ponds as required. It is always preferable to keep the small fish in the upper end of the series and the larger fish below. With

FIG. 5.25. LARGE RAINBOW TROUT FARM WITH FEED BEING DISPENSED BY FEEDING CART SUSPENDED FROM OVERHEAD RAILS, IDAHO

FIG. 5.26. TRACTOR–PULLED AUTOMATIC FEEDER FOR TROUT

proper management planning, it is possible to keep most of the ponds full at all times with fish of varying ages. This provides fish for planting every month of the year. If fish do not move from pond to pond, it is possible to have screens only at every third or fourth pond.

Waste material from the fish accumulates in the ponds. As the fish grow larger and become more active the waste material drifts down to the lower end of the concrete raceway type rearing ponds and accumulates against the checkboards. A piece of corrugated fiberglass mounted

FIG. 5.27. AUTOMATIC FEEDING OF TROUT USING TRUCK

on 2 vertical pieces of 5 × 10 cm (2 × 4 in.) wooden posts placed against the upstream side of the checkboards so the fiberglass will be 9 cm (3 ½ in.) above the pond bottom and extend above the water surface will force all water to pass under the fiberglass and up between it and the checkboards. This current will move the accumulated material from pond to pond which will virtually make the ponds self-cleaning of all solid waste material that reaches the lower end of the pond. The waste material will now move down in an earthen pond as well as it will in a concrete pond.

Advantages of Concrete Ponds.—The many advantages of the concrete raceway type rearing ponds over earthen ponds are summarized: (1) no weed growth or bank erosion; (2) no structural maintenance; (3) swim-up fry can be put directly into ponds from incubators without excessive loss, and the need for hatchery troughs is eliminated; (4) good circulation; (5) good control for chemical flush treatment of disease; (6) use of mechanical crowder for cleaning ponds, grading, and herding fish together for loading tank trucks or pumping into other ponds through a pipe; (7) good accessibility for tank trucks, fish pump, and mechanical pellet blower; (8) easily adaptable for use of automatic feeder; (9) compact unit in small areas to simplify working conditions; (10) self-cleaning devices aid in moving solid waste material from the ponds; (11) smaller area simplifies bird control problem; (12) seines are not required for catching and moving fish; and (13) simplify overall management.

This all adds up to mechanizing many of the hatchery operations with a great saving in labor which reduces the unit cost of producing fish.

Salmon Rearing

After the salmon fry are stocked in rearing troughs or silos, they are reared in similar fashion to rainbow trout until smolting. Under natural conditions, the smolt stage or migration stage occurs from 1 to 2 years after hatching. However, in Washington and Oregon, the eggs hatch in January or February and the fry are raised in water temperatures of 10°–15°C (50°–60°F) recirculated fresh heated water. Because of the water temperatures, feeding, and other environmental factors the fish smolt in 4–5 months and in June–August are acclimatized slowly to sea water and can be transferred to floating seawater pens or ponds on land into which sea water is pumped. Of course, as shown by coho salmon runs in Lake Michigan, the smolt can continue to be grown in fresh water.

During the so-called smolt stage, physiological changes make it possible for the fish to live in sea water at 34–35 ppm salt content. The skin colors change from blue-brown-green to a shade of silver. This takes place between 15 and 40 g sizes (½ and 1 ½ oz).

Temperatures and Growth

No single factor has as much effect on the development of eggs or growth of the fish as temperature. Rainbow trout reared in water at a constant 15.5°C (60°F) grow at about 2.5 cm (1 in.) per month. At 7.4°C (45°F), growth is only about 0.6 cm (¼ in.) per month.

Temperature unit calculations can be used to compute growth. This is expressed as:

$$TU = T_1 - T_2 \times 1 \text{ month}$$

where T_1 is water temperature
T_2 is 3.7°C (38.6°F) or zero growth water

Using this formula with 10.9°C (51.6°F) water, the following can be computed:

$$TU = 10.9 - 3.7 \times 1 \text{ month } (51.6 - 38.6 \times 1 \text{ month})$$
$$TU = 7.2 \times 1 \ (13 \times 1)$$
$$TU = 7.2 \ (13)$$

Since 11.7 TU's (21 TU's in °F) are required per 2.5 cm (1 in.) of growth, and we start with 8.75 cm (3 ½ in.) fish and desire 28.75 cm (11.5 in.) fish, we need 20 cm (8 in.) of growth. This requires 93.3 TU's (168 TU's in °F). Dividing 93.3 by 7.2, we get 12.9 months of time required. (In °F, divide 168 TU's by 13 to arrive at 12.9 months.)

FIG. 5.29. SILOS FOR REARING SALMON FRY TO SMOLTS, WASHINGTON

There is some discrepancy in recommended water temperatures. In general, rainbow spawners should not be held at water temperatures exceeding 12.8°C (56°F) for at least 6 months before spawning. After spawning, the eggs are acclimatized at the hatchery. Temperatures of 8° to 13°C (46° to 50°F) are recommended, although waters as warm as 13° to 14°C (55° to 58°F) have been used. Rearing water temperatures can be between 10° and 20°C (50° and 68°F). However, the authors have successfully fed rainbow trout at 22°C (72°F) for short periods of time. It is generally agreed that yearling and adult rainbow can withstand temperatures for short periods of time up to 25.5°C (78°F) without feeding, and lethal temperatures are somewhat higher.

Hatching temperatures for coho salmon are similar but optimum growth is experienced at about 15°C (59°F). Salmon are not as tolerant of warmer waters as are rainbow trout.

Dissolved oxygen levels found in water are a function of temperatures, altitude, stocking rates, volume of feed fed, and related factors. In general, the higher the dissolved O_2 levels are toward saturation, the better the health of fish and the better the feeding activity and growth. Ellis *et al.* (1946) stated that the lowest safe level for trout is about 5 ppm and 7 ppm is preferable. Ellis stated that 10 to 11 ppm is best for trout and they may show discomfort when dissolved O_2 is less than 7.8.

Grading Fish

It has been said that a hungry trout 7.5 cm (3 in.) long will devour a trout 3.75 cm (1 ½ in.) long, providing he can catch it, and that a 15 cm (6 in.) trout will eat a 7.5 cm (3 in.) trout, and so on up the line, until the word cannibalism becomes frightening. There just isn't any question that large trout will eat small trout. Both trout and salmon are handled in comparatively large numbers when reared in troughs or ponds. Therefore, any lot of fish of fair numbers must consist of the offspring of several females. This, plus the fact that all of the eggs from a single female are not always of the same size, and the fact that the larger fish in any group are better able to compete for food than the smaller fish, explains some of the reasons why trout and salmon grow irregularly in size.

Grading of fish is necessary to get good growth, reduce cannibalism, prevent competition between fish of smaller size with their larger kin, and obtain fish of the correct size to meet management requirements. Furthermore, the total weight of fish in any one group can be more accurately determined for computing the amount of food to feed in percentage of body weight if the fish are of a nearly even size. Grading equipment has been designed which will grade fish to a number of sizes in one operation. Grading in more sizes than is necessary, however, is time consuming and normally grading one group of fish into three sizes, small, medium, and large, is sufficient.

Many types of fish grading devices are currently in use for segregating the various sizes of fish from the time they hatch until they are ready for release in the stream. Each of these devices has been developed to operate under a certain set of conditions to accomplish a desired objective. All of them are efficient to some degree and can be adapted to almost all situations, with varying degrees of efficiency.

If fish are raised in hatchery troughs to the fingerling size, they are generally graded before putting them into rearing ponds. In most situations, the fish are graded one or two times while they are in the rearing ponds. It is general practice to grade fish before planting which can be done very easily with the fish pump.

Fish graders most commonly used in California are the Wilco adjustable fish grader or modification, and the fish pump grader, the latter being the more efficient.

Wilco Adjustable Grader.—The Wilco adjustable fish screen or grader was developed by Wilco Products, 327 Burnett Ave. North, Renton, Washington 98055. It can be used to grade fish in a concrete pond or the concrete trunk of an earthen pond.

The grader consists of oval shaped vertical bars in a frame 154 cm long,

75 cm high, and 3.75 cm thick (61 ¾ in. long, 30 in. high, and 1 ½ in. thick). Openings between the vertical bars can be adjusted by turning the adjusting knob, located near the end of the screen, to the desired setting and locking it by tightening a thumb nut. The openings can be varied from 1.56 mm (1/16 in.) minimum to 20.3 mm (13/16 in.) maximum. Larger screens or graders are available by special order.

The grader is placed in the upstream end of the pond trunk of earthen ponds. The fish are pulled into the trunk behind the grader with a seine and crowded against the grader with a push rack. A hand push rack can be used on ponds with concrete walls much more efficiently than using a seine. As the fish become crowded between the racks, those small enough to pass through the grader swim into the pond above. Those unable to pass through remain in the pond behind the grader where they can be temporarily held for planting or other purposes.

FIG. 5.30. MECHANICAL TROUT SIZING SEPARATOR

The flow and water depth of the pond should be held at normal operating levels during the grading operation. The time required to grade a pond of fish varies from only a few minutes to 2 hr depending somewhat on local conditions. Several graders can be operated at one time in the same pond series should the need arise.

FIG. 5.31. MECHANICAL TROUT SIZING SEPARATOR IN ACTION

FIG. 5.32. SIZED TROUT FROM MECHANICAL SEPARATOR
BEING TRANSPORTED BY PIPE TO DIFFERENT RACEWAYS

The advantage of the Wilco grader is that it is almost automatic. The only operation required is that of selecting the size of spacing, setting the grader in place, and forcing the fish in behind it. The attendant can then continue with other duties and within a short time the fish have graded themselves with a minimum of handling.

There is a serious disadvantage to this method of grading fish in that from 10 to 25% of the small fish do not pass out through the vertical bar graders. This method is definitely not as effective as other methods of grading fish although it is less time consuming.

Fish Pump.—The 15 cm (6 in.) fish pump has many uses as a labor saving piece of equipment in grading fish, moving fish between ponds, and loading fish into tank trucks.

The pump unit is a Paco Horizontal Non-clog Pump, Type NCH, directly connected with a gas engine. It has a single port, non-clog impeller with the blade edge well rounded to avoid catching trash or, in this case, injuring fish. This type of pump is used in conveying fruits, vegetables, and food products in food processing plants. It is manufactured and distributed by the Pacific Pumping Company in Oakland and Los Angeles.

The pump is mounted on a trailer that can be moved to the ponds where it is to be used. It has a reinforced suction hose attached to a pickup box that is placed in a section of the pond from which the fish are to be removed. The pickup box has a gate that is perforated so it will draw water at all times that the pump is running and draw fish only when it is opened. The fish pass through the pump up to the separator tower and a grader box equipped with an adjustable Neilsen grader that can be set from 3.12 mm (⅛ in.) minimum to 25 mm (1 in.) maximum spacings which slope downward so the fish will slide into a chute and drop into a tank truck. The water and small fish drop through the parallel bars and pass through a hose back into a pond. The separator tower can be raised or lowered hydraulically to accommodate different size tank trucks. This pump can load fish into a tank truck in a small fraction of the time required by any other method. It is only necessary to be sure the fish are crowded into the area of the suction hose and pickup box.

Fish can be moved from one pond to another by removing the parallel bar grader so all water and fish will pass through the discharge hose and be piped to any pond within a reasonable distance.

Fish can be graded by pumping them into the parallel bar grader with proper spacings for the size fish desired. The graders can be changed easily to grade different size fish. The smaller fish pass through the grader and discharge hose with the water and can be piped to a pond in the nearby areas. The larger fish can be loaded into a tank truck and hauled to another pond or run through another pipe to another pond by supplying water to this outlet.

The fish pump grader does a more thorough job of removing the small fish than can be accomplished with the Wilco adjustable grader method previously described. Small fish do not have much chance of getting into a tank truck load of fish for planting if the grader in the pump is adjusted properly. This is the most rapid and efficient method of grading catchable and subcatchable size fish.

This fish pump and fish grader have been assembled by Neilsen Metal Industries, Inc., 3501 Portland Road NE, Salem, Oregon 97303.

A 7.5 cm (3 in.) fish pump has been designed for use with fingerlings. It is the same principle as the 15 cm (6 in.) pump but has a lighter suction attachment like a vacuum sweeper so it can be moved around easily to catch the smaller fish.

AMERICAN EEL (ANGUILLA ROSTRATA)

In any part of the world where eels are consumed they are a luxury food. This is due to the declining catch of wild eels on a worldwide basis and the expense of culturing eels. In the United States, the wild eel catch has not yet achieved its maximum sustainable yield from wild stocks and culturing is in its infancy. There are probably fewer than 50 eel farms in the country. Irrespective of present lack of importance, the authors feel that the future of eel culturing is bright and therefore have included them in this book.

Existing production practices in the United States have borrowed techniques from the other countries, particularly Japan. Hence, culturing techniques and the following discourse rely heavily upon the techniques of other countries, which have been tested, refined and improved by more than 100 years of culturing.

Culturing Practices

In recent years, attempts to hatch eel eggs have been successful. However, efforts to raise the young larvae for the first few months have not been successful. Hence, the industry relies entirely on the catch of "wild" elvers.

The American eel, as other eels, is a catadromus species. This means that it is spawned in the ocean, migrates to fresh water for growth, and after obtaining sexual maturity, migrates back to the sea to spawn. It spawns in the Sargasso Sea and some 6 to 12 months later is distributed, apparently at random, on the American mainland throughout the Gulf of Mexico and along the east coast north to Greenland. Because of the immense distance involved, the elvers may be caught as early as October or November in Florida to as late as August in northern Canada. There are several reasons for this: (1) areas near the Sargasso Sea receive elvers earlier, and (2) the elvers are not attracted to the river or creek mouths until the water temperatures warm up to 8° to 10°C (48° to 50°F). Prior to coming into fresh or brackish water, they undergo metamorphosis from leptocephalae at sea and become transparent elvers about 5 to 6 cm (2 ½ to 3 in.) long. They enter the river mouths mainly on spring tides or about at two week intervals. They swim and drift with the incoming current. When the tide turns they wiggle into the bottom mud and wait for the next incoming tide after darkness.

They swim only at night and keep next to the banks. They usually swim on the surface. Because of their habits, most fishermen have never seen an elver after a lifetime of fishing. While all elvers and eels are sensitive to sunlight, they are attracted by lights at night. They are captured by dip netting along the banks after being attracted by a bright light, by trolling, by special traps at obstructions in the stream, and larger nets across small streams.

The elvers are very delicate and are not touched by hand. In small quantities they can be placed in a box lined with wet muslin and taken to a holding tank within a few hours. If the night is cold the box should be covered. They can be transported in large lots in aerated tanks, in polyethylene bags with water or oxygen or in polystyrene trays. There will be about 4000 to 6000 elvers per kg (1800 to 2700 per lb).

When the elvers reach the eel farm, they are placed in elver tanks, usually in a greenhouse or inside. The tank water is maintained at 25°C (77°F). This heated water permits them to get off to a fast start. This first tank is usually circular, about 5 m (16 ft) in diameter and 60 cm (2 ft) deep. Water is sprayed on the surface so that the water in the tank

FIG. 5.33. EEL RECIRCULATION POND SHOWING CATCH BASIN AND FEEDING STATION

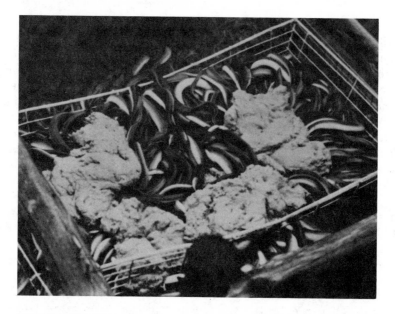

FIG. 5.34. FEEDING EEL

whirls. Drainage is from a pipe in the center. The elvers are kept in this tank about one month. The water is maintained crystal clear and well oxygenated. For the first two weeks, minced earthworms or bivalve flesh is fed. The next two weeks they are gradually shifted to minced tubifex or similar worms, and mackerel, sardines, or anchovy. For the fifth to tenth week, minced fish flesh or artificial feeds are fed. They are fed nearly continually. The food is lowered into the water on trays and uneaten food is removed. Before feeding the fish it is desirable to dip them into boiling water for about 1 min to soften the skin. After the first month the elvers have metamorphosed into black eels and are transferred to fingerling ponds which may still be inside and heated, depending on outside water temperatures. Until the eels are about 20 g in weight they will try to leave the pond on rainy nights so it is necessary to have a lip on the elver and fingerling tanks or ponds. When pond water outside gets over 23°C (74°F), the fingerlings can be moved outside. After the tenth week they can be gradually shifted to an artificial diet or one of all trash fish. The trash fish are fed by passing a line through their eyes so that the heads and backbones can be removed from the feeding place after feeding. Some producers after the first 4 weeks shift entirely to a fish meal mash diet containing 55 to 60% protein. The more mature fish have a mash diet containing 43% protein. A typical diet contains:

White fish meal	70%
Potato starch	20%
Yeast	5%
Vitamin mixture	1%
Salt	1%
Minerals	1%
Other	2%

After about 18 to 24 months the fish reach about 200 g (7 oz) sizes and are ready for market. Because of extreme variations in individual growth, which may vary by 200 times, eels are sorted and harvested at various times throughout the years.

In the United States, many of the young eels captured are already black eels and can be stocked directly into fingerling tanks instead of elver tanks.

Total survival over the two year period will range from 25 to 90%. The fish are fed only if the water temperatures are over 15°C (60°F). Above 30° to 32°C (88° to 90°F), oxygen levels in the water are low and supplemented oxygenation must be used and feeding levels reduced.

Stocking is from 20 to 40 eels of 10 to 20 g per m^2 (1 to 2 eels per ft^2) or about 3000 kg per ha (3000 lb per acre). Density can ultimately be 12 to 14 tons per ha (4.8 to 5.7 tons per acre).

Rickards et al. (1979), reporting on raising eels in North Carolina, followed a slightly different feeding program. The newly captured elvers were placed in each indoor tank. For the first four days the feed consisted of 86.5% deboned fish, 12% earthworms, and water blended to a thick liquid. Table salt (1.5%) was added gradually to the mixture to give it the consistency of stiff paste. Feeding was at a 5% rate at the bottom of the tank. After 4–5 days the earthworms were omitted and the food tray was gradually raised over a few days to the surface. After the initial feeding period they were fed twice a day, the feeding rate was at 10% for one week and increased to 20% by the fourth week. After 25–30 days of the deboned fish diet, the elvers were gradually converted to a diet of 75% fish meal (mostly menhaden) 25% alpha-corn starch, trace amounts of tetracycline (3 mg per 150 g of feed) and a vitamin pre-mix of 10 mg per 150 g of food. Tetracycline has not been registered for use in fish diets, so mention herein does not imply endorsement of use. This new diet gradually replaced the old diet at the rate of 10% per day until the new blend was 100% fish meal-starch. This was fed at 8% per day until the young black eels were moved to outside ponds. After water temperatures were at least 18°C (65°F), the same diet of fish meal-starch was continued. Generally the eels were fed twice a day when the waters were above 18°C (65°F), once a day at 12° to 18°C (56° to 65°F), every other day at 10° to 12°C (50° to 54°F) and not at all below 10°C (50°F).

BAIT AND GOLDFISH

There are four species of minnows or fish used as bait minnows. These are the golden shiner (*Notemigonus crysoleucas*), fathead minnow (*Pimephales promelas*), bluntnose minnow (*Pimephales notatus*), and white or common sucker *(Catostomus commersonii)*. Goldfish (*Carassius auratus*) (with over 100 varieties) are often used for bait fish. Because the minnows and goldfish have similar spawning and rearing techniques, they are included together in this section.

FIG. 5.35. AIR PHOTO OF TYPICAL BAIT OR GOLDFISH FARM

Stocking of Spawners

Shiners should be stocked in brooder ponds at the rate of 1235 to 2470 per surface ha (500 to 1000 per surface acre [Altman and Irwin, undated]). Prather *et al.* (1953) state that the stocking rates of brood goldfish depend on methods of rearing. They advise with Method I (wild or free spawning) to stock after the danger of frost is past at the rate of 494 to 741 per ha (200 to 300 per acre). With Method II, the egg transfer method, and with Method III, the fry transfer method, they advise stocking at 7410 to 49,400 per ha (3000 to 20,000 per acre).

Prather *et al.* (1953) state that brood fatheads should be stocked at the rate of 2470 per ha (1000 per surface acre) between February 15 and May 1. This is for use using Method I (wild or free spawning); no mention is made of Methods II and III. Altman and Irwin (undated) state that

FIG. 5.36. AERATING
PUMPED WATER

bluntnose minnows can be stocked at somewhat higher rates.

The male fathead minnow is typically larger than the female, so only the larger fish should not be selected for spawning. Research at Colorado State University has shown that a sex ratio of 5 females to 1 male results in higher production. Goldfish reproduce well with approximately equal numbers of sexes.

Propagation

Cultural methods for golden shiners and goldfish are similar. Both species are termed "vegetation spawners" which prefer to deposit their eggs in shallow water, either on living plants or on artificial material, such as mats made of Spanish moss, hay, or straw.

During the spawning season, it is advisable to keep plant fertilization at a minimum. Spawning is usually curtailed when water becomes overly abundant with plant and animal life. If this should occur, rapid addition of fresh water should be made. Any shortage of natural foods is compensated for by the provision of supplemental feed.

Three general methods of propagating these species are used.

Method 1: Wild or Free Spawning Method.—Brood stock is placed in ponds containing suitable natural vegetation or artificial spawning material. It is a common practice to drop the water level in the pond during early spring to permit the growth of grass along the shoreline. When the ponds are refilled, the grass provides spawning sites. Aquatic plants also make excellent spawning materials. If plant growth is absent or scarce, hay or straw may be placed in shallow areas. Such material should be anchored so that it will not drift with wave action. Mats of Spanish moss may be used if available. These should be removed and washed when spawning ceases and hatching is completed.

Should spawning activity diminish, it may be stimulated by rapidly raising the level of the pond.

Egg laying, hatching, and growing of young to marketable size occur in the same ponds. Adults are left in the pond with the young-of-the-year. It should be noted that parasite-laden brood stock will therefore transmit the infestation to their offspring.

Method 2: Egg Transfer Method.—In this method, the brood stock is kept in ponds at a high stocking density, perhaps 450 to 560 kg per ha (400 to 500 lb per acre). It is vital that the pond be kept completely free of any natural vegetation to prevent egg deposition. Leaves and roots of marginal plants must also be prevented from entering the water along the shoreline.

When eggs are desired, specially constructed spawning mats are placed in the pond. If necessary, the pond level may be raised quickly to stimulate spawning.

Mats are constructed by placing Spanish moss between woven wire. Steel welded wire (5 × 10 cm or 2 × 4 in. mesh) is folded so that each mat will be approximately 50 × 120 cm (20 × 48 in.) in size. Selected wires are usually removed from the top side making the mesh size 10 × 10 cm (4 × 4 in.). Hog rings are used to hold the top and bottom together.

Spawning mats are placed in shallow water areas with one side of the mat at the edge of the pond. Several mats should be placed in groups in an end-to-end arrangement along the shoreline. Mats are ready for transfer when an even coverage with eggs is apparent. They should not be left in the ponds until eggs become so abundant that they touch one another. Excessive numbers of eggs on a mat encourage the growth of saprophytic fungi which may spread over developing eggs.

The number of mats to be placed in the brood stock pond is dependent on the spawning action of the fish. Only as many mats should be provided as the fish can fill with eggs in a 12 hr period. As spawning action increases, the number of mats used should be increased to prevent overloading with eggs. Care should be taken to avoid placing too many mats in a pond as it is vital that the fish utilize the spawning sites efficiently.

Method 3: Fry Transfer Method.—In this technique, fry produced by either of the methods described above are trapped, counted, and transferred to rearing ponds. Fry to be transferred are produced in hatching ponds that have been purposely overstocked with eggs. Baited lift-traps (Wascho and Clark 1951) constructed of fine mesh are used to remove fry from the hatching pond. This process should be done early in the morning or on a cloudy day to prevent injury to fry. Numbers of fry can be determined accurately and rigid control can be exercised over the number of fry stocked per hectare (acre) in rearing ponds. When properly applied, this method provides uniform production and optimal use of all pond areas. Pond space is conserved early in the growing season so that subsaleable fish left from winter storage may be grown to market size before ponds are needed for the current year's crop.

Cultural methods for fatheads and bluntnose minnows differ from those of golden shiners and goldfish largely because of a significant difference in their spawning habits. Fathead and bluntnose culture is restricted to use of either the free spawning method or the fry transfer method. A favorite method of fish culturists utilizes the fry transfer method. Spawning ponds are heavily stocked with brood stock and adequate spawning sites must be provided. Rocks, pieces of tile, bricks, or boards can be used to supplement existing sites. Some operators staple pieces of old lumber cut in 10 × 30 cm sections (4 × 12 in. sections) to a wire anchored parallel to shore in shallow water. Boards are spaced approximately 30 cm (12 in.) apart.

It would be difficult, if not impossible, to suggest which of these methods is best suited for your operation. The method that is ideal for one hatchery may be inadequate for another. Often a combination of two or more of these methods is advisable. Whenever possible, the egg transfer method or the fry transfer method should be used since these allow greater precision when stocking rearing ponds. Ideally, the farmer hopes to achieve a population of 494,000 or more fry per ha (200,000 or more per acre).

Estimating the number of egg-laden mats that should be used is a difficult task and requires considerable experience. Dependent on the number of eggs per mat, the number of mats stocked per ha may vary from 124 to 185 (50 to 75 per acre). If egg production is slow, a farmer may double or even triple the number of mats transferred. When spawning activity is very light, most of the mats should be removed. However, it is a good practice to have at least one mat in each spawning area. This will help to prevent fish from seeking less desirable areas in which to spawn.

Counting fry is a relatively easy task when compared to estimating the number resulting from the hatching of eggs on a mat. Determinations are made by first counting the number of fry in 3 cl (1 oz) liquid measure. The count is then multiplied by the number of centiliters or ounces of transferred fry. One hundred cl equals 1 liter (32 liquid oz equals 1 qt). A bucket graduated in liters (quarts) is used to transfer fry. For example, if there are 50 fry per cl, the volume of fry needed to yield a stocking of 494,000 is approximately 100 liters. (If there are 200 fry per oz then 31.25 qt are needed to stock 200,000 per acre.

Spawning

Golden Shiner (Notemigonus crysoleucas).—The golden shiner prefers lakes, but can be found in shallow waters of streams throughout the central and eastern United States. The females grow faster and larger than males. In the southern U.S., this species becomes sexually mature at 1 year of age or about 6 cm (2 ½ in.). Some individuals have been known to live up to 8 years and obtain a length of over 25 cm (10 in.).

Golden shiners start spawning (in Arkansas) when water temperatures reach 22°C (72°F), and continue spawning through June. It is reported that in Alabama, spawning begins when the temperature reaches 20°C (68°F), and continues throughout the summer with as many as 4 spawns. The adhesive eggs are deposited on vegetation at up to about 27°C

FIG. 5.37. ONE TYPE OF SPAWNING CONTAINER

(80°F). Eggs are emitted indiscriminately in areas where plant life exists. While living plants appear to be preferred, shiners will spawn on detritus or plant roots. Eggs are adhesive and cling to the surfaces of plants, debris or rocks. No protection or care is given to eggs or fry by the adults. In fact, the adults may become predators on their own fry.

FIG. 5.38.   CHECKING SPAWNING MATS

The golden shiner is omnivorous and will eat any plant or animal life that is small enough to enter its mouth. Artificial foods are readily accepted. The fish are quite delicate and extreme care needs to be exercised when they are handled in hot weather.

Fathead Minnow (Pimephales promelas).—This fish does not often grow to more than 8 cm (3 in.). The males grow larger than the females. Many anglers know this fish as the blackhead minnow because of the dark coloration on the heads of males during spawning season.

Sexual maturity is reached in about one year. Spawning commences when the water temperature nears 18°C (65°F) and continues through June.

Altman and Irwin (undated) report that spawning takes place between 18° and 21°C (64° and 70°F). Flickinger (1971) reports that spawning ceases above 29°C (85°F). Prather et al. (1953) reported that spawning began between 15.5° and 16.5°C (60° and 62°F) and continued to as late as October 1. The female deposits adhesive eggs on old lumber, tile, rocks, concrete, metal, or any object that provides space for the activity of the male. The average incubation period is about 5 days and mortality of the adults is very high after the spring spawning period (Markus

1934). Females spawn repeatedly; one was reported to have produced over 4000 offspring from 12 spawns in an 11 week period. The spawning female releases a small number of eggs at a time. After fertilization, the male picks up the eggs and places them in the nest site. Eggs are attached to the underside of any object in shallow water. This process is repeated until a nest is completed. A male may use eggs from several females to complete the nest which he then guards until fry emerge.

Like the shiner, the fathead feeds on aquatic organisms small enough to be eaten. Artificial feed is readily accepted.

With care the fathead can be handled during warm weather. It is, therefore, a popular bait fish. The species has one serious disadvantage in that many adult males die after spawning.

Bluntnose Minnow (Pimephales notatus).—The bluntnose minnow resembles the fathead minnow in appearance and in breeding and spawning habits, but in general seems to be less prolific and will withstand less crowding in minnow containers (Dobie et al. 1948).

The bluntnose has diversified food habits with no change of food preference throughout life except for size of food particles. It feeds upon plant and animal plankton both on the pond bottom and in open water. Vegetable matter consisting of blue-green algae (unicellular and small filamentous forms), and animal organisms composed of Caldocera, Ostracods, Cyclops, and midge larvae were found to be part of the diet of bluntnose minnows (Kraatz [undated]).

Breeding behavior is similar to that of the fathead minnow. Spawning occurs when water temperatures reach 21°C (70°F) and on objects like those listed for fatheads. The incubation period is from 2 to 5 days.

Brooder ponds should be stocked at a somewhat higher rate with bluntnose minnows than with fatheads, but the rearing ponds may be stocked at a comparable rate.

Goldfish (Carassius auratus).—Goldfish need no description as this species is familiar to most people. Their color may vary from white through black. Some fishermen prefer red or gold individuals; others like brown or bronze colored fish.

Goldfish normally start spawning when water temperatures reach 15.5°C (60°F). Spawning habits of this species closely parallel those of golden shiners. A female may deposit 2000 to 4000 eggs during each of several spawns.

At water temperatures of about 15.5°C (60°F) eggs usually hatch in 6 to 7 days. Only 3 days are required when temperatures are about 26.6°C (80°F).

Food habits of goldfish are very similar to those of the shiner and they grow well on artificial feed.

FIG. 5.39. GOLDFISH FRY
AFTER HATCHING

By careful selection, growers can readily control the color and shape of goldfish. This has important applications in bait fish culture. If a brown color is preferred, brood stock should be so selected. If an increase in the number of red individuals is desired, early-coloring fingerlings should be selected from the rearing ponds for potential brood stock. Such fish should be taken from the population immediately, grown in a separate pond, and ultimately used as brood stock. Body contour may be improved in a like manner if necessary. Generally, a slim-bodied variety is preferred for bait. If any of the brood stock is undesirable in shape (heavy, thick-bodied) it should be removed before spawning occurs.

The goldfish is a hardy species which attains a large size if not overcrowded. It is often plagued by many parasites and diseases which do not infest shiners or fathead minnows.

Goldfish have been blamed for destroying game fish habitats in a manner similar to carp if they are released in natural waters. Consequently, the sale of this species is banned in certain states.

While not as widely accepted as the golden shiner because of their sluggishness on a hook, goldfish make an ideal trotline bait fish.

Rearing Ponds

Fry should be trapped or seined from the shallow water of the brooder ponds and transferred to rearing ponds when they have reached 2 to 3 cm (¾ to 1 in.) in length. Fry should be stocked at a rate of 247,000 to

494,000 per ha (100,000 to 200,000 per surface acre) of rearing pond water depending upon the fertility of the water and the amount of supplementary feeding to be done.

Rearing ponds, like brooder ponds, should be freed of parasites, diseases, and predators before minnows are stocked. If the ponds have remained dry for some time treatment may not be necessary, but ponds that do not become dry over winter should be poisoned before stocking. Five percent rotenone power applied at the rate of 3.7 to 5.5 g per ha-m (1 ½ lb per acre-ft) and thoroughly mixed in the pond water should destroy predatory fish present. Calcium hypochlorite is an effective agent for the control of crayfish and some other forms.

Put 15 to 20 cm (6 to 8 in.) of water in the deep end of the rearing pond and stock with the desired numbers of minnow fry, fertilize if necessary, then fill slowly by increasing the water depth 2.5 cm (1 in.) per day. Most operators expect to produce saleable minnows in 5 months, but it seems possible to reduce this time to as little as 1 month if proper growth can be maintained. Supplemental feeding and/or fertilizing may increase minnow production and growth, but caution in their application is essential.

Minnows too small for sale as bait in the fall or minnows kept for brood stock can be stocked in overwinter holding ponds. Minnows stocked in these ponds may be crowded somewhat more than in ponds during the summer season. Some supplemental feeding will probably be desirable to carry the minnows through the winter and to promote growth of the small ones to prepare them for early market.

To have fish of the right size at times of market demand, vary the time of spawning, the density of fish in the nursery pond, or the density of fish in the rearing pond. Growth of shiners may be rapid at 185,000 fish per ha (75,000 per surface acre), moderate at 370,000 per ha (150,000 per acre) and slow at 494,000 per ha (200,000 per acre).

The mortality experienced from stocking shiners and goldfish fry in rearing ponds is not known, but fry 2.5 cm (1 in.) long of fathead minnows may have a 20 to 25% mortality. To produce saleable young-of-the-year fathead minnows for the fall and winter markets, low densities (247,000 fish per ha or 100,000 per surface acre) must be used to obtain maximum growth. Although weight of individual fish increases rapidly at low densities, total weight does not approach carrying capacity because there are so few fish. Consequently, pond space is used inefficiently, but sometimes this procedure is necessary to meet market demand. To produce saleable yearlings for the late spring and early summer markets, medium densities (around 500,000 per ha or about 200,000 fish per surface acre) would be the best choice. At medium densities growth of individual fish is slower, but total weight increases rapidly, and pond space is used more efficiently. Higher densities (up to 750,000 fish per ha or 300,000 fish per surface acre) are used to produce saleable fish later in the second summer. Also by stocking earlier or later the first summer, saleable fish will be available correspondingly earlier or later the second summer. Size of fingerlings at time of stocking also affects the length of time necessary to produce a saleable fish. Fathead minnows averaging about 2200 fish per kg (1000 fish per lb) will easily reach saleable size in one growing season. However, fathead minnows averaging more than 17,600 fish per g (8000 fish per lb) will not reach saleable size in one growing season unless low densities are used. Because saleable and maximum sizes are nearly the same for the fathead minnow, growth starts to slow as the fish approach saleable size. Optimum harvest time comes during this period of change from accelerated to decelerated growth because of low rearing costs associated with rapid growth.

Similar results from density of stocking occur for shiners, bluntheads, suckers, or goldfish. Experience over time will indicate the general stocking rates for each species, depending on spawning time, density, fertilization, and supplementary feeding which will permit the individual to have the proper sized fish for his markets at different seasons of the year.

The rate of growth of the young fathead under normal conditions is shown as follows:

Number of Days	Length of Fish (cm)	(in.)
10	0.51	0.20
20	1.02	0.40
30	1.47	0.58
40	1.93	0.76
50	2.49	0.98
60	2.92	1.15
70	3.43	1.35
80	3.89	1.53
90	4.45	1.75
100	4.95	1.95
110	5.46	2.15
120	5.89	2.32

Fathead production is usually expressed in kg (lb), and wholesale minnow sales are usually in liters (gal.). Table 5.7 presents the number of fatheads of various sizes per unit of weight and per unit of volume.

TABLE 5.7. RELATIONSHIP BETWEEN FATHEAD MINNOW LENGTHS, WEIGHT PER KG (LB), AND NUMBER PER LITER (GAL.)

Length of Fish		Number of Fish		Number of Fish	
cm	in.	per kg	per lb	per liter	per gal.
2.54	1.0	5,720	2,600	5,495	20,800
3.81	1.5	1,628	740	1,563	5,920
4.08	2.0	660	300	634	2,400
6.35	2.5	330	150	317	1,200
7.62	3.0	185	84	178	672
8.89	3.5	114	52	110	416
10.16	4.0	75	34	72	272

[1] 1 gal. = 8 lb.

The rate of growth of golden shiners in ponds is as follows:

Number of Days	Length (cm)	(in.)
10	0.76	0.3
20	1.53	0.6
30	2.29	0.9
40	3.05	1.2
50	3.81	1.5
60	4.57	1.8
70	5.33	2.1

Golden shiner production is usually expressed in kg (lb) and wholesale minnow sales are usually in liters (gal.). Table 5.8 presents the number of golden shiners of various sizes per unit of weight and per unit of volume.

TABLE 5.8. RELATIONSHIP BETWEEN GOLDEN SHINER LENGTHS, WEIGHT PER KG (LB) AND NUMBER PER LITER (GAL.)

Length of Fish		Number of Fish		Number of Fish	
cm	in.	per kg	per lb	per liter	per gal.
2.54	1.0	9,350	4,250	8,982	34,000
3.81	1.5	2,464	1,120	2,367	8,960
4.08	2.0	946	430	909	3,440
6.35	2.5	473	215	454	1,720
7.62	3.0	260	118	249	944
8.89	3.5	156	71	150	568
10.16	4.0	103	47	99	376
11.43	4.5	70	32	68	256
12.70	5.0	51	23	49	184
13.97	5.5	35	16	34	128
15.24	6.0	26	12	25	96

[1]1 gal. = 8 lb.

Even though propagation of the shiner requires more expertise than other bait fish and stocking rates of rearing ponds and total numbers are less per hectare (acre) of water (due to larger sizes), it is the most important bait fish. It also requires higher temperatures and thus has a shorter growing season. It has up to 10,000 eggs per female but an ovarium parasite often reduces this number greatly. It is also not as hardy and requires careful handling. It has the same diseases as the fathead plus *Plistophora* and occasionally *Ichthyophthirius*. The fathead doesn't winter as well and small fish (more than 12,200 per kg or 5000 per lb) do not survive well. Postspawning mortality may range up to 90%.

Rearing culture is usually conducted in one of two ways: either an intensive or an extensive type of farming is applied. The culture method used by an operator will depend on the nature and size of his ponds, the availability of water, the experience of the fish culturist, and the size of the operation. Intensive culture permits a greater utilization of land area but demands more water per ha (acre) than the extensive method. When adequate land is available but water is limited or difficult to distribute, an extensive type of farming is usually used. Questions concerning the type of farming to be employed should be resolved before a hatchery site is purchased.

Under intensive farming, growth of the fish is based primarily on supplemental feed. Feeding is usually done at the rate of 3 to 5% of the body weight of the fish per day but will need readjustment according to the size of the fish. Amounts of feed should be reduced if an excess is observed on the ponds 15 to 20 min after the fish are fed. Fry may require more time to "clean up" their feed but should be fed more often if labor permits.

Ponds are flushed continually with fresh water. Production levels increase with amount of water added.

Well-managed ponds under intensive culture will yield a high production of uniformly sized fish. Size control is possible. Average production figures for the intensive culture of bait fish vary from 450 to 1010 kg per ha (400 to 900 lb per acre). Greater production may be achieved in individual ponds.

Extensive fish culture utilizes only enough water to fill the pond and maintain water level. Supplement feed is used sparingly and is replaced by natural food acquired through fertilization.

Under extensive culture, large pond sizes prevent precise control of the fish population. Since most of the growth is dependent upon natural foods, size variations may be extreme within ponds. Often, a large fraction of the fish may be too large to command optimum prices. Production figures also vary greatly. Experienced producers consider 280 kg per ha (250 lb per acre) as average.

The choice between intensive culture and extensive culture often depends on the economic importance of bait fish farming to the operator. Intensive culture requires considerably more time than the other method. Where other types of farming operations are also involved, the extensive culture may be preferred. Consideration should be given to the amount of permanent and seasonal help available.

Experienced fish culturists differ concerning the ideal size of ponds. Generally, small ponds of 2 ha or less (5 acres or less) are used in intensive culture. Where the egg transfer method is used, it is advisable to complete stocking of a particular pond within 3 to 5 days to help acquire uniformity of fish size. This is usually impossible in large ponds.

A comparison of the merits of intensive versus extensive rearing methods is in order.

The advantages of intensive culture are:

(1) It permits greater control of size of fish produced; therefore, manipulation of population
(2) It requires less land area
(3) Harvesting is simplified
(4) Control of vegetation and disease is easier
(5) Partial harvesting of marketable fish can be employed to provide space for growth of the subsaleable individuals.

Disadvantages include:

(1) More water and more feed are used per ha or acre than in the extensive type of culture
(2) More labor is required
(3) Increased danger from oxygen depletion
(4) Epizootics of parasites and diseases are more common in the crowded situations of intensive farming.

Extensive farming has the following advantages:

(1) Less water per ha or acre is required
(2) Less labor is required per ha or acre of ponds
(3) Dangers from oxygen depletions are minimized
(4) Less investment in feed.

Disadvantages include:

(1) Control of weeds and diseases is often difficult and always costly
(2) The operator has less control of size of fish produced
(3) Large land areas are required
(4) Partial harvesting is less effective as a means of increasing and/ or controlling production

(5) Harvesting is usually more difficult and often may be limited to cool weather.

Water Level

Maintenance of a suitable water level in a pond is essential for the proper growth and reproduction of minnows. Brooder ponds, especially during the breeding season, should be maintained at a constant water level.

Evaporation and seepage of pond waters will be determining factors in the amount of water which must be kept flowing into the pond to maintain the desired water level. Seepage of water has an added undesirable feature in that the fertility of the pond in the form of nutrients can be carried away or otherwise made inaccessible.

Minnows seem to feed better and grow faster when the water level of a pond is rising. Because of this fact, certain management procedures which make use of a rising water level are beneficial.

Manipulation of the water level of a rearing pond is sometimes beneficial when the minnows are not feeding or growing properly. Minnows can often be induced to resume feeding and growth by suddenly decreasing the water level (by draining part of the water) and then filling the pond again by slowing adding the water until the pond level is brought back to normal.

Supplemental Feeding

"At least twice as many fish can be raised per acre of water by artificial feeding" (Dobie *et al.* 1948).

The process of supplemental feeding is one which should be undertaken with extreme caution by the beginner in the minnow rearing business. An overabundance of food added to the water can bring about an oxygen depletion and resulting loss in minnows.

Supplemental feeding, particularly during the hot summer months, calls for close surveillance of the ponds to make certain that the amount of food being fed does not exceed the amount being utilized by the minnows.

Day by day weather, water temperature, minnow size, and a great many other factors exert a controlling influence on the amount of food which may be fed with safety to the minnows in a pond.

It is a safer practice to begin the supplemental feeding program by administering only a small amount of food. This amount may be increased in a few days if the fish are consuming all that is being placed in the pond. Periodic increases in the amount of food supplied can be introduced until the maximum amount of food that a pond can safely take is reached.

The prepared mixture of food is generally scattered daily at the same time and at the same place on the surface of the water along the shallow end of the pond.

A few hours after feeding or even the next morning an inspection of the pond area where the food was placed will reveal the degree of food utilization by the minnows. If all the food is being eaten the amount may be increased. Cloudy days or days with a minimum of sunshine can upset the feeding program. It may be necessary to reduce or suspend feeding for a few days.

Food formulas vary widely, depending a great deal on the availability and cost of materials to be used. Some hatcherymen are supplementing regular food formula by the addition of some of the antibiotics which are now present in livestock feed. However, no published data on amounts or types being used are available at present.

There can be numerous diets which could be adequate. A few suggestions follow.

Naturally occurring foods in ponds are not sufficient to produce maximum fish growth, but they are necessary to prevent dietary deficiencies from commercially prepared food. Dr. Walter H. Hastings, nutritionist for the Fish Farming Experimental Station, Stuttgart, Arkansas, advises these formulas for fish foods:

Starter Feed

Ingredients	lb	kg
Soybean flour	200	440
Blood flour or feather flour	100−200	220− 440
Oat flour or ground wheat	800−900	1760−1980
Distiller solubles	100−200	220− 440
Delactose whey or skim milk	200	440
Gluten meal	100−200	220− 440
Fish meal	200	440
Vitamin concentrate	20	44
Minerals	50	110

Grower Feed

Ingredients	lb	kg
Soybean flour	200	440
Blood flour	100	220
Skim milk	100	220

Ingredients	lb	kg
Alfalfa meal or reground pellets	250	550
Fish meal	200	440
Poultry by-products meal	150	330
Ground wheat	950	2090
Vitamin concentrate	20	44
Minerals	30	66

Ingredients in the starter feed should go through a micropulverizer to ensure that the particle size is small enough for young fish to ingest. About 1 kg of food per ha (1 lb of food per acre) should be scattered around the entire pond twice daily in order to train newly hatched fry to eat commercially prepared food. As the fry get accustomed to feeding in about three weeks, the feed may be broadcast downwind along one side of the pond. The grower formula is designed primarily for feeding larger fish, and it should be fed in pellet form since greater utilization is attained with pelleted food. If it is impractical to make small pellets, larger pellets may be made and then reground to a small particle size. Most broiler (chicken) rations are also adequate for pond fish if the texture is suitable.

Fish should be fed 2 to 3% of their body weight per day and thus the amount of food needed will increase as the fish grow. A feeding schedule based on food conversion and mortality rates may be computed and used to increase the amount fed. Another method is to weigh and to count a sample of fish every two weeks to determine growth. If fast growth is needed, adjust the feed allotment daily, feed more than 3% body weight per day, and feed more than once a day. When large quantities of food are used, pollution can occur from a waste food and fecal deposition. Such buildup sometimes causes oxygen depletion resulting in heavy loss of fish. Up to 45 kg of food per surface ha (up to 40 lb of food per surface acre) per day may be fed if the ponds can be flushed with a larger volume of water when problems arise.

Prather et al. (1953) used different feeds for goldfish and fathead minnows but did not give a feed for shiners. For goldfish they advised feeds having a protein content of 30 to 40% (Tables 5.9 and 5.10) Since at that time there was little difference in the costs of feeds used in the experiments, they recommended the one that gave the highest production.

Feeds for goldfish production listed in order of decreasing value were:

(1) Mixture of 22.7 kg (50 lb) hog supplement and 22.7 kg (50 lb) poultry laying mash
(2) Soybean meal or cake

TABLE 5.9. PRODUCTION OF GOLDFISH ATTAINED IN 161 DAYS BY USE OF SUPPLEMENTAL FEEDS

| Treatments[1] | Goldfish Production per 0.4 ha (1 Acre) | | | | Feed Required to Produce a kg or lb of Goldfish | |
| | Total[2] | | Due to Supplemental Feeding | | | |
	kg	lb	kg	lb	kg	lb
Fertilizer only	128.9	283.6	—	—	—	—
Fert. + wheat shorts (17% protein)	471.0	1036.1	342.0	752.5	5.09	5.09
Fert. + poultry laying mash (20% protein)	462.2	937.7	297.3	654.1	5.85	5.85
Fert. + poultry laying mash + hog supplement (30% protein)	648.2	1426.1	519.3	1142.5	3.35	3.35
Fert. + soybean meal (41% protein)	585.8	1288.7	456.9	1005.1	3.81	3.81

[1]Ponds stocked at rate of 26,000 goldfish per 0.4 ha (1 acre); fertilizer treatment, 6 applications of 90 kg (200 lb) each of 8-8-2 per 0.4 ha (1 acre); supplemental feed, 1723 kg (3828 lb) per 0.4 ha (1 acre) or 26,000 goldfish per 0.4 ha with 6 applications of 90.9 kg each of 8-8-2 per 0.4 ha and 1740 kg of feed per 0.4 ha.
[2]Average results of 2 experiments run at different times.

TABLE 5.10. PRODUCTION OF GOLDFISH ATTAINED IN 175 DAYS BY THE USE OF SUPPLEMENTAL FEEDS

| Treatments[1] | Goldfish Production per 0.4 ha (1 Acre) | | | | Feed Required to Produce a kg or lb of Goldfish | |
| | Total[2] | | Due to Supplemental Feeding | | | |
	kg	lb	kg	lb	kg	lb
Fertilizer only	139.0	305.9	—	—	—	—
Fert. + wheat shorts (17% protein)	385.9	848.9	246.8	543.0	6.27	6.27
Fert. + cottonseed meal (36% protein)	405.0	890.9	265.9	585.0	5.82	5.82
Fert. + red dog flour (18% protein)	412.0	906.5	273.0	600.6	5.67	5.67
Fert. + poultry laying mash (20% protein)	512.5	1127.5	373.5	821.6	4.14	4.14

[1]Ponds stocked at the rate of 26,000 per 0.4 ha (1 acre); fertilizer treatment, 4 applications of 90 kg (200 lb) each of 8-8-2 per 0.4 ha (1 acre); supplemental feed, 1533 kg (3406 lb) per 0.4 ha (1 acre) or 26,000 fish per 0.4 ha, 4 applications of 90.9 kg each of 8-8-2 per 0.4 ha and 1548 kg per 0.4 ha of feed.
[2]Average of 3 replications.

(3) Peanut meal
(4) Poultry laying mash
    Wheat shorts
    Red dog flour
    Cottonseed meal

No significant difference in goldfish production from these feeds. Select the feed with the lowest cost.

It should be noted that although the cost of producing 1 kg (1 lb) of goldfish was cheapest where soybean meal was used, the highest weight of goldfish per ha (acre) was obtained with a mixture of laying mash and hog supplement.

For best results the materials used in supplemental feeding should be fed daily at about the same hour and at the same location. After several days the fish become accustomed to regular feeding periods and congregate near the areas where they are fed much the same as do chickens and hogs.

It is not necessary to mix water with feeds. The dry materials are usually thrown by hand into shallow water along one side of the pond much the same as in fertilization.

Supplemental feeding rates for young goldfish in rearing ponds (Table 5.11) appear to be satisfactory where there is little or no overflow of water.

TABLE 5.11. FEEDING SCHEDULE AT 3 DIFFERENT RATES

| Age of Goldfish in Months | Feed per 0.4 ha (1 Acre) per Day | | | | | |
| | Rate I | | Rate II | | Rate III | |
	kg	lb	kg	lb	kg	lb
0−1 (April)	4.5	10	4.5	10	4.5	10
1−2 (May)	4.5	10	4.5	10	4.5	10
2−3 (June)	4.5	10	9.1	20	13.6	30
3−4 (July)	4.5	10	9.1	20	13.6	30
4−5 (Aug)	4.5	10	9.1	20	13.6	30
5−6 (Sept.)	4.5	10	9.1	20	13.6	30
6−7 (Oct.)	4.5	10	9.1	20	13.6	30
7−8 (Nov.)	4.5	10	9.1	20	13.6	30
8−9 (Dec.)	4.5	10	9.1	20	13.6	30
Fish production per 0.4 ha (1 acre) from fertilization + feed	818	1800	955	2100	1023	2250

Recently hatched goldfish fry are quite small and, therefore, need only small quantities of supplemental feed in addition to the tiny plants and animals upon which they normally feed. Although many hatcherymen feed recently hatched fry every 2 to 3 hr for the first several days, such frequent feeding is not necessary since the young fish have abundant quantities of natural food present in fertilized ponds. A higher rate of feeding is obviously required as the minnows increase in size.

The rates of feeding shown in Table 5.11 can be fed safely in shallow ponds with no danger of fish being killed. Higher rates of feeding can be used. The fisheries station at Auburn, Alabama, has fed 22.7 kg (50 lb) of feed per 0.40 ha (1 acre) per day safely in shallow ponds with no overflow water without killing fish; commercial producers occasionally use up to 45.5 kg (100 lb) of feed per day where water is available to flush ponds in event of low oxygen. In hot weather, ponds should be inspected frequently where heavier rates of feeding than those listed are used since the decomposition of feces, uneaten food, and dead plankton organisms may cause the supply of oxygen to become dangerously low.

The same supplemental feeds were fed to fathead minnows as for goldfish. The production rates are shown in Table 5.12.

TABLE 5.12. THE EFFECT OF VARIOUS RATES OF FEEDING ON FATHEAD PRODUCTION IN FERTILIZED PONDS

Treatments[1]	Feeding Period	Gain over Fertilization Only kg	lb	Total Production per 0.4 ha (1 acre) kg	lb
Fertilizer only	—	—	—	67.3	148.0
Fertilizer + 1250 kg (2750 lb) poultry laying mash	Nov. 1–Feb. 18	163.8	360.4	231.1	508.4
Fertilizer + 1318 kg (2900 lb) soybean cake	Oct. 1 –Feb. 25	325.2	715.5	392.5	863.5
Fertilizer + 1709 kg (3760 lb) soybean cake	Oct. 1–April 4	474.0	1042.8	541.3	1190.8

[1]Fertilized Feb. 3 to Sept. 8, with 727 kg (1600 lb) 6-8-4 and 18 kg (40 lb) nitrate of soda.

Prather *et al.* (1953) had difficulties in feeding shiners during the summer. They stated that in experiments at Auburn, Alabama, a kill from oxygen depletion occurred in all ponds where shiners were fed in the summer. Consequently, in this region golden shiners should not be fed during this period. A low rate of feeding, not to exceed 11 kg of soybean cake or meal per ha per day (10 lb per acre per day) can be used in the winter without too much danger of loss from oxygen depletion.

While feeding cannot be used in the summer, fertilization can be used safely to increase production when a pond is stocked properly. From 30,000 to 235,000 shiners weighing 421 to 674 kg (375 to 600 lb) can be raised per ha per year. If 674 kg per ha (600 lb per acre) are produced, the maximum number that can be raised to an average length of 7.5 cm (3 in.) is 28,000 per ha (70,000 per acre).

Using Prather's procedures the average production was 185,250 minnows and 561 kg per ha (75,000 minnows and 500 lb per acre). Of these, 80 to 90% were 7.5 cm (3 in.) or larger.

It is recommended that changing the ingredients of a minnow food should be gradual. By gradual substitutions, changes can be made.

If fish suddenly stop feeding, trouble is developing. Disease and low oxygen are the two most common reasons for an abrupt cessation of feeding.

Eliminating Undesirable Fish

It is important that all fish be removed from hatchery ponds before they are restocked since any remaining fish will seriously reduce production in the succeeding crop. Usually time does not allow these ponds to be completely dried, so that poisons must be used. The poisons recommended are powdered derris, powdered cube, or emulsifiable rotenone each containing at least 5% rotenone. Any of these poisons used at the rate of 11 kg per ha-m (3 lb for each acre-ft) of water and thoroughly mixed with the water in the pond will kill most fish usually present in hatcheries if the water temperature is above 15.6°C (60°F). Where the pond is treated with chlorinated lime or hydrated lime as a disinfectant, it is not usually necessary to poison as the lime will kill the undesirable fish.

Emulsifiable rotenone gives the best results and is the easiest to handle since it is a liquid and requires no mixing before application to the pond. The price of emulsifiable rotenone is higher than that of powdered derris or cube which requires preliminary mixing with water. Emulsifiable rotenone is the most effective of the three in controlling top minnows when it is thoroughly stirred into the pond water. When the powdered forms are used very often some of the top minnows are not killed because it is difficult to mix the poison with the thin upper layer of water even when the pond is stirred by an outboard motor.

The procedure for poisoning a pond is as follows:

(1) Partially refill the pond so that all wet places are covered with water.
(2) Carefully estimate the area and average depth of the water to be treated.
(3) From the surface area and average depth find the amount of poison necessary in Table 5.13. The figures in the table represent grams (ounces) by weight if a powder is used, or fluid centiliters (ounces) if emulsifiable rotenone is used.
(4) Weigh or measure the amount of poison required. If powdered derris or cube is used it should be mixed with a small amount of water to form a paste. Then the paste should be diluted with additional water sufficiently to permit even dispersion of the poison. The emulsifiable rotenone may be diluted with water before application.
(5) Place an outboard motor mounted on a sawhorse in the pond. Gradually pour the diluted poison into the "prop wash" of motor, stirring until the poison is evenly distributed. By moving the motor the pond can be treated from several positions, giving more thor-

ough coverage. This is desirable when the surface area of the pond treated is over 1000 m² (¼ acre).

(6) After the pond is completely stirred, wait 24 to 48 hr or more, then drain it completely.

(7) The pond can then be completely refilled and is ready for use. Care should be taken that higher concentrations of poison than those recommended not be used and that the poison that remains after draining be sufficiently diluted so that it will not kill fish. Allow at least a week or more for the rotenone to detoxify before stocking.

TABLE 5.13. AMOUNT OF CUBE, DERRIS, OR EMULSIFIABLE ROTENONE REQUIRED TO KILL FISH IN PONDS OF VARIOUS SIZES AND DEPTHS

| Surface Area | | \multicolumn{6}{c}{Average Depth of Water} | | | | | |
m²	Acres	15 cm cl	6 in. oz	30 cm cl	12 in. oz	45 cm cl	18 in. oz
400	¹⁄₁₀	8.9	3	14.8	5	23.7	8
800	⅕	14.8	5	29.6	10	44.4	15
1000	¼	17.8	6	35.5	12	53.3	18
1200	³⁄₁₀	23.7	8	44.4	15	65.1	22
1600	⅖	29.6	10	59.2	20	85.8	29
2000	½	35.5	12	71.0	24	106.6	36
2400	⅗	44.4	15	85.8	29	130.2	44
3000	¾	53.3	18	106.6	36	159.8	54

Harvesting

When only a small portion of the minnows in a pond are to be sold immediately, the pond can be seined or trapped. However, draining is the only method for harvesting the entire pond, although seining is most commonly used.

Before draining or seining there must be facilities to haul, grade, hold or dispose of the minnows harvested.

The ponds can be partially drained, and the minnows concentrated in either the holding area or catch basin. If they are concentrated for more than one hour, fresh well-oxygenated water should be passed through the catch basin. The minnows are harvested using a minnow seine or dip net and transferred to tanks or drums of water.

When minnows are brought from ponds to holding tanks, they should be placed in water that is only slightly cooler than the summer pond water. Then a cold water flush should be maintained. After fish have had a day to recover from the stress of harvesting, they should be graded into sizes.

Regardless of harvesting method, excessive handling at one time should be avoided. A good rule of thumb is harvest one day, grade the following

FIG. 5.40. ONE METHOD OF HARVESTING USES A HARVESTING BASIN

FIG. 5.41. HARVESTING BAIT FISH BY NET IN A POND

day, and then transport the third day. If water temperatures are above 26.7°C (80°F), fish are difficult to handle without their suffering considerable mortality. If possible, fish should be harvested in the coolness of early morning or on cloudy days.

The most common method of harvesting bait fish is seining. In small ponds the seine should span the entire width of the pond. However, a seine haul can start anywhere along the length of the pond depending upon how many fish are needed. Spanning large ponds is impractical.

FIG. 5.42. THINNING OUT THE BAIT FISH POPULA-TION BY PARTIAL HAR-VESTING USING A TRAP NET

Baiting fish into one corner of a large pond and subsequently entrapping them with a seine pulled diagonally across the corner is effective, especially with fathead minnows.

The best all-around mesh size for a seine is 14.8 mm ($\frac{3}{16}$ in.). However, subsaleable fathead minnows will become caught by their gill covers in this mesh. Mesh of $\frac{1}{8}$ in. reduces gilling, but it is more expensive and picks up more debris. Some minnow farmers use 4.6 mm ($\frac{1}{4}$ in.) mesh for medium and large golden shiners. Woven mesh produces less injury than knotted mesh. Another requirement for a seine is sufficient length and width to allow the bag to form properly. To estimate proper size one should have a seine that is approximately 50% longer than the width of the pond and twice as wide as the average depth of the pond. For example, if the pond is 30.5 m wide and 1.2 m deep (100 ft wide and 4 ft deep), the seine should be 45.7 m long and 2.4 m wide (150 ft long and 8 ft wide). In ponds with soft mud bottoms, seines equipped with double-braided, manila rope lead lines are excellent for harvesting fish. This type of seine does not readily pick up crayfish, which seriously injure fish when trapped in a net.

The fish can be counted or graded for size when harvesting small lots or they can be transferred ungraded to holding tanks. If they are stressed during harvesting they should be held in tanks for 24 hr to allow the fish to overcome harvesting shock prior to grading.

The fish can be graded with mechanical fish graders, counted by hand, or weighed. If counted by hand, a sample should be counted and weighed and then, based on weights and number, the remainder can be weighed.

Fatheads are graded much the same way as goldfish but instead of several size groups they are separated into those large enough for bait and those that are too small. Marketable size will depend on the markets but bait size is usually 5 cm (2 in.) or more. Smaller minnows can be discarded or restocked to grow to marketable sizes.

The fish to be restocked for further growth are stocked at the rate of 247,000 per ha (100,000 per acre) and grown out as long as there is at least a month between the time of stocking and the normal spawning period. Since fatheads are not as hardy as goldfish they must be handled with greater care and speed, and since fatheads cannot stand crowding the use of oxygen is recommended when holding or transporting them. This is also true with shiners which are even less hardy than fatheads. Shiners are also separated into two size groups, those of marketable size and those too small. Marketable size is usually 7.5 cm (3 in.) or larger. The smaller fish can be discarded or restocked.

FIG. 5.43. HAULING BAIT FISH TO A PACKING HOUSE FOR SEPARATION BY SIZE

White or Common Sucker (Catostomus commersonii)[1]

This fish is widely distributed in the United States east of the Great

---

[1]This section is taken from Dobie *et al.* (1956) pages 69–78. The treatise is one of the few dealing with the white sucker. The information has been updated to include the latest techniques.

FIG. 5.44. HOLDING TANK FOR GOLDFISH

FIG. 5.45. SCREENS AND BASKETS FOR SIZING BAIT FISH OR GOLD-FISH

Plains from northern Canada south to Arkansas and Georgia. It prefers clearwater lakes and streams. The sucker runs upstream to spawn early in the spring. Temperatures of 13.8° to 20.0°C (57° to 68°F) are best for hatching eggs. In this temperature range the incubation period is 5–7

FIG. 5.46. AUTOMATIC SEPARATION MACHINE FOR SIZING BAIT FISH

days. Nearly 50,000 eggs have been taken from only one female.

The sucker has diversified feeding habits. It seems to feed on any food that may appear in the water. A study of 1080 suckers from Minnesota natural ponds shows the average food content to be Cladocera, 30.6%; copepods, 17%; ostracods, 2.4%; chironomid larvae, 26.4%; miscellaneous insects, 1.5%; rotifers, 10%; protozoans, 0.8%; nematodes, 0.6%; and miscellaneous organisms, 10.7%. This list suggests that the planktonic crustaceans are the preferred food of the sucker, but a closer study reveals that chironomid larvae are eaten whenever they are available irrespective of crustacean abundance. Very small suckers prefer small organisms but can exist on larger forms when necessary.

The common, or white, sucker is a popular minnow for propagation because it is easy to raise in large numbers, grows rapidly, is very hardy in the minnow pail, and is preferred by fishermen as a bait for walleyed pike. Suckers are raised more cheaply in natural ponds than in artificial because the sucker needs a large amount of growing space that can be provided more cheaply in natural ponds.

Production.—The white sucker is naturally a fish of clear waters, so ponds for its production must be selected more carefully than those used for other bait species. Experience has shown that the following points are important in choosing sucker ponds:

(1) Ponds of moderate fertility usually produce the most suckers. Sterile ponds do not produce enough food for the fish and very fertile ones often produce enough algae to cause summer kill. Any pond that becomes pea soup-green should not be stocked with suckers because production will be very small. If the pond is over 10 ft deep and the algal bloom is moderately heavy, the by-products of algal decomposition will be dispersed widely enough to make a fair sucker production possible.

(2) Ponds with large populations of chironomid fly larvae, or bloodworms, in the bottom muds will produce good sucker crops more consistently year after year than ponds that do not have an ample supply of these larvae.

(3) The texture of the pond soil is very important. Ponds with loam and sandy loam soils produce best, peat and peat loam ponds average, and silt and clay loam soils poorly. The pond soil is important in its effect on water fertility and the production of chironomid fly larvae.

(4) Ponds with heavy, mosslike growths of filamentous algae over the bottom do not produce good crops of suckers. One Minnesota pond always produced large sucker crops until the filamentous algae got started and covered the entire bottom. Since then production has been almost zero. This is possibly because filamentous algae decompose readily and produce toxic ammonia just as the waterbloom does, and because the algal mat on the bottom may interfere with the feeding activities of the sucker.

Collecting the Eggs.—Suckers and minnows that spawn in running water are usually stripped and the eggs are hatched in jars. Taking eggs from the sucker and fertilizing them is not difficult, but considerable strength is required. The sucker not only is large, but it also is one of the most active and powerful fish for its size native to our waters. On the upper Mississippi and its tributaries, suckers literally swarm during May and June over the shallow, rocky stream bottoms in swift water, as well as along the rocky, wind-driven shores of many of the northern lakes.

The fish in these spawning runs are caught with seines or traps and are sorted. The ripe males and females are carefully put into separate tubs of water and the unripe fish are released for another day. If those selected for stripping do not give their eggs and milt freely under light pressure, with the thumb and forefinger moved downward over the abdomen toward the vent, they should also be released. Eggs forced from the fish by heavy pressure will not prove fertile. The males mature somewhat earlier in the season than the females, and the bulk of them may have moved higher upstream than the point at which the bulk of the females are taken, resulting in a local scarcity of males. Both sexes would be available, however, if the fish were caught as they ran up the stream and were put into a suitable holding pond until needed for stripping operations. In any event, eggs should not be taken unless a ripe male is immediately available for fertilizing them.

The female is held over a dampened pan into which the eggs are expressed. Immediately after the eggs are taken, a male is stripped of his sperm; the milt and eggs are thoroughly mixed by gently swirling the pan. Four or five pairs of fish may be stripped into one pan providing each batch of eggs and milt is thoroughly mixed immediately after stripping. After a lapse of 2 or 3 min, water may be slowly added to the pan and the stirring continued at intervals by rocking the pan gently to and fro, swirling the water.

The milt can now be washed out by frequent changes of water. If the eggs have a tendency to stick together in clumps, a cup of muck or cornstarch of the consistency of bean soup should be added as the eggs are stirred. The muck or cornstarch is then washed out with the milt.

After being washed, the green fertilized sucker eggs are transferred to a tub to harden. The tub is placed in cold creek water and is shaded from the sun. Periodically the eggs are stirred gently and the water is changed. After 2 hr the eggs are hard enough to withstand the rigors of transportation to the hatchery.

Artificial Hatching.—At the hatchery, the eggs are transferred to Meehan hatching jars. Usually 2 or 3 liters or quarts of eggs are placed in each jar, and the water is adjusted so the eggs are in constant but gentle movement throughout the lower portion of the jar. For best results, the water should contain sufficient dissolved oxygen for the eggs, but should be free of air bubbles because the bubbles adhere to the eggs and carry them up and out of the jar. Length of hatching time depends on temperature of the water. Eggs will hatch in 1 to 6 days in water warmer than 12.3°C (65°F), in 10 to 15 days in water of 10° to 15.6°C (50° to 60°F), and not at all in water colder than 10°C (50°F). Minnesota hatchery operators prefer to start the eggs at 10°C (50°F) to prevent clumping, and then increase the temperature to 12.8° to 15.6°C (55° to 60°F) for hatching, if possible. In some states, the water from lakes and streams reaches optimum temperature in time to be used in the hatching battery.

Fortunately, sucker fry stay in the jars after hatching, and do not swim out with the water until they are about 5 to 10 days old. Consequently, the fry can be held in the jars until they are free swimming, and are not put in the pond until they are strong enough to search for food. Because the suckers stay in the jar and settle to the bottom when the water is turned off, it is very easy to determine the number on hand and the number to be used in each pond. The fry can be poured into a glass measure graduated in milliliters (ounces) and measured after they settle. Counts made in Minnesota indicate that there are 2720 5-day-old sucker fry per 28 g (1 oz). Suckers grow rapidly in ponds. In 60 days they average 7.1 cm (2.8 in.) and may reach a length of 8.9 cm (3.5 in.) in that time.

Stocking the Ponds with Fry.—While it is important for the operator to know how many fry per ha or acre to stock in each pond, there is little exact information on the subject. Some years ago the Minnesota Bureau of Fisheries set an arbitrary figure of 40,000 fry per 0.4 ha or acre, because that was a fair division of the fry available for distribution. Experience has since shown that this stocking rate is satisfactory. Recent studies indicate that certain ponds tend to produce only so many fish no matter how many are stocked. Of course, understocking produces fewer fish, but overstocking wastes fry. If the supply of fry is limited, the dealer should experiment until he knows the optimum stocking rate for each of his ponds. For example, when a Minnesota pond was stocked with 69,000 fry to 0.4 ha or acre of water, the survival was 17%; when 51,000 fry were stocked, the survival rate increased to 27%; and when 37,000 fry were stocked, the survival reached a peak of 40%. The number of fish produced per ha during the 3 years was 29,650, 34,600, and 37,100 fish of pike-bait size (12,000, 14,000, and 15,000 per acre). The survival rate for suckers in all Minnesota ponds under observation during these 3 years averaged 22% and reached a high of 50%.

Producers who are confronted with seasonal markets, which require fish of acceptable size, may find that it is not advisable to produce the maximum number of suckers in a pond. Studies of natural sucker ponds in Minnesota have shown a very definite relation between the number of fish produced in a pond and the size the fish will be in 60 days. As the number produced in the pond is dependent on the number of fry stocked, the producer must decide before stocking time the size of fish he wishes to raise and stock accordingly. The following table shows the relation between the number of fish produced in some Minnesota sucker ponds and the size of the fish at 60 days.

Number of Fish Produced per 0.4 ha or Acre	Average Length at 60 Days (cm)	Average Length at 60 Days (in.)
19,000	5.0	2.0
8,000	6.3	2.5
4,000	7.5	3.0

The rate of growth of white suckers under "normal" conditions is as follows.

Number of Days	Length of Fish (cm)	(in.)
10	1.8	0.7
20	2.8	1.1
30	3.8	1.5
40	5.1	2.0
50	6.1	2.4
60	7.1	2.8

Sucker pond production is usually expressed in kg (lb) and wholesale sales are usually in liters (gal.). Table 5.14 presents the number of suckers of various sizes per kg or lb and per liter or gal.

These values may not hold true in other areas, so each dealer will have to study his ponds and determine the prevailing relationship. Of course, the stocking rates necessary to produce certain numbers of fish will vary with each pond according to the survival rate that exists in that pond.

TABLE 5.14. RELATIONSHIP BETWEEN WHITE SUCKER LENGTHS, WEIGHT PER KG (LB) AND NUMBER PER LITER (GAL.)

Length of Fish cm	in.	Number of Fish per kg	per lb	Number of Fish per liter	per gal.[1]
2.54	1.0	9,350	4,250	8,982	34,000
3.81	1.5	2,464	1,120	2,367	8,960
4.08	2.0	968	440	930	3,520
6.35	2.5	484	220	465	1,760
7.62	3.0	260	118	249	944
8.89	3.5	154	70	148	560
10.16	4.0	101	46	97	368
11.43	4.5	68	31	66	248
12.70	5.0	48	22	46	176
13.97	5.5	35	16	34	128
15.24	6.0	26	12	25	96
16.51	6.5	22	10	21	80
17.78	7.0	18	8	17	64

[1] 1 gal. = 8 lb.

The dealer who knows these relationships for his ponds will be able to stock some ponds lightly to produce bait for midsummer. Ponds stocked moderately will produce the same size bait for late summer, and those stocked heavily will produce small fish that can be held over the winter for the early summer season in the following year.

Operators that do not know the stocking requirements of their ponds well enough to build a graduated series of populations that will produce pike-bait sized suckers during the entire fishing season can adjust the pond populations by moving fish from one pond to another. By a system of periodic test nettings, the dealer can determine the growth rate of the fish in each pond. By moving the fish from one pond to another, he can release some populations for faster growth and can crowd others for slower growth.

In actual practice, the operator tries to obtain the desired minnow population in each pond by regulating the stocking rate and then compensates for errors in judgment and seasonal variations in the survival rate by moving fish from one pond to another. When this program is used in conjunction with an overwintering pond, a year-round supply of pike-bait sized suckers can be produced.

Fertilizing Sucker Ponds.—As most natural ponds produce enough water fleas to feed all the suckers the pond will hold, fertilization is usually not necessary and should be avoided whenever possible. A number of Minnesota ponds have been operated for years without fertilization and are still producing good crops of suckers. Most ponds are fertilized or partially fertilized through runoff waters from cultivated fields. Suckers seem to grow faster and more consistently when feeding on chironomid fly larvae than when feeding on water fleas. If the fish in a pond are growing very slowly, the pond should be fertilized with manure to increase the number of chironomids in the bottom muds.

Commercial inorganic fertilizer should be used sparingly on northern natural ponds because the phosphorus tends to produce heavy algal blooms that may result in fish kills. In very fertile ponds, the control of algae with copper sulfate may be more important than fertilization.

Harvesting.—Natural ponds can be harvested most efficiently with a large seine that is set out in a semicircle from a boat and pulled in slowly to a good landing beach. When the net reaches the shore, it is bagged and moved quickly to deeper water so the minnows will not smother or choke on silt. The minnows are transferred to a floating live box as soon as possible, and all turtles, salamanders, and crayfish are thrown out. If the minnows are uniform in size and are large enough for pike bait, they are loaded into a tank of fresh water and hauled to the bait shop. If the haul produces large numbers of undersized minnows, the fish are put in a slat grader and the small ones returned to the pond for further growth. The pond is then reseined at periodic intervals until further hauls are not practical. The minnows that have been missed can be trapped under the ice during the early part of the winter.

The time of harvest for each pond will depend on the seasonal market and the size of the fish being raised. In Minnesota, the sucker harvest starts during the last week of July and continues until September. The only ponds that have not been harvested by that time are those producing fish to be held over the winter. In areas where winter spearing is allowed, the cash return from a poor pond can be improved by holding the fish until October or November, when they can be sold as decoys for spearing. Higher prices can be obtained for them at that time if they were sold as pike-bait sized minnows in the summer.

While the production of sucker ponds varies with pond conditions and pond management methods, the average production of sucker ponds in Minnesota was 10,000 fish per 0.4 ha (1 acre) for 26 pond-seasons, with a high of 25,000 and a low of 1500. The average was 75 kg (165 lb) per 0.4 ha (1 acre), with a high of 223 kg (490 lb) and a low of 3 kg (6 lb). These production figures are far below the yield goals set in some publications on minnow propagation, but the ponds still are considered very practical. The cost of operation was low and the margin of profit was high.

In most sucker ponds, the weight of fish produced can be greatly increased by cropping and grading the fish periodically. On the average, the production of Minnesota sucker ponds was increased 75% by cropping. When the ponds were cropped twice in a season, the weight increase was only 5 or 6%, but when the ponds were cropped 6 to 8 times, the weight increase was as high as 140%. This means that if the producer harvests all of the minnows from a pond the first day he seines, large numbers of the suckers will be small and will have to be sold as crappie bait. If the dealer removes only the pike-bait size each time and allows the small fish to stay in the pond and grow, he will be able to sell the entire production as pike bait. The number of times a pond can be cropped will depend on the cost of seining and the value of pike bait versus crappie bait.

Holding Suckers over Winter.—In the northern states, where 1 m (3 ft) of snow and −29°C (−20°F) temperatures are common, propagation ponds are not practical for overwintering suckers. The fish must be seined or trapped and moved to holding ponds supplied with running water. River water ponds are preferable to those fed with spring water, because the river water supplies some natural food and the minnows are in better condition in the spring. In general, no supplemental feeds are fed in the winter. However, most ponds are aerated to keep ice from covering the surface.

Suckers held over winter in holding ponds will be thin the following spring. The best way to fatten them is to place the fish in shallow natural ponds that have a good supply of natural foods. Fish that are about 6.25 cm (2.5 in.) long will soon grow to pike-size bait and those that are 8.75 cm (3.5 in.) will increase to bass-bait size by June or July.

Other bait minnows which have been or might be cultured are:

(1) Creek Chub or Horned Dace (*Semotilus atromaculatus*)

(2) Pearl Dace or Leatherback Minnow *(Margariscus margatita)*
(3) Hornyhead Chub or Redtail Chub *(Nocomis biguttatus)*
(4) River Chub *(Hybopis micropogon)*
(5) Blacknose Dace or Slicker *(Rhinichthys atralulus)*
(6) Longnose Dace *(Rhinichthys cataractae)*
(7) Finescale Dace or Rainbow Chub *(Pfrille neogaea)*
(8) Northern Redbelly Dace, Yellowbelly Dace or Leatherback *(Chro-somus eos)*
(9) Southern Redbelly Dace *(Chrosomus erythrogaster)*
(10) Emerald Shiner *(Notropis atherinoides)*
(11) Common Shiner or Skipjack *(Notropis cornutus)*
(12) Spotfish Shiner or Blue Minnow *(Notropis spilopterus)*
(13) Brassy or Grass Minnow *(Hybognathus hankinsoni)*
(14) Stoneroller or Racehorse Chub *(Campostoma anomalum)*
(15) Western Mud Minnow *(Umbra limi)*

Holding Facilities

Hold facilities are usually in the form of tanks made from poured concrete or cement blocks. A suitable size is 1 m × 7.3 m × 66 cm (3 ft × 24 ft × 2 ft) with dividers at 2.4 m (8 ft) intervals and a water depth of 50 cm (18 in.). If raceways are constructed in pairs, the inner wall can be shared to reduce construction cost. Cold water at 13° to 16°C (55° to 60°F) from groundwater or other source is essential for holding facilities during the summer months. Because water from these sources is often void of dissolved oxygen but high in free carbon dioxide, aerating towers, baffles, or spraying devices are needed to increase dissolved oxygen content and to reduce free carbon dioxide. Electrically driven agitators are usually placed in the holding vats to maintain high oxygen levels. Water recirculation with the water spraying on the surface of the vat is also used. A cold water supply for holding minnows should have a flow capacity of 4 liters per min for each 400 liters (1 gpm for each 100 gal.) capacity of the holding tanks. Since holding tanks require large quantities of water, the tanks should be located near the water source. With a flush rate of 4 liters per min for each 400 liters (1 gpm for each 100 gal.) capacity, 1 kg of fish in 45.6 liters (1 lb of fish in 5 gal.) of water may be held satisfactorily. When fish are crowded, free carbon dioxide increases, ammonia and other metabolic wastes build up, and dissolved oxygen can be depleted. Flushing and aeration are important in ensuring the well-being of the fish. Cool temperatures are important because the water's holding capacity for dissolved oxygen is higher and the metabolic rate of the fish is lower. Some fish farmers like a roof over the holding tanks to provide shade. Unshaded tanks develop algal growths, and the water

FIG. 5.47.   GOLDFISH IN HOLDING TANKS READY FOR SHIPMENT

temperatures may become dangerously high during the day. Other fish farmers report healthier fish in unshaded tanks. Climate and length of time fish are to be held will help reconcile these conflicting viewpoints on shade.

After 24 hr in the holding tanks the fish can be graded.

The preferred market size varies a little with locality. The sizes listed in Table 5.15 are acceptable in central and northern states, but in the south the sizes will be shifted down. That is, a "medium" shiner in the north will be a "small" shiner in the south.

TABLE 5.15.   MARKET SIZES OF MINNOWS

Grader Spacing (in.)[1]	Size Classification	Approx. Total Length		Approx. No. of Fish	
		cm	in.	per kg	per lb
	Fathead Minnow				
< 13/14	Subsaleable	—	—	—	—
> 13/64 but <15/64	Small	4.4	1.75	770	350
> 15/54 but <17/64	Medium	5.6	2.25	550	250
> 17/64	Large	6.9	2.75	385	175
	Golden Shiner				
< 12/64	Subsaleable	—	—	—	—
> 12/64 but <14/64	Small	4.4	1.75	715	325
> 14/64 but <16/64	Medium	6.3	2.50	440	200
> 16/64 but <18/64	Large	7.5	3.00	275	125
> 18/64	Jumbo	—	—	—	—

< Fish pass through that spacing.
> Fish are retained by that spacing.
[1] 1 in. = 2.5 cm.

Transportation

Most minnow farmers produce too many fish to do all their own retailing. Hence, many or most sales are at wholesale prices to fish dealers and fishing camps. This means that minnows may be transported hundreds of miles. This calls for special handling and specialized equipment.

Since the condition of the minnows at harvest is very important to the success with which they may be transported without experiencing heavy losses, the fish are usually "hardened"; this is done by reducing the feeding rate in the rearing ponds during the last few weeks. This may reduce weight slightly but produces a tougher minnow. After the fish are harvested and placed in holding tanks they should not be fed for at least 24 hr before transporting. This allows the body wastes to be passed so that they do not foul the water during shipment. If the fish are in poor condition at this time, they should be transferred back to a pond to recover.

Transportation should be conducted on cool days or at night, but this is not always possible. Short trips of not more than 3 hr duration can be accomplished by the use of simple equipment, while trips of 300 to 500 km (200 to 300 mi.) or more will require more specialized equipment.

Containers for Short Hauls.—There are many types of containers that

FIG. 5.48. SELECTED BAIT FISH READY FOR SHIPMENT TO WHOLESALER

suffice for transporting minnows over short distances. It would be advisable for the beginner to the minnow business to observe some of the containers in use at other hatcheries before choosing one. Several of the types in use will be mentioned here but the list is by no means complete.

(1) Large canvas bags based on the principle of drinking water containers as used by our armed forces will suffice for short hauls. Nine to twelve heavy canvas bags about 0.9 m deep by 0.45 m (36 in. deep by 18 in.) in diameter can be suspended by the top, open to the air and be carried in a pickup truck. Their advantages include simplicity, low cost, little injury to the minnows, little splashing, and the elimination of supplemental aeration. The water is oxygenated by absorption through the canvas.

(2) Some dealers use barrels or drums as containers. In most cases oxygen is supplied to the water by means of a compressor or by oxygen cylinders. Wooden barrels are preferable in that they tend to stay cooler than metal barrels.

(3) Tanks of various sizes, shapes, and materials are also used extensively in the transportation of minnows. Usually tanks are constructed to make use of agitators, circulating pumps, oxygen cylinders, ice, or a combination of these.

(4) Cylindrical strainer type containers are useful for carrying assorted sizes of minnows when only a small number are to be transported.

Long Distance Transportation.—A bait farm dealer who plans to transport minnows over long distances should visit several large bait or fish hatcheries and observe the methods and equipment being used before investing money or planning equipment.

Many types and kinds of tanks are used to transport bait for long distances. Nearly all methods involve the use of some device for aeration or agitation to ensure an adequate oxygen supply for the fish and the use of some form of cooling device. Also some successful shippers use chemicals to partly anesthetize the fish to lower their activity and oxygen demand and also use additional devices to remove the carbon dioxide and other organic wastes. No special tank for long distance transportation will be discussed. However, plans for tanks to be carried on trucks might be obtained by contacting your State Game and Fish Department. It is felt that one can get a better idea of what is needed by visiting several of the larger fish or bait farms.

The safe transportation of minnows involves the following factors:

(1) Minnow size and container size

(2) Care of minnows

(3) Distance to be transported

(4) Season of the year

(5) Temperature of the water

(6) Oxygen supply

(7) Waste products

(8) Fatigue from continued splashings

All of these factors are more or less integrated and one should take all of them into consideration when planning the transportation of minnows.

The kinds and sizes of minnows to be transported must be considered in determining the number of minnows to be carried in a certain sized container. Fathead minnows and bluntnose minnows will withstand more crowding than will the golden shiners. Generally speaking, the smaller the fish the less weight in grams (lb) that can be carried per liter (gal.) of water (Schaeperclaus 1933).

The kind and size of containers being used will also be a determining factor in the number of minnows to be safely carried. Generally, by cooling with ice and aeration by pumps, one can safely carry 1000 5 cm (2 ½ in.) minnows to each 38 liters (10 gal.) of water. Experience will indicate the number which can be transported safely (Altman and Irwin [undated]).

In general, 454 to 908 g (1 to 2 lb) of minnows can be placed per 3.8 liters (1 gal.) of water in sealed, plastic bags containing oxygen and water for medium hauls. For short deliveries, up to 2.3 kg (5 lb) of minnows per 3.8 liters (1 gal.) of water can be used. The sealed bags should be kept out of direct sunlight so that the water doesn't get too hot.

With mechanical agitation, 454 g of fish per 3.8 liters (1 lb per gal.) of water can be transported in 15.5°C (60°F) water. Adding pure oxygen into the water will permit doubling the volume of fish per unit of water. If transportation goes over 24 hr, filtering or changing the water may be necessary.

Care of Minnows.—Harvesting and handling methods prior to and during the journey will have an effect on the condition of the minnows on arrival at their destination.

A hardening or tempering process will be of value in conditioning the minnows for a journey. This process of hardening or tempering consists of gradually raising or lowering the temperature of the water in which the minnows are held until a suitable temperature is reached.

"Water in non-aerated tanks should be kept at 18.3°C (65°F) or lower .... A minnow should not be subjected to more than a 10° [5.5°C] change of temperature unless the change is very gradual. Proper tempering requires twenty minutes for every 10° change" (Dobie *et al.* 1948).

The sudden shock encountered when minnows are changed from one water temperature to another is often fatal. This shock is greater or has more lethal effect on the minnows when the change is from a colder to a warmer water. This factor should be kept in mind when unloading the minnows at their destination and the tempering process again should be used to condition the minnows to the temperature of the water in which they are to be placed.

The risk of loss will increase as the distance to be traveled or the time spent en route increases. Therefore, long trips call for more specialized equipment and greater precautions en route.

Season.—The season of the year has a decided effect on the number of minnows per container that may be safely transported. Fall, winter, and early spring months are generally considered the best months for safe transportation of minnows. However, bait farm dealers cannot always dispose of their crops during these months and must transport some minnows during the hotter months. The number of minnows per container must then be reduced to ensure maximum safety during the trip.

"In Alabama the transportation of fatheads is confined to the winter and spring since they may not be handled safely in summer . . . . The transportation of golden shiners is limited to the cooler months of the year. They are never shipped via railway express" (Prather *et al.* 1953).

Temperature.—Most experienced bait haulers agree that keeping the temperature of the water at 18.3°C (65°F) or below prevents the occurrence of excessive losses of minnows during transportation. The minnows are less active and use less oxygen.

The most common method of lowering the temperature of the water in the hardening or tempering process and holding it at the desired temperature is that of adding ice to the water. Make certain that the temperature of the water is not lowered faster than 5.5°C (10°F) in 20 min because the sudden shock will cause a high mortality of minnows.

Oxygen.—An adequate supply of oxygen for the minnows must be maintained at all times. The supply of oxygen during transportation is maintained by several methods or combinations of methods.

(1) *Agitation.* Agitation of the water by means of an electrically driven impeller, which is connected to the power supply of the vehicle, is one means of mixing or stirring the water and increasing the oxygen supply. This method is suitable for short trips with a small number of minnows. Several types of agitators are on the market, or a suitable one may be constructed by the dealer.

(2) *Circulation.* Another method of aeration is by means of a circulating pump that takes water from the bottom of the tank and sprays

it into the top of the tank, thus absorbing oxygen as the spray passes through the air.

When transporting golden shiners the spray should not fall directly on the water but should be baffled because shiners have a tendency to jump, causing many injuries (Prather *et al.* 1953).

"When tanks are aerated with running water a continuous flow of not less than 3.8 liters (1 gal.) per minute for each 95 liters (25 gal.) of water in the tanks should be maintained. The water should reach the tank from pressure jets placed well above the water level. Each tank should have a minimum of two pressure jets and at least one jet for every 95 liters (25 gal.) of water in the tank" (Dobie *et al.* 1948).

Many modifications of the water circulating type system are used in the transportation of minnows. Some haulers state that by the addition of a venturi (a particular type of pressure nozzle) in the discharge pipe of the water circulating pump the hauling capacity of a tank has been increased tenfold (Feast and Hagie 1948).

Oxygen in either pure or atmospheric form is used where circulation of water is not employed and sometimes in combination with the water circulation process. The water can be oxygenated from air compressors or from metal cylinders of pure oxygen. When oxygen is supplied by a compressor or a cylinder, it is necessary to have tubes fitted with valves and an atomizing device to carry, control, and disperse the oxygen. Carborundum aerator stones or a perforated oxygen release tube which produces small bubbles of oxygen as it enters the water are commonly used as atomizing devices. When using oxygen under pressure, an oxygen regulator valve and gauge of the type used by welders are necessary to control the pressure and to determine the amount being used.

Only a small amount of oxygen entering the tank is necessary since somewhere around 3 ppm is the minimum quantity. Amounts greater than 10 ppm are a waste and can be dangerous.

In the transportation of golden shiners in Alabama, oxygen alone (without water circulation) was recommended as the best method for trips lasting 12 hr or less. "With this method, up to 7500 shiners 10 to 12.5 cm (4 to 5 in.) in length can be hauled per 378 liters (100 gal.) of water without injury if the temperature is 18.3°C (65°F) or lower" (Prather *et al.* 1953).

Transportation Methods Employing Chemicals.—Experiments with and recent developments in the use of hypnotic drugs have shown that possibilities exist for their application to fishery practices, especially during transportation and handling.

Most of the drugs, when dissolved in water, reduce the respiration rate and the activity of minnows, thus permitting an increased number of fish per volume of water.

Several drugs that have been tried are mentioned but no specific recommendations for their use are given.

Thiouracil when used on stonerollers and on bigmouth shiners at the rate of 9 grains per 15,000 cc of water (388 ppm) reduced the oxygen consumption about 20% and seemed to increase the fish's ability to withstand lower concentrations of dissolved oxygen (Osborn 1951).

Urethane (ethyl carbamate) decreased the oxygen consumption of fish under its influence and has been used in transporting fish during hot weather. In a 0.5% solution redfin shiners, small common suckers, and rainbow darters could be left for 18 min without injury (0.5% solution equals 19 g [⅔ oz], dissolved in 3.8 liters [1 gal.] water). The cost of 19 g is approximately $0.16 (Gerking 1949). A weaker solution which would cause decreased activity and decreased respiration but not completely anesthetize the fish would be convenient for transportation.

Sodium amytal and sodium seconal have been tested by employees of the California Department of Fish and Game in experiments conducted in the transportation of trout and some other fish. In general, the results of the tests indicate that the carrying capacity of containers can be approximately tripled by the use of drugs.

These two drugs tend to lose their strength in water over 10°C (50°F); therefore, when using the drugs, the water should be kept at 10°C (50°F) or lower. In most cases a pressurized aerating system was utilized while the fish were en route. The combined method of drugs and aeration increased the number and weight of the fish that could be handled per volume of water. The cost of the drugs is approximately ½ cent per grain and would probably be cheaper in bulk packages (Reese 1953).

Further experiments in the use of drugs will probably disclose a formula "which will be acceptable for general use in concentrating fishes or minnows for transportation" (Altman and Irwin [undated]).

TROPICALS[2]

The commercial raising of ornamental (tropical) fish had its start in Florida—on the east coast in 1926 with a "Miami" style operation—and on the west coast in 1931 with a "Tampa" style operation. Because of the nature of the soil, the lower part of the state has very few areas that can be used to dig dirt ponds. The coral formations seep and water levels can't be maintained in a practical way. This area uses large concrete vats to raise fish and this is the "Miami style." The middle of the state and particularly the area of Tampa-Lakeland-Bradenton has a high water table and soil that holds water on a year-round basis. Fish are raised in

[2]By Ross Socolof.

dirt ponds and this is the "Tampa style."

The vast majority of the ornamental fish raised in the United States can be raised only in Florida. Still, there are a few fair-sized greenhouse operations scattered throughout the country and one or two small farms raise fish outdoors in ponds near the Mojave Desert's Salton Sea. Some people take advantage of the 4 or 5 months that fish can be kept outside unprotected north of mid-Florida and raise or size ornamental fish during these warm months. This amounts to a fractional part of the whole, and most certainly better than 95% of the ornamental fish produced in the USA are raised in Florida.

The only exceptions to this are goldfish and koi (hybrid colored carp). These are cold water ornamentals and are mainly raised in the colder parts of the country.

The ornamental fish industry raises livebearing fish and egg laying fish. Both are equally sophisticated enterprises. The notion that a few fish can be put into a pond or concrete vat and basically left untended to propagate prolifically for easy money happens as often as lead turns to gold. As a business, it has been very good for many people for many years. However, it has been good only for those people who have dedicated themselves to hard work and constant study. In recent years the nature of fish farmers as individuals has changed. More and more well trained ichthyologists and business oriented people are entering the industry in Florida.

Solid statistics are difficult to find. With a small margin allowed for error, the following numbers are correct.

Ornamental fish are the largest single air freight item out of Florida. Between 15,000 and 20,000 boxes of live ornamental fish are shipped by air each week from Florida.

There are more than 200 full time ornamental fish farming businesses in Florida. There are at least another 100 smaller part time operations. The Florida Tropical Fish Farmers Associatation has 131 members.

There are more fish farmers raising fish to sell to other fish farmers than there are fish farmers shipping fish out of the state. This means that many people specialize in raising a few varieties. They sell these fish to other farmers who raise part of their fish and buy more so that they can offer the retail stores and wholesale fish jobbers a wide variety of product.

There are approximately 20,000 dirt ponds currently being used to raise ornamental fish in Florida. The average pond is 30.5 × 7.6 × 1.8 m (100 × 25 × 6 ft). These are on some 2430 to 3240 ha (6000 to 8000 acres) of land.

The total value of the retail dollars generated each year from the ornamental fish shipped from Florida to be sold in the United States is over $75 million and approaches 1 million boxes.

Before going any further, we will look at the basic product. First we will discuss the livebearing fish, then the egg laying fish. After these, details on operating and managing an ornamental fish business in Florida will be presented.

Livebearing Fish

These miniature fish give birth to living young. This phenomenon started the keeping of tropical fish. In 1908 the first livebearing fish were offered for sale in the United States. These were nondescript, undesirable fish by today's standards. The first fish was *Gambusia affinis* (our common mosquito fish). This species was followed shortly by (imported from Germany) *Girardinus, Xiphophorus*, and *Poecilia* species. In the United States immediately after World War I the hobby started to boom and the boom hasn't yet subsided. As the years passed more colorful hybrids were created and the wild, uninteresting, and drab livebearers lost favor. The hobby concentrated on the various *Poecilia* and *Xiphophorus* species. These two genera include all of the fish produced commercially in large quantities. They are best known as guppies, mollies, swordtails, and platies. With the exception of the common or wild type green swordtail and guppy, all of today's volume production of ornamental livebearing fish are hybrids which we will discuss individually.

Guppies.—Guppies are also known as "millions fish" for the obvious reason that they are as prolific as aquatic rabbits. A full grown male is slightly more than 2.5 cm (1 in.) in length and a female seldom exceeds 5.0 cm (2 in.). A mature female delivers as many as 200 young and records of almost 300 living young are often reported. My experience indicates that under proper conditions it will take only 3 weeks for a female to have successive broods of young. I have observed that under proper conditions a newborn baby will reproduce itself in 6 to 8 weeks. The factors that will speed up this process (optimum) would include proper temperature (about 28°C or 83°F), plenty of room, extensive photoperiod (midsummer), and, most important, a varied and ever present supply of live food. In certain areas the hybrids must be treated differently and other results are to be anticipated. One very important factor a breeder must always be conscious of is the realization that one impregnation will suffice for 5 or more broods. The females store sperm bundles from one impregnation for future fertilizations without the need of a male. This is also true of most other livebearing fish. Guppies mature so fast that to breed properly the young must be isolated at birth and raised separately until ready for selective breeding.

Hybrid Guppies.—Breeding hybrids is difficult because most of the desirable traits are found in the recessive genes. This gives a first brood that will all look alike. The desirable traits will be masked. The siblings must be bred to each òther or back to the father. The F_2 generation will show results. The first hybrid strain established was a gold or blond guppy. Most of today's hybrids can be had with either a normal or a gold body color.

In the mid 1930s the hybridization of guppies started seriously and after a short hiatus during World War II has progressed to a spectacular art today. There is an International Fancy Guppy Associatiation that sponsors well attended shows. Each year larger, better shaped, and more colorful fish appear on the scene. The guppy characteristics that have been highly developed are size, color, and tail shapes. In recent years the delta tail guppies have proven most popular, mostly because of the huge tail size and the exacting symmetry possible. This appears as a basic equilateral triangular shape with the apex of the triangle at the caudal peduncle. Other strains have been developed with pintails, veiltails, single and double swordtails, scarftails, etc. A recent characteristic that has been developed is males and females with the area posterior to the eyes a solid black. This combined with a vivid red tail makes a spectacular animal. Another very desirable character is the snakeskin or cobra skinned pattern on the body and tail. These also are now produced in quantity. Albino guppies are available, but have never been popular.

From the preceding, it should be obvious that the production of fancy guppies is not a simple accomplishment nor will it ever be. It is best to consider production of the large volume guppy items, which are common (wild type), gold body and colored small tail guppies (known as mixed fancy), separately from the fancy hybrids. These three are produced in dirt ponds in very large quantities. They are sold at low prices and they sell well. Top quality hybrids are seldom raised to adults in dirt ponds, the major drawback being the huge tails that slow the fish down for predators. As a result few perfect fish are produced. The time expended in the culling process for top quality fish makes dirt pond culture impractical. Most of these fish are started in dirt ponds and removed when young, graded for sex, and raised to adulthood in concrete vats, wooden vats, or glass aquariums. While Florida produces a large number of fancy guppies, the largest number is produced in Singapore. This is a labor intensive process and the cheap labor available in Singapore has made it difficult for Florida to compete. Higher air freight rates have now started a reversal of the balance and it would seem that in the coming years the production of fancy guppies will accelerate in Florida.

Guppies have a reputation for survival in poor environments. They also are reputed to be able to withstand severe stress from temperature, handling, etc. This is true to a certain extent with the wild type fish, but the reverse is true with the hybrids. They suffer quicker and more severely from temperature drops than any other livebearer. In a cold winter they are the first livebearers to die. The intensive inbreeding over the years has weakened the fish. The foreign fish are constantly subjected to a variety of antibiotics which have resulted in fish that carry antibiotic resistant strains of bacteria. They do well until stressed, and they are stressed in shipping. It requires a knowledgeable and vigilant fish keeper to keep these fish healthy.

Using 380 liter (100 gal.) or larger concrete vats inside to raise the fancy hybrid fish to maturity is the most practical system. Water must be partially changed daily. A constant (open) flow system adding water in small quantities works best. Environment must be kept clean. Live food is highly recommended. Newly hatched brine shrimp nauplii are best. *Cyclops*, rotifers, *Daphnia*, midge larvae, white worms, and similar foods are easily cultured and work. They live until eaten which is not true of brine shrimp. The advantage of brine shrimp nauplii is safety. No predators can be introduced. The disadvantage is the high cost of the eggs.

Mollies.—The fish grown today are all hybrids with the exception of wild brackish to freshwater forms that run from the middle coast of the United States around the Gulf of Mexico south to Spanish Honduras. The wild fish are similar in appearance and not easily differentiated at first glance. Some are collected wild and resold, but most of these are cultured in dirt ponds. These are *Poecilia latipinna, P. velifera*, and *P. petensis*. All other mollies are mutations from these (and several other species such as *P. sphenops, P. orri*, etc.). The sailfin black molly is a standard and remains one of the more important items in the industry. This fish was developed in the early 1930s by William Sternke from wild black mottled mutations found in the wild mollies *(P. latipinna)* in Florida. The strain today is "perma-black." Perma-blacks have almost eliminated the problem the fish has of fading into a black-grey pattern during the colder months. The color will return as the weather warms. The large sailfin possessed by adult males is spectacular.

Mollies and most other livebearing fish produce fast-maturing and late-maturing males. Some fish can be sexed as early as 3 weeks after birth as the gonopodium develops. A large percentage of the apparent females are late developing males, and these fish can appear as females for as long as 9 months before turning into developed males. These late developing males are the spectacular large sailfin specimens. This explains two questions that have nagged fish culturists for years. It was accepted by all of the people in the industry that in the early fall many females turned into males. What actually happens is that the unsexed fish develop into males in the fall. The other explanation this information provides is the apparent lack of large males. Sailfin black mollies are black forms of the wild green sailfin molly *(P. latipinna)*.

Dr. Harry Grier has demonstrated and proven the effect photoperiods have on the development of ova by the females. The eggs simply will not develop unless the fish are exposed to a photoperiod of 13 hr or more. This causes zero production during the midwinter months with the first young appearing as the days lengthen in early spring. The babies born in March develop into large males in the August−November period.

The regular sized black sailfin molly is 1 of the 3 or 4 largest volume items produced and sold. A recent study showed that some 3000 dirt ponds are devoted to black molly production in Florida. In order of importance these other mollies are cultured in dirt ponds: black yucatans (a black form of *P. sphenops*), black lyretail (hybrid *P. latipinna*), sphenops (a hybrid *P. sphenops*), marble sailfin (hybrid *P. latipinna*), chocolate (a hybrid *P. latipinna*). In the last few years the production of both albino and albino lyretail mollies has moved to Miami and they are mostly cultured in vats. Several new and exciting hybrid mollies have been developed in recent years by both the industry in Florida and Dr. Joanne Norton, a geneticist in Ames, Iowa. These include many new crosses highlighting burnished golds and reds with various trade names such as starburst, goldfire, goldust, copper topaz, and opal. One other mutation that exists and could be interesting commercially is a veiltail molly, but this mutation has been difficult to produce commercially.

Handling and diseases are discussed later, but there is a problem particularly prevalent in black sailfin mollies and other hybrids of *P. latipinna*. This is a condition called "shimmies." The fish, if handled roughly, stressed by cold, or not given an opportunity to acclimate slowly in holding tanks, develop "shimmies." It is not a disease. The word describes the condition, and proper handling will eliminate it. It is most common in the winter months and as a result many buyers switch to black yucatan mollies during this time of the year as they seldom develop this problem.

Swordtails.—*Xiphophorus helleri* is the wild green swordtail of the aquarium industry. Since its introduction about 70 years ago it has been a large volume item. Several other species have infrequently been seen, but *X. helleri* is the fish that is the basis for all swordtail strains in the industry today. Platy fish are discussed later. They are mentioned at this point as they are a swordless species of the genus *Xiphophorus*. They cross back and forth so that platy/swordtail hybrids are important.

The most important commercial swordtails are the bright reds and

FIG. 5.49. DIGGING A TROPICAL FISH POND

FIG. 5.50. TROPICAL FISH POND COVERED FOR COLD WEATHER

these are first and foremost the velvet or blood red fish. These sell in great numbers. They are an almost solid velvet red color with only (at times) a trace of white on the ventral surface below the gill plates. Wagtail livebearers are those fish with an all black tail and black dorsal fin. A velvet wagtail swordtail has the solid red body with a black tail and black dorsal fin. The third of the popular bright red strains is the velvet tuxedo swordtail. This fish has ideally about 40% of its body black and this will be on the ventral part of the body. It is also the most

FIG. 5.51. WIRE BASKET TRAP FOR TROPICAL FISH

FIG. 5.52. PLASTIC TRAP FOR TROPICAL FISH

difficult to raise as the strain has always been difficult to set. More labor in culling is needed. This cuts down production and increases the cost. A fourth highly developed red fish is the so-called Berlin swordtail.This velvet red fish has a black blotched body. Unfortunately the strain develops melanistic sarcomas, and its production in recent years has declined.

A brick red swordtail strain preceded the bright velvet or blood red strains. These are very prolific fish that are produced commercially and are sold in varying amounts including the following strains: green tuxedo, green wagtails, brick wagtail, brick tuxedo, all black, all gold, golden tuxedo, golden wagtail, and painted or freckled. This last is a black blotched fish on a mostly brick red body.

Albino swordtails exist and sell well. There are 3 fin mutations that have been combined with the color varieties just described. These are a beautiful lyretail, a high broad dorsal fin, and a veiltail. The first is by far the most important. The high dorsal (known in the trade as the Simpson sword) has been combined with the lyretail. The lyretail genes carry along with them genes giving an elongated gonopodium. It is not possible for mature fish to reproduce themselves so normal gonopodium male fish must be used to raise lyretails. These males should be those carrying the hidden genes for the lyretail rather than identical looking males without the lyretail genes. The veiltail sword has little commercial interest.

The albino forms have been crossed with the lyretails and high dorsal types. All of these are raised "Miami style" in vats. There is a hybrid known as a red-eyed red sword which is an albino form with red eyes, but still keeps a deep red body coloring. To raise good swordtails one must cull out the runty males or one will suffer shortly with undersized fish. A good fish should be at least 3 in. in length exclusive of the tail spike past the caudal fin.

Platy or Moon Fish.—This is the fourth of the 5 basic commercial livebearers; it is *Xiphophorus maculatus*. It was first introduced before World War I and in the early years was often referred to as a moon fish. The caudal peduncle in most wild strains has a definite moon-shaped crescent.

There are various races of wild platyfish and this has helped develop the many varieties. There are wild fish with a great amount of black pigmentation and others with metallic sheens. As this is a *Xiphophorus* and the swordtails are also, it could be expected that they would easily cross. This is true and today's swordtail and platy hybrids have been crossed back and forth many times.

The fifth and last of the basic livebearers is also a *Xiphophorus* (*variatus*) and all three hybridize easily. Platies grow to a maximum size of 7.5 cm (3 in.) but this is unusual. A good commercial fish will be approximately 3.5 cm (1 ½ in.) with the females slightly larger. Regardless of the strain the male platyfish will take on more color. A red wagtail platy is colored identically regardless of the sex, but with a bit of training it becomes apparent that the males have a slightly deeper color. This is true for all color strains of platies.

The only metallic platy strain raised today is the blue platy. It is a very prolific fish. The strain was one of the earliest and is firmly set. This makes for easy production and as a result it is the lowest priced platy fish produced.

Sales in recent years have not been as brisk as in the past. The bright red varieties are the best sellers. There are four, and in order of importance and sales they are: first, the red wag platy. This fish is all red with a black tail and black dorsal and anal fins. A good strain should be at least 95% true with gold and gold wag platies as the culls. Very rich deep mahogany red strains have been developed. These have never been prolific enough to prove economical although they are more attractive.

A very similar fish and the second most important platy strain is the all red platy. Color is the same as the wagtails but is an overall red with no moon patterns, pigment spots, or metallic scales.

The third bright red platy strain is the red crescent platy. The color is not as good as the pure reds, and it has the typical black crescent marking on the caudal peduncle. The advantage of this strain is that it is much more cold resistant. As a result many are sold during the colder months.

The fourth bright red platy is a red tuxedo platy. This fish tends to remain small, and is not as popular as the others. This last strain has also been further refined. A wagtail has been combined with the tuxedo pattern to give a red tuxedo wagtail, but it is a very difficult strain to set. Sales have never been good enough to justify the higher price which is necessary for the fish to be commercially important. The result is that very few people raise them today. However, it is a very attractive fish.

Albino platies exist but are rare, unattractive and generally difficult to produce. They have no commercial importance.

The (Simpson swordtail) hi-fin has been put on all of the platies. Most of these are raised "Miami style" in concrete vats. The strain is difficult but a higher price can be gotten (in spite of the ruthless culling necessary) and they can be produced in enough quantity to make them commercially worthwhile.

No one has put the lyretail on platies as yet. It might be worthwhile, but my experience indicates that it would not prove profitable. However, I am sure it will be available shortly.

There are several strains of platies that are unique to platies and do not have counterparts in the swordtails. It is true that with a little work any platy pattern can be put on a swordtail and the same is true in reverse. The fact that they are two very differently shaped fish results in some patterns doing better in one shape as against the other shape. This results in unique patterns on platies.

Bleeding heart platies and milk and ink platies are two strains that fit the preceding description. The red fish and black blotches (similar to the Berlin swordtail strain) don't have the cancer problem in platies and so we have many so-called calico or hybrid platies in this color pattern. This with the addition of a calico wag strain nicely fills a niche for this attractive color pattern.

Before the original wagtail was developed there was an intermediate platy strain which showed a black border on the dorsal and ventral edge of the caudal fin. This was called first a comet variety and later a victory variety as some similarity exists to the "V for Victory" hand signal used during World War II. They are still produced but not in important numbers.

There also are a solid gold, a gold crescent, a gold wagtail, a gold tuxedo, and a red saddle gold platy (dorsal fin and below are red) that all sell profitably. There is no brick red strain of platy fish as in swordtails. The red crescent is the closest and it is a better red.

A salt and pepper platy is still produced. It has a variety of background colors and the body is covered with small black spots. The more red in the strain the better they sell. In recent years the red strains with black blotches (calico) have taken a great part of this market. They still sell but only because of a much lower price and the strain will probably disappear in the next few years.

Platy Variatus.—This is the third of the *Xiphophorus* genus and the species is *variatus*. There were 2 distinct wild races that developed into the sunset variatus and the rainbow variatus. These were from the 2 original wild races. The body shape of variatus is midway between that of platies and swordtails. A saleable *variatus* will average 4 to 5 cm (1 ¾ to 2 in.), with the female always growing larger. Maximum size would be slightly over 7.5 cm (3 in.).

Variatus have proven the most attractive of the 3 *Xiphophorus* when the hi-fin mutation is applied. This is because of the striking contrasts. Strains with a bright yellow dorsal fin on an all black or bluish body are dramatic. *Variatus* are extremely hardy, and can take more cold than the other 2 members of this genus. In addition to the hi-fin strains, little if anything has been done with other fin mutations, so we may shortly see lyretail or veiltail varieties. The most popular and largest selling *variatus* today is the marigold strain. This is basically an orange strain although some better strains are being developed that will have a color closer to red without impairing production. These are known as ruby red marigolds. There are strains of marigold wagtails and marigold tuxedo variatus, the latter being the more popular.

A black bodied fish has been around for many years and there is a wide difference in the available strains, the ideal being the nubian variatus strain, which has a black body with a red tail and also a bright yellow dorsal fin. It takes a long time for the colors to fully develop and as production is slow the price is high. More prolific black strains have little of the beauty of the nubian, but can be produced in larger quantities. They are commercially valuable because of the lower price.

The original sunset strain still sells very well and is a beautiful fish. It is very prolific and the biggest problem is to size the fish before they have overproduced and become stunted. The other original strain of redtailed rainbow variatus is also prolific. It can be beautiful, but in recent years its popularity has declined.

There are some new varieties just appearing on the market that should prove to be important items if they can be produced in economical quantities. The most important are strains incorporating the metallic colors from the blue platy strain and the blotching from the calico platy strains.

We have already discussed the other 4 species of important commercial livebearers and this last concludes the list. The 5 species (guppies, mollies, swordtails, platies, and platy variatus) are the cornerstone of the ornamental fish farming industry. It is interesting to realize that these 5 wild forms have developed into at least 300 hybrid forms. Most are still available. More will come along, and some have or will disappear. The number should remain close to the 300 varieties existing today.

There is one old world livebearing fish that is commercially important. This is the halfbeak livebearer from in and around Southeast Asia. It sells very well and grows to about 10 cm (4 in.), with the males not getting more than half that length. They are more cannibalistic than any of the other livebearers. Their young are also larger than would be expected. There is also a golden form which is worthwhile to raise.

Mollies will cross with guppies. Platies, swordtails, and variatus will cross with each other. There are a number of wild species of livebearing fish with limited commercial markets. Few are being raised today, but some would prove profitable. The best possibilities would include *Alfaro cultratus, Poecilia melanzona, Belonesox belizanus, Phallichthys amates,* and various *Anableps,* etc.

With the exception of the halfbeaks, almost all of the other livebearers come from Mexico and Central America with a few from the United States and South America.

Pond Breeding of Livebearing Fish

Constructing the Farm.—Ponds should be at least 1.5 m (5 ft) deep. An average pond is 24 to 30 m long (80 to 100 ft), and it should be dug about

6.1 m (20 ft) wide. Ideal terrain for Florida ornamental fish culture is low and wet. This type of hectarage (acreage) can usually be bought at lower prices than higher ground. As Florida has a definite wet season (July through September), care must be taken during construction to be certain one does not have future flooding problems. When the pond is dug, all of the earth that is removed to make the pond is carefully spread around the perimeter of the pond. This in effect will raise the ground level 0.8 m (2½ ft). This (normally) is sufficient to dike the ponds safely. An unusual low spot is cured by digging a pond at that site deeper and/or larger to obtain more fill to raise the ground level.

A common error that can easily be avoided is attempting to crowd too many ponds to the hectare or acre. There is always normal erosion of fresh pond banks. A 6.1 m (20 ft) wide pond will settle out to about 7.6 m (25 ft) by the time the banks become knit tightly with grass. If one loses 1.5 m (5 ft) on each side to erosion, one has in reality lost 3 m (10 ft) between 2 ponds. If one had decided on a 6.1 cm (20 ft) road between ponds, one suddenly has a serious problem. There is only 3 m (10 ft) remaining of the 6.1 m (20 ft) planned. Therefore, leave at least 9.1 m (30 ft) between ponds. This way there is always enough room to drive around the pool.

The first pools were dug in 1931 by Albert Greenberg for his Everglades Water Gardens, just east of Tampa, Florida. These first ponds were dug by hand with no equipment used except a shovel and a wheelbarrow. The next ponds were dug using mule power. A mule was used to pull a drag or sledge type contraption out of the pond with the earth.

Today pools are constructed using a combination bulldozer and dragline or only a dragline. The bulldozer can get the pond started fast, but must shortly stop as the water table will seep into the cut and mire the equipment. During the wet season the water table is almost to the surface (at times above it). During the most severe drought it is never less than 1.2 to 1.4 m (4 to 4½ ft) deep. This means that the ponds never dry out themselves. It also means they can never be dried out. The advantages far outweigh the disadvantages.

The dragline that is used to dig ponds should have at least a 68 cm (¾ yd) bucket. A 136 cm (1½ yd) bucket works better if the operator is skilled. When constructed the ponds should be dug 1 ft down and then 1 ft out. This is known as a 1:1 ratio and it provides a good sloping bank.

After the ponds are dug the dirt that has been removed is graded out using a surveyor's transit to be sure there is even ground. Most land has a natural slope and this should be maintained provided it is kept even and minimal. The goal is to allow the drainage ditches to empty themselves. A drainage ditch should be provided after every 2 rows of ponds and is used to move the water from the ponds when a pond is pumped.

Ponds should be dug only between October and March. Rainfall is at a minimum during this period. It will give an opportunity to establish the pond area with the minimum amount of erosion. Grass seed should be spread as soon as possible.

As soon as the ponds have been completed and the ground graded a water system must be installed. Deep wells are necessary to obtain sufficient water. Turbine pumps are the only practical type. The well should be 20 cm or more (8 in. or more) if one is constructing 100 or more ponds. One can expect to get at least 1137 liters (300 gal.) of water a minute. The water is needed to maintain water levels during drought periods.

More important is the use of water during extreme cold periods during the winter months. The water which usually comes up from 150 to 240 m (500 to 800 ft) will have a constant temperature. It will usually be close to 22.2°C (72°F) and it does not vary at any time. The deeper the well the warmer the water. Running this warm water into the ponds during extreme cold gives the fish an area to collect in to avoid mortality.

The well should be located so that a minimum amount of pipe is needed and where sections can be turned off by the use of valves at the well. Pipe of 15 cm (6 in.) should leave the well. This will reduce into 10 cm (4 in.) pipe at every row. Pipe should be set 45 cm (18 in.) below the roads and run down the middle. The outfall from the 10 cm (4 in.) main lines into the ponds should be 1.25 cm (½ in.) pipe.

There are several types of suitable plastic irrigation pipes available. The pipe used should be strong enough to prevent crushing. Plastic as a material is very suitable and is easily repaired when leaks occur. Never cover the water system until water has run for several days. When sure that all leaks have been corrected, cover it.

It is difficult to maintain an even pressure and volume of water in an entire system. End caps can be placed over the 1.25 cm (½ in.) outfall pipe in the ponds. If left intact the pond is taken out of the system as no water can run into it. If holes of various size are drilled in the end cap, the water flow will be restricted. This makes it possible to balance the flow throughout the system.

Stocking the Ponds.—Newly dug ponds can be used 3 days after construction with no additional treatment. Great care must be used in obtaining suitable stock of the livebearer strains. The better the strain, the less labor involved in culling. Fish should be examined closely before being stocked. Runts or questionable appearing fish must be discarded. Go with your first instinct when stocking. If you start to question a fish, discard it and disregard your second thoughts. The number of fish used to produce a crop varies with the species. If possible get as many females as possible. The following amounts for initial stocking are suggested, with the expectation of a saleable crop 6 or 7 months later. Stocking should be done in March as this is the ideal time of the year. Stocking will continue year round once one is in full production. Most of the important stocking is done in early spring. Starting stocks of 300 guppies, 400 platies and swordtails, and 200 variatus work out well. If one intends to stock a total of 5 ponds with red platies, it makes good sense to put the entire 2000 red platies into 1 pond and let them spawn. Once these fry are large enough to handle, they are culled and the remainder are used to stock the other 4 ponds. This will give a big jump in cleaning up a new strain.

This basic procedure should be followed until satisfied with the purity of the strains. Never, under any circumstances, return unculled fish to a pond. Time must be found to remove the culls. This will prevent the strain in the pool from deteriorating fast. I have many times stocked ponds with only 25 to 50 fish. These were the best fish of the particular strain I was making or had obtained. It always has paid and there are no short cuts.

It is impractical for anyone engaged in a normal sized ornamental fish farming operation to attempt to breed and spawn all strains inside. This is done by using virgin females and selected males. One can work inside on one or two projects at a time. The obtaining of virgin babies from the ponds and the stocking of these fish after culling is practical. It is time consuming and takes experience, but it gives good results. Experience comes with time.

Many fish develop adult colors different from their juvenile coloration. This takes time to learn. Mollies, swordtails, and platies will result in 2 distinct types of males. There is a small fast developing male and a much larger late developing male. For many years it was (and to a great extent still is) believed that some mollies and swordtails would change sex in the early fall. What actually was happening was that a large unsexed fish (that looked like a female) developed into a late developing male. If one will discard the small runty (early developed) males when stocking, one will improve the overall adult size of the strain.

Sizing and Feeding.—An overabundance of fish will retard growth. Keeping production and sales reasonably close eliminates this problem. When one has too many good sized fish and a quantity of younger fish in a pool with no immediate prospect of sales, something must be done. If nothing is done one will have a pool of fish of small size and the larger fish will be lost. The cure is to move the fish. Split them into 2 newly prepared pools. If this isn't possible then they should be culled and the oversupply discarded. Good management will prevent this from happening as at best it is costly and time consuming.

Production can be kept down and fish sized by utilizing a predator fish to eat the young. Several small fish are used. The very best, from my experience, is *Fundulus chrysotus. Chrysotus* means golden ear. This native Florida cyprinodont has a flashy golden spot on the operculum. Males are easily distinguished from females. They will pond breed and there also is a limited market for the fish.The trick is to set one pond aside to produce the fish. Take *only* the males and use about 300 in a large pond. Fundulus will not be able to eat or injure the large fish; they will eat only the juveniles and this will limit production in a natural way.

Care must be taken not to spread fundulus around the farm as they are difficult to eliminate. Their eggs develop slowly and often will not be destroyed when a pool is pumped and limed. A second pumping after 15 days becomes necessary to be sure they are eliminated. Maximum length is about 3 in.

The feeding of the ponds is very important; the fertilizing of the ponds is also very important. Feeding and fertilizing are interrelated. They complement each other if done properly and can impair production if done improperly. Almost all of the pond water in Florida is hard and alkaline. This means that the proper fertilizer should be used. As all water is different it is best to experiment until you find the types and amounts of fertilizers that work best in your water. I said at the start that feeding and fertilizing are interrelated. The meaning of this when feeding is that uneaten food will act as a fertilizer. Some farmers seldom if ever fertilize their ponds but rely on the fish food to do it for them. This has never worked for me and I question the practice.

No experimental work has ever been done on the feeding requirements of pond raised ornamental fish. Feeding practices have evolved from individual experiences. Every farmer has a different concept. Until the time comes when solid nutritional research work is done, feeding these fish will remain a hit or miss method. The most popular feeding formula is composed of 2 parts of bolted oats and 1 part Maine Concentrate. Maine Concentrate is a high protein, high quality fish meal. I use this as the basis for my feeding program. My exact formula is as follows: I mix up 682 kg (1500 lb) at a time. It is composed of 227 kg (500 lb) of Maine Concentrate, 136 kg (300 lb) of 45% meat scraps, 227 kg (500 lb) of oat flour and 91 kg (200 lb) of wheat mids. Before this is mixed, I add 19 liters (5 gal.) of fish oil, 454 g (1 lb) of trace minerals, and 1702 g (3¾ lb) of a vitamin unit.

The most important thing about pool feeding is when and how much. When feeding it is best to feed late in the day. Feeding early can interfere with trapping. If you know you are going to trap a pool the next day, the pool should not be fed. Feeding once a day is sufficient. If you can feed smaller quantities more often it will be beneficial. This seldom is practical. Normally, fish are fed 5 or 6 days a week. During the coldest days of the year feeding should be stopped as the fish will not eat. During the very hot summer months fish often will not feed actively and feeding should drop down to 3 times a week. Using 272 g (8 oz) of feed is sufficient for the average pond. Overfeeding can act like a time bomb. A pool can foul out overnight and without warning. Feeding fish when they aren't eating is dangerous. This is particularly true in the winter. When the water warms the excess food will foul the pond.

Success in quickly raising certain varieties of fish to a large size depends on feeding. This is particularly true of kissers and other gouramies. Proper feeding is both an art and a science. Fertilizing should initially be done when the pond is pumped. I find it makes no difference if I fertilize immediately or wait for a day or two. I pump, lime and fertilize at the same time. A combination of inorganic fertilizer and organic fertilizer is recommended. I use both but apply inorganic first.

There is a wide variety of available materials to use as inorganic fertilizers. Best results from my experience are obtained using cottonseed meal. Alfalfa meal and soybean flour are other organics that are occasionally used. Some farmers add one of these organic fertilizers to their feeding mix, but this is usually a very small percentage of the mix.

Fertilizing should be done on a scheduled basis. Once a week or 3 times a month is about right. Every pool should be examined visually and fertilized when needed. Amounts vary depending on the condition of the pond. The trick is to touch up the ponds before the bloom disappears. If one can properly feed the planktonic bloom with the fertilizing program, one will achieve maximum results.

Weed Control.—The ideal situation is to put a bloom on the pond and keep it. This cuts down the available light and retards or prevents the growth of higher forms of aquatic plants. The closer one can come to this the better one's results will be. It can not be done at all times, but one should keep trying. The nature of the soil has a lot to do with the ease in putting color on a pool as well as with keeping a bloom on a pond.

Different areas on the same farm will react completely differently to identical feeding and fertilizing programs. When you get a heavy aquatic growth, it is necessary to remove as much of it as you can by hand. It means using a long handled rake. A heavy leaded line dragged through the ponds will break loose most of the rooted plants and make it easier to remove them. It is hard work, but it is too dangerous to try to kill a heavy weed growth. Raking out weeds is a necessary job that all too often is avoided.

Chemicals have become increasingly expensive and the less used the better, for many reasons. An important reason is the potential danger to the crop. Chemicals react with the temperature. Unfortunately they re-

FIG. 5.53. TROPICAL FISH PRODUCTION PONDS

FIG. 5.54. TROPICAL FISH FARM SHOWING HATCHERY BUILDINGS

act most when it is hottest. This is also the problem time insofar as weeds are concerned. Aquatic weeds grow faster during the hottest part of the year. This is the reason why it is important to prepare the pond before treating by removing as much of the organic plant material as one can.

The most dangerous aquatic plant is *Hydrilla verticillata.* This plant is

FIG. 5.55.   INDOOR TROPICAL FISH HOLDING AREA AND HATCHING TANK

FIG. 5.56.   4-IN. PUMP USED ON TROPICAL FISH FARM

widespread throughout the state of Florida. Care must be taken that it is not accidentally introduced as it is very difficult to eliminate. It is easy to kill back but it produces a resting winter bud called a turion which will produce a new plant at a later date. Proper and continuous treatment will eliminate *Hydrilla*. One application will not do it.

Water hyacinth, water lettuce, *Salvinia*, duckweed, *Azolla*, and *Riccia* are the floating plants one will have to contend with. Water hyacinth and water lettuce are not easily introduced, but they are helpful in culturing certain fish, as they provide shade and hiding. When it is time to remove them, the best method is by hand. Duckweed is the most common of the floating plant plagues. All can be handled the same way. They are spread on the feet of wading birds, the backs of turtles, and often by careless workers. They are also spread on traps and seines that are not clean. If care is taken, this problem is negligible. A fine spray of kerosene will kill duckweed. This is the cheapest method, but unfortunately it seldom will get all the plants so it is not recommended. Diquat works best on an overall basis as it is not dangerous to the fish. Karmex can be used but it is very tricky and can easily kill fish.

Regardless of the toxicity of the chemical, what kills most fish in weed control is fast fouling of the pond with the fish dying from lack of oxygen. Weed control is most important in the summer and unfortunately most dangerous at that time of the year. All of the other aquatic plants can be killed with any number of chemicals. The most important are Karmex, Cutrine, unchelated copper sulphate, Dowpon and Casaron.

Emergent plants such as cattails can be sprayed; they can also be killed back with kerosene. This method is not the best but is the cheapest. It is hard work but effective. It should be used when there is not a serious problem. If it has gotten away from you (and time is important), then use Dowpon with a sticker. It will kill completely. Check the weather before spraying; it if rains the day you spray, you will have to repeat it.

Experimental work has been done recently with aniline dyes, the objective being to darken the water artificially to retard light and slow down photosynthesis. The theory is sound and shows promise. I have attempted this using potassium permanganate with mixed results. The grass carp or white amur (*Ctenopharyngodon idella*) is a very effective biological weed control. One good sized fish in a pond will control the plants. They do not eat fish and can be removed easily to another pond with a wide meshed seine. However, they have been banned in Florida for some years. At present there is serious talk about their being used with proper permitting in impoundments of less than 10.1 ha (25 acres). If this materializes, it will be a very large financial aid to ornamental fish farming.

Harvesting.—There are two basic types of traps used to collect fish. One is a plastic minnow trap that has a wide funnel for the fish to enter and a removable rear door to pour the fish out into the sorting container.

This trap must be tended. It is used early in the day and will usually be ready to be pulled within 1 hr. A minimum of 2 plastic traps are set in a pond. This is done to reduce the chance of the fish not trapping. Care must be taken. Although the traps are open to the available oxygen in the water, they can be overcrowded to the point where mortality occurs. Labyrinth fish such as the paradise fish and gouramies must be able to reach the surface to breathe. A trap set completely under the water will cause the trapped fish to drown. Everyone does this once and then never does it again. During the summer the water near the surface can become terribly hot. Traps set too high will not trap fish. Dry dog or cat food works well as bait in the traps.

The second type of trap is known as a heart trap. The trap is similar to a heart in shape. It is a wire basket with the indenture in the heart being the opening through which the fish are trapped. Once inside they remain there. The shape keeps them circling and few will escape. This trap is open at the top to the atmosphere and is usually set the evening before the fish are needed. The fish are in the trap ready for harvesting early the following day. If a pool is ready to be pumped out for restocking, it should be trapped heavily and quickly. Up to 90% of the fish will usually trap out within 48 hr. The key word in that last sentence is "usually." One can get some nasty surprises when one has a pool 75% emptied and suddenly realizes there are a tremendous number of fish remaining in the pond. When that happens, use a seine and move the fish out fast. If there is no place to move them at the moment, let the pool fill up again and go through the pumping procedure at a later date.

A good practice is to prepare a pool for pumping. When satisfied that the salvageable fish have been trapped out, another step can be added before pumping. If there are no weeds, the fish can be killed. If there are weeds, the fish and the plants can both be killed. This eliminates a great deal of time when pumping the pond. Hydrothol is an effective poisoning agent and will kill everything. Rotenone can be used if there are no plants to be killed.

Pumping.—As a general rule a pond must be reset (pumped and re-stocked) every year. Some ponds can be kept going for 3 and more years when a good colony of fish is established and reproducing. This will usually be something like rosy barbs or common guppies. Some pools that are used for sizing can be used 2 and 3 times in 1 year.

Pumping is done with gasoline powered water pumps. Two pumps are used. The largest, which is usually a 10 cm (4 in.) pump of the "mud hog" or "trash" type, removes the water. The second "wash" pump has the water constricted to increase the pressure and is used to clean. The pond's bottom can be reshaped and all of the accumulated detritus removed. I washed a pond every other pumping for years but am now convinced that the extra time required to wash every time pumped is worthwhile in increased production. Properly managed, pumping can be done while other work is also being done. The person pumping can, while the pump is operating (for at least 1 hr), pull traps, set traps, or do any other jobs needed. With the exception of seining the pond after it is ⅔ empty, to remove whatever fish are left, one person can do the job.

When a pool is pumped, it reduces the water level in the ponds in the immediate area. This canbe helpful if pumping more than one pond in an area. It will save some time as 30 cm (1 ft) or more of the water has been removed through the water table. Two or three ponds can be pumped in a day.

When the pond has been emptied and washed, it is limed. Hydrated lime comes in 22.7 kg (50 lb) sacks. Two bags will do the job on the average pond. It is broadcast. With a little practice the powdered lime can be made to evenly cover the exposed bottom. Protective goggles should be used as lime is a bad irritant to the eyes. If one's back is always kept to the wind, no problems can be anticipated. However, everyone will probably throw lime into the wind one time, but then never again. The emptied pond will fill itself with water within 3 days. Fertilize the ponds after liming them. The pond should be ready for fish in 15 to 20 days. Often a pond will bloom and clear before it can be stocked. In that case, touch it up with fertilizer. There is nothing more self-defeating in fish farming than stocking fish in a sterile pond.

Predators and Control.—The list of predatory animals is long. Experience indicates that most of the most serious predators are not recognized, and a number of animals considered to be serious predators really do more good than harm.

With the exception of *Homo sapiens* there is only one serious mammalian predator, the otter. Their numbers in Florida are greatly reduced due to the encroachment of civilization, but if one should be in an area where otters are abundant, they can do a great amount of damage.

There have been many instances of "fish rustling." Several years ago a group of organized thieves was apprehended and brought to trial and convicted. Since that time, I have heard little about mysterious disappearances of fish. When fish are stolen it is done at night. It is a good idea to mark ponds with numbers. Never use signs to designate the contents. There is no point in helping a potential thief!

Avian predators can be very serious. Occasionally sea gulls and terns come in, but these are normally sea birds. The herons, larger egrets, and kingfishers are the most serious. At certain times of the year birds such as *Anhinga,* cormorants, and various fish-eating ducks move in for short periods of time. All of these can do much damage. A few dogs on the farm will help with this and other predator problems. A clean farm with a minimum amount of brush discourages birds and in the long run is the most effective deterrent. Carbide shotgun machines that can be adjusted to periodically (every 2 to 4 min) shoot a loud blank do no good. They work well the first day or two and then the birds come closer and closer. After 3 or 4 days they are worthless for anything except migratory birds. At certain times it makes sense to patrol and scare birds after hours. During drought periods when water is not readily available for the water birds, they will come in larger numbers. Dawn and dusk are the worst times.

Reptiles are not serious predators. The ones to focus attention on first are alligators, which are increasing in numbers all the time. Small animals (less than 0.9 m or 3 ft) can and do consume small fish. *Natrix taxispilota,* the brown water snake, and others of this genus do eat fish. This particular snake is easily mistaken for a cottonmouth moccasin, which is seldom seen. Caution should be exercised. Most of the turtles are vegetarian. The only fish eaters are the softshell, mud, musk, and snapping turtles. The softshell is the most common. These and the other 3 fish-eating turtles can and will bite badly. While all of these reptiles are predatory, they actually eat more predators than fish. They eat aquatic insects, each other, frogs, tadpoles, crawfish and young birds. Taking an overview, I am of the opinion that they do more good than harm.

Amphibians are a problem. The larger frogs are the only serious ones. *Amphiuma means* is seldom encountered. It is a large salamander that can exceed 3 ft in length. It has a pair of tiny forelegs and no rear legs. It is known as a "congo eel," "ditch eel," or "siren," and is seen rarely. Contrary to common opinion, frogs probably do more good than harm. They eat a variety of aquatic predators and adult dragonflies that have predatory aquatic larvae. The tadpoles, however, are a serious problem. They compete with the fish for food. They can seriously crowd and stress fish in a trap. Eggs when seen should be removed from the ponds. The tadpoles when trapped should be thrown out and onto the ground. The only fish I know that eats them is *Astronotus ocellatus.* Some can be used as food for this fish. No one has yet found a selective toxicant for tadpoles. They cannot be drowned as they can absorb oxygen through their skin while under water if they cannot reach the surface for air.

Aquatic invertebrates are the most serious predators. Fortunately there are controls today that if properly utilized can control the problem. Larval forms of terrestrial insects are bad predators. Some of the worst are the water tiger (*Dytiscus*), water dragon (larval forms of dragonflies, darning needles, mosquito hawks, etc.), and damselfly larvae. Water boatmen (Corixidae) and back swimmers (Notonectidae), *Argulus, Hydra*, crawfish, leeches, and adult diving beetles are only a partial list. Those that come to the surface to breathe, such as the water boatman and back swimmers, can be killed cheaply and effectively by the use of kerosene. One liter (1 qt) is usually sufficient. The liter (1 qt) of kerosene is evenly spread as tightly as possible along the edge of the pond against the wind so the kerosene will spread with the wind. A gentle and slight wind is sufficient. The bugs die on contact, and the kerosene evaporates within hours. The insects will fly back in again so this control work must be done periodically. Fuel oil or anything heavier should not be used as it will not all evaporate.

The other more expensive and broader control chemical is called Baytex 4 and is an emulsifiable insecticide. It kills almost all invertebrates except some worms and snails and kills all of the important predators. It has helped the industry more than any single tool since its use in ornamental fish farming was discovered. It will also kill fish and so cannot be used casually. The hotter the weather, the more effective Baytex becomes. A treatment of 14 g (½ oz) does not sound like very much for an average sized pond, but it will do the job. Baytex should be kept cool and in the dark when not in use.

The use of copper sulphate is recommended to eradicate snails. Snails themselves cause little if any problem, but they are intermediate hosts for such disfiguring parasites as *Clinostonum marginatum*, the grub worm. Copper sulphate has been used for many years as an aquatic weed control tool. Many people are under the impression that continued use over a period of time will reduce the fish's fertility, but I do not agree.

Spawning the Egg Laying Fish

This section breaks down immediately into those fish that breed themselves in ponds (which we will discuss first) and those fish that must be bred indoors and grown out in ponds.

Colonies of self-perpetuating egg layers of a limited variety can be established. The ponds must be maintained insofar as weed control, feeding, fertilizing, etc., are concerned, but basically left untended, some ornamental egg layers will reproduce themselves. Various species in family groups will be discussed.

Characins.—As most of these fish come from soft acid waters, few reproduce themselves naturally in Florida waters. However, some will. Those reasonably sure of pond breeding include *H. scholzei*, the blackline tetra; *H. bifasciatus*, the yellow tetra; and *H. caudovittatus*, the Buenos Aires tetra. Others do pond breed, but when it is a hit or miss situation, it does not make sense to tie up a pond on this kind of proposition. It is better to learn the art and science of breeding indoors. This will tremendously increase the chances of obtaining consistent inventories. A good general water temperature range is 24° to 28°C (76° to 83°F).

Cyprinidae.—The most important member is the zebra danio *(B. rerio)*. This species plus the other small danios will usually pool breed. They are easy to breed indoors and that is the best method. The reasoning is to keep the fish uniform. Pond breeding can give too many mixed sizes. The giant danio *(D. malabaricus)* will do well in a pond, and this method is recommended. Flying barbs *(E. malayensis)* will pond breed. This member is in disfavor currently and does not sell. None of the labeos have bred to my knowledge. Several of the *Barbus* and *Puntius* species are easily pond bred. The one most highly recommended is the rosy barb *(B. conchonius)*. The important barbs all occasionally pond breed, but not often enough to be meaningful. White clouds *(T. albonubes)* have potential as a pond bred fish. Results are very mixed, but several people (not including the author) have been able to commercially pond breed these fish.

Cobitidae.—*Misqurnus anguillicaudatus,* the Japanese weather fish, will pond breed. It will also move overland in wet weather. I know it is bred in Michigan in outdoor ponds, but not in Florida, and I would not recommend it. A *semicinctus,* the kuhlii loach, has been bred a very few times and has possibilities. None of the other loaches have been pond bred.

Callichthyidae and Other Catfish Families.—*H. plecostomus* is bred in dirt ponds in ever increasing numbers and is a very important item. Some of the other catfish are being bred now but are still in the experimental stage. They may become important.

Cyprinodontidae.—Little has been done with the top minnows, but experimentally there have been some encouraging results. In smaller controlled ponds, worthwhile results could be anticipated.

The local Florida cyprinodonts such as *Fundulus chrysotus, Cyprinodon variegatus, Chriopeops goodei,* and *Jordinella floridae* can be caught wild. It is much more economical to let them pond breed and be easily available. *Oryzias latipes,* the medaka, will pool breed, and there is a golden strain which does sell better. These fish were so cheap for so many years that the industry stopped breeding them, but today's higher freight costs for medaka (from Japan) will make breeding them a profitable venture.

Atherinidae.—Several of the Australian rainbow fish are being bred and in large numbers. *M. maccullochi* and *M. nigrans* are the most popular. There are some very beautiful newly available atherinids from New Guinea which will pool breed and would be very popular.

Anabantidae.—These are the gouramies and paradise fish. Almost all will pool breed, but they are unpredictable and unprofitable for most breeders. The ones that will pond breed but also can be easily bred inside should be bred inside. The mixed sizes and breeding activities retard harvesting and selling to a large extent. The more difficult breeding inside actually ends up easier and cheaper.

Several species are difficult to reproduce indoors, and these do best outside. The most important are *T. pectoralis*, the snakeskin gourami, and *T. microlepis*, the moonlight gourami.

Centropomidae.—*Chanda ranga*, the glassfish, is the most important. All of the fish produced in Florida are produced naturally. They can be bred indoors with much difficulty but can be raised more economically in ponds.

Cichlidae.—This comprises the largest list, but it is easier to list the important exceptions. Many other species are hit or miss, but generally pool breeding can be accomplished for those cichlids coming from hard and alkaline waters. The important soft water cichlids are not bred. These include all of the *Appistogramma* (the most important is *A. ramirizi*), *P. pulcher*, the kribensis, *N. nudiceps* and other Stanley Pool African cichlids, *A. maroni*, the keyhole cichlid, *S. discus, Uaru, Crenicichla, Crenicara*, and *P. scalare*. This last is at times bred, but not in an economical manner. The hard alkaline Central American *Cichlasoma*, almost all of the *Aequidens, Etroplus* from Asia, and the African Rift Lake fish from Lake Malawi will all breed in Florida waters. The one important exception are the beautiful Lake Tanganykia cichlids. Lake Tanganyika water is extremely hard and alkaline, but the mineral content varies radically from Florida water. To date, these fish have not been bred with any degree of success in outdoor ponds in Florida.

Eleotridae.—The Australian purple striped gudgeon *(Mogurnda mugurnda)* can be pool bred. A few of the dormitators can be bred. These latter are not really commercial fish.

Gobiidae.—Several species have been bred. Not enough has been done with these fish as yet to make any definite statements. I suspect several species can be bred on a commercial basis..

Mormyridae, Pantodontidae, Osteoglossidae, Notopteridae.—Mormyridae, Pantodontidae, Osteoglossidae, Notopteridae, and many other families of fish that are important in the ornamental fish industry have never been bred or seldom bred. There is a great amount of potential in these fish.

Using the same formula as in the previous section, the indoor breeding of the various types of egg laying fish will be discussed. The average breeding tank is 38 liters (10 gal.) or less. Most breeders use a 19 liter (5 gal.) aquarium for the smaller characins, gouramies, paradise fish and Siamese fighting fish. Established breeding pairs of large fish such as *Discus* and angels are kept in 57 liter (15 gal.) tanks and larger. Breeding of some of the bigger characins and cichlids is accomplished in 114 liter (30 gal.) tanks and larger.

In order to breed indoors commercially it must be done on a large scale. Enough fry must be produced to utilize a large empty dirt pond. Raising several thousand fish of normal value in a full sized dirt pond is self-defeating and not profitable. One must set up sufficient breeding tanks for each species. This means a minimum of 60 tanks of an item that will spawn heavily and a success ratio of 90% or better can be anticipated. The most successful commercial breeders seldom if ever set up fewer than 200 tanks at a time for each pond. The notable exception to this is with medium and larger gouramies. They spawn heavily and 40 or 50 pairs set up at one time are sufficient.

Breeding Egg Laying Fish

Characins.—All are bred similarly. Best results will come from using the following procedures. Breeders are kept separately (by sex) until used. Use them every two weeks. Keep the breeders fed with a variety of food and keep them in large tanks. Use small numbers of fish in each aquarium and set them up in pairs. Most characins will breed within 24 hr. At least three days should pass before separating the pairs that do not spawn. One exception to this is *Mettynis*. They can breed after being set up for 1 week or even 10 days. Most characins do best set up in pairs. Good results can also be had using trios (2 males to 1 female). Breeders should be removed after spawning. There is no need to panic as egg eating among characins has been overemphasized for years. Spawning will occur in the morning in most cases. The breeders can be removed in the afternoon.

Eggs are semi-adhesive. The skeleton of the Spanish moss plant is an ideal egg receptacle and is widely used. Most characin fry should spend several weeks inside before being put into ponds.

Cyprinidae.—The danios should be bred in groups. They are voracious egg eaters. The eggs are nonadhesive and a breeding trap is mandatory. This trap can be made out of hardware cloth or plastic netting. Both

work equally well. Two inches of water depth in the trap is recommended. Groups of 20 to 40 fish work well, and a ratio of 2 males to 1
female is needed for optimum results. Eggs hatch quickly. As with all
fish, the temperature determines the actual speed of hatching. The fry
will spend a day on the bottom and a day plastered to the side walls before free swimming. They can be placed directly into prepared ponds.

Barbs should be handled identically to the characins. The fry are kept
inside for at least 2 weeks before being put into ponds. They breed best
in pairs. The most important barbs commercially are sumatra, cherry,
clown, filimentosa, *Arulius, Lateristriga,* gold, and checkerboard. Two
of the genus *Rasbora* are bred commercially. These are *Trilineata,* the
scissortail rasbora, and *Borapetensis,* the brilliant rasbora. They are
bred in pairs. White clouds are usually colony bred indoors but can also
be set up in pairs or trios. White clouds do not eat their eggs or their
young. Colonies can be skimmed on a daily basis and the baby white
clouds removed for grow-out at an even rate. There are several important cyprinodont hybrids. The best are various sumatra barbs.

Callichthyidae.—The largest volume and the most important is the
genus *Corydoras.* The culture of these fish is accelerating. Currently
Corydoras aeneus and *paleatus* are being bred. These fish are also
available in albino forms. They are the only members being raised
indoors in large quantity. The total, while large, constitutes less than 5%
of the demand. Many more of these fish plus some 50 other ornamental
catfish can be raised. Progress is being made by many ornamental fish
farmers in this direction. Corydoras are erratic spawners. Each breeder
must first learn how they wil breed best for him. Once this is established
a regular and consistent production can be anticipated. They can be
spawned in pairs or can be raised in colonies, and parents will not eat
either their eggs or their young. Best results are obtained by separating
sexes until spawning. Water temperature, cleanliness, and photoperiods
are the critical factors in successful breeding.

Cyprinodontidae.—A few breeders specialize in raising members of the
genus known in the trade as "killifish." These include *Nothobranchius,
Cynolebias, Aplocheilus, Epiplatys, Pterolebias, Aphyosemion,* and
others. They are all delicate, specialist fish. There is a good market for
them, but first the breeder must establish his identity among the buyers.
This is different from most ornamental fish as both buyers and sellers
of these fish are specialists, and the number of each is limited. At present
there are more buyers than breeders so a good market awaits breeders
who will include some of these varieties in their production schedules.

Atherinidae.—The most important item at the moment is the *Celebes* rainbowfish. This is a fish being bred by only a few breeders with a large demand awaiting the product. It will increase manyfold in the future. There are other similar fish not yet available. When breeding stock becomes available they will also be bred.

Anabantidae.—This is a very important family of fish in the ornamental fish farming industry. Most should be bred indoors. The market is very large and quality fish year after year sell out. One of the best known anabantids is *Betta splendens*, the Siamese fighting fish. Only unsexed and female fish are bred in quantity in Florida. The large males require a very high labor cost. Most large males are imported from the Far East. The other two major groups are the paradise fish and the gouramies. The paradise fish are probably the oldest aquarium fish kept ornamentally except for the goldfish. There are a red, blue, black, and albino paradise fish. These are all *Macropodus opercularis* although there is some question as to whether or not the black is a species and not a mutation (it is often called *M. concolor*). Based on its size and behavior, I think it is a different species, although I have no real taxonomic basis for the statement.

The gouramies include *Helostoma temminicki* and *rudolfi,* the kissing gouramies, the former known as the pink kisser and the latter as the green kisser. Members of the genus *Colisa* are important. The most important are *fasciata* the giant, *chuna* the honey, *labiosa* the thick lipped, and *lalia* the dwarf. The genus *Trichogaster* is also very important. This includes *leeri* the pearl, and *trichopterus* the blue or three spot. There are also important mutations of *trichopterus.* These all are bred and sold in large numbers. The opaline, the gold, the platinum, and the white lace are all attractive. In 1978 a golden form of *T. leeri* appeared on the market and promises to be a popular fish.

There are lesser known Anabantidae that are bred infrequently. A limited market is available for these various *Bettas, Trichopsis, Ctenops,* etc.

All of the Anabantidae breed similarly, the only difference being the extent of the bubble nest. In some species there is a complete absence of the nest. Eggs are light enough to float. All should be bred in pairs except for the kissing gouramies. Kissers do best with 2 males to 1 female or can be set up using 3 males and 2 females. Container size is important. Spawns range from averages of 100 eggs *(B. splendens, C. pumilus)* to 30,000 or more *(H. temminicki).* Aquariums of 8 liters (2 gal.) are large enough for *Bettas,* dwarf gourami, *chuna* gourami, and similar sized fish. The 7.5 to 10 cm (3 to 5 in.) species do best in 30 to 38 liter (8 to 10 gal.) aquaria. The large fish are bred in containers of 114 liters or more (30 gal. or more) to the pair. Often kissing gouramies are set up in wooden or plastic vats of approximately 378 liters (100 gal.). I recommend this practice. Males tend the nest. The bettas are the only males that will eat the fry. The eggs hatch and swim free in approximately 48 hr at 27°C (80°F). With bettas, females can and should be removed after spawning. Males can drive hard so the breeding set up should include floating material to help anchor the nest. A 3 in. piece of styrofoam does fine plus a piece of crockery or plastic to give the female a place to hide from the male. Many females will be lost if this is not done. The males are intense breeders and sometimes kill the females.

It is not important to keep the kissing gourami sexes separate prior to spawning, but they do best if rested for 3 weeks before each spawning attempt. Males should be removed as soon as juveniles swim free. Fry will not be eaten using this timing. Bettas, being the most aggressive, require the closest attention while breeding. Conditioning is done by keeping both male and female bettas separated. Males must also be kept separate from each other. The practice is to keep them in 1 liter (1 qt) jars. There are still many questions unanswered as to the importance of the bubble nests. All of the nest building fish will occasionally release and fertilize eggs with no bubble nest prepared, and babies hatch normally. Africa has 4 or 5 outstanding Anabantidae species of the genus *Ctenepoma*. These include *acutirostre, ansorgei, fasciolatum,* and *oxyrhynchus.* They have all been bred occasionally, but not commercially. A large market awaits the breeder who can produce them commercially.

Cichlidae.—In recent years the keeping of cichlids has become very popular. This was stimulated first by the discovery of the beautiful red, yellow, white, and pied polymorphs of *Cichlasoma labiatum* and *citronellum* in Nicaragua. Then a few Rift Lake fish were exported from Lake Tanganyika in the late 1950s from Zaire (then Belgian Congo). The civil war in that country stopped the export for more than 10 years. In the mid 1960s fish started coming in from Lake Malawi and Lake Tanganyika. They are almost all mouthbreeders. The colors rival the most beautiful of the marine coral reef fish. Many of these can be pool bred, but many have not been as yet. The industry in Florida is raising more and more of these each year. The other important cichlids that are being produced indoors are the wild and golden forms of *A. ramirizi, P. pulcher,* the kribensis, *N. nudiceps, P. scalare,* the angelfish, and *S. discus.*

The market for angelfish has been insatiable since its introduction after World War I. Many color and fin mutations have developed and all sell. These include first the veiltail varieties. This long sweeping caudal fin has also been bred into every color variety. The normal angelfish remains the best seller. The other varieties are black lace, marble, dusky, all gold, all black, zebra, half black, albino, and pied. In addition, there are varieties that mostly exist in the imagination of the breeder. These include

FIG. 5.57. LIMING PUMPED OUT TROPICAL FISH POND

FIG. 5.58. TROPICAL FISH POND ONE DAY AFTER PUMPING AND LIMING

so called red, blue, brown, and lavender angelfish. They will come in time but are not here yet.

S. discus is still one of the most expensive and popular of the cichlids.

They have unique breeding requirements that make quantity reproduction difficult. The babies when they swim free eat as a first food the heavy slime secretion on the parent's bodies. The parents are notorious egg eaters, which complicates the problem. Babies are and can be raised away from the parents, but this is an arduous and time consuming process. It is also difficult to do in a commercial establishment. Parents will usually eat several spawns before settling down to raise their fry, but there are no patterns. These fish are worthwhile to pursue as the market is never satisfied and prices are always high.

Cichlids raised commercially are either mouthbreeders or substrate spawners. The mouthbreeders do best in colonies. One male is used with 5 to 8 females. The eggs are brooded for as long as 3 weeks and can be brooded by either the male or the female depending on the species. Once procedures are established, a breeding routine is easily maintained. Most other cichlids lay adhesive eggs on a hard surface. Some prefer to spawn on a horizontal surface (rams, kribensis, etc.) and others spawn on a vertical surface (angelfish, discus, etc.). Once you think a pattern has been set, they will fool you and spawn on airlines, bottom or side glass, on top of a pipe instead of inside it, or most anything to shake your complacency. A high ratio of success is attained by removing the material and affixed eggs from the parents to hatch separately and raise separately. The notable exception is the discus. Water from the spawning tank must be identical to the water the fertilized eggs are placed in for hatching. Temperature must be the same. Live food is essential to condition the breeders and keep them breeding. Mated pairs should periodically be rested to get maximum yield from the fish. Angelfish, discus, and others will breed on a weekly basis, but this cannot be allowed to continue for long periods of time. If it is, one will not get maximum utilization from the fish. Pairs must be split up and rested 2 or 3 times a year. Several varieties of cichlids including *Astronotus ocellatus* the oscar, *C. severum* the deacon fish, and *C. festivum* are produced best by letting the fish spawn in dirt ponds on carefully placed pieces of spawning material. This material (flat pieces of slate work well) is checked daily for spawns. The spawns are taken from the pond into the hatchery and hatched. They can be raised inside until saleable or after 3 or 4 weeks returned for grow-out to other ponds. There have not been many cichlid mutations excepting the angelfish. The important ones are the golden severum, red and tiger striped oscars, pink convicts *(C. nigrofasciatum)*, and the red chromide.

Breeding Stock

When obtaining oviparous fish to use for fish breeding stock, there are very important points that must be considered. Do not use mature fish. There are several good reasons to avoid mature fish. (1) One cannot tell the fish's age. (2) One cannot tell if these fish have been used as breeders and are now " over the hill." Both of these are good reasons but the last is the most important reason. (3) Fish raised in one's own water have a much better chance of breeding than fish that have been raised in other waters. (4) In addition to this, it is mandatory for best results that the first spawns gotten from any new breeders be saved for replacement breeders as they will make superior breeders.

Always back up stock with what might look like superfluous breeding stock. Have available additional potential breeders over and above those apparently needed, so these fish will be available when the breeders suffer from an unforeseen disaster. If possible, and I strongly recommend it, keep duplicate and triplicate sets of breeders. Keep them in different ponds, vats or tanks. This precaution will pay for itself over and over again. Having insufficient, inadequate, or partially grown breeders can lose an entire season for the breeder.

When breeding stock is needed the most (early spring), it is most difficult to obtain. When buying breeding stock, buy it locally (in Florida). Avoid wild fish, which are extremely difficult to breed. Avoid foreign-produced egg layers, which are usually available from Southeast Asia, as they are badly inbred. Also, the careless use of excessive amounts of antibiotics by breeders in the Far East has caused many strains of pathogens that are resistant to normally effective antibiotics.

It is an effective precaution to keep the breeding stock in the hatchery building remote from any possible contamination. I recommend the equivalent of a permanent quarantine area for the breeding stock. Fish jump and nets move diseases, as do hands and gloves. If the fish you buy for resale are isolated, you will eliminate most danger of contaminating your own stock. Keep your breeders uncrowded, clean, and fed a variety of food, and you will have a minimum number of problems.

Diseases—Prevention and Control

This is a very important part of fish farming. Fortunately, a great amount of work has been done on the subject. The information is available in great detail and must be read. Suggestions for further readings on diseases are given in Chapter 7. The ability to learn to identify the particular pathogen is very important. It necessitates a knowledge of the proper use of the microscope and also a basic knowledge of the anatomy of a fish. The equipment needed to identify diseases is not expensive. An inexpensive microscope, a simple dissecting kit, and some good hand lenses are recommended. The microscope will also be needed to check water samples for infusorians, etc.

The first rule in fish health control is prevention. Regardless of whether the fish entering the holding facilities come from Singapore or a pond 9 m (30 ft) from the hatchery door, they must be handled identically. The fish are stressed. Stress is what really causes most disease problems. Prevent the pathogen from taking over while the fish recovers (from the stress) in the holding tanks and most problems disappear. This is done by using a variety of chemicals and drugs. The assumption is made that one does not see a problem. One will seldom immediately see a problem as few shippers will send obviously sick fish. If one does see a specific problem, treat specifically, and if not, treat preventively.

There are several combinations of chemicals and drugs that are effective in preventing diseases. One neglected tool is methylene blue. I had heard for years that its only use was to make an aquarist feel good when he colored the water. In actual fact it is very useful. It inhibits the growth of bacteria. I am told (and it works for me) that methylene blue makes toxic chemicals like acriflavine (tripaflavine) less toxic. Anything that stimulates slime secretion is an effective tool. I use Epsom salts (magnesium chloride) at the rate of 14.8 ml (1 level tbsp) to each 12 liters (3 gal.) of water. Sea salt or rock salt is more generally used. They are both effective. Sodium sulphathiazole has been used with good results for many years. One thing that must be understood when using "sulpha" is that it will drive up the pH. This does nothing to Florida pond raised fish, but it can cause disasters if used on fish kept in neutral or lower pH's. The best alternative is tetracycline hydochloride. It is inexpensive and does nothing meaningful to the water quality. Use at the rate of one 250 mg capsule to each 27 liters (7 gal.) of water. Another effective compound is copper sulphate. A supersaturated solution is made. The liquid is effective at 1 drop to each 11.4 liters (3 gal.) of water. Formaldehyde is another useful tool that works extremely well alone or in combination with other drugs (particularly malachite green). Use formaldehyde at 2 drops to 4 liters (1 gal.). All of these chemicals can and are used to prevent disease. I have refrained from being overly specific, as what works for me may not work for you. Various people in the industry use somewhat different combinations and evidently have satisfactory results. Experience teaches not to correct a problem one doesn't have. Don't change what works. Correct only the problems which are there.

Fish are medicated on arrival and should not be fed for 48 hr. They should be left undisturbed as much as possible. At the end of this quarantine period, they can either be moved to clean water or the holding aquarium can be siphoned down and half the water removed before refilling. The latter is easier on the fish. The fish are then carefully examined for problems. You may have to re-treat as before, or treat specifically.

One generally unrecognized fact is that all fish must be taught to eat in captivity. The fish have never seen the type of food being presented (generally), and they usually have never been fed in a totally artificial environment. It is very easy to overfeed. Start feeding small amounts to get a few of the fish eating. Feed 2 or 3 times and they will then quickly get the idea. One gross overfeeding of newly arrived fish can create a problem fast.

Specific treatments for specific diseases are important to know. External bacterial damage to the fish is known in the trade as "fin rot," "tail rot," and "mouth fungus." This is one of the most common and serious diseases. The preventive medication program will eliminate most of this problem. Heavy rains, which stress the fish in the ponds during the summer, can cause bacterial problems in the ponds which can make this problem more serious during the summer. A combination of methylene blue, acriflavine, and liquid copper, which is made up in advance in stock solutions that we call "cocktail," is used along with an antibiotic and works well. It is not the only treatment, but the one used (with slight variations) most often.

ICH (*Ichthyophthirius multifiliis*) spreads fast and is often out of control before it is recognized. It is a protozoan. Vigilance is the key. Experience will automatically direct attention to a tank with fish that have an altered behavior problem. One can, in time, and with little error, recognize this disease before it is obvious. There are many treatments. The most effective is malachite green. The warmer the water, the quicker the life cycle of the protozoan, and the quicker the cure. Care should be taken to quarantine the affected aquarium until cured. Velvet or *Oodinium* is another similar and smaller protozoan parasite. It kills slowly, is quite common, and also cures best with malachite green. A particularly stubborn strain has been coming in from the Far East in recent years.

There are a number of protozoans and flukes that attack the gills and ultimately will kill fish. Fish riding the surface can be suffering from oxygen starvation caused by the gills not being able to function normally. Other things will cause fish to ride the surface (toxic wastes and low oxygen supply are most common) so the exact cause must be determined. Formaldehyde at the rate of 2 drops/4 liters (1 gal.) will do a fast and complete job of killing most gill parasites. Formaldehyde will take oxygen out of the water so if the problem is low oxygen and not gill congestion, formaldehyde can compound the problem.

There are many pathogens to be aware of and to know how to handle. These include parasitic copepods like anchor worms; parasitic crustaceans like *Argulus*; other types of worms, both internal and external; dropsy; true fungus diseases; etc. Many excellent books are available and should be included in one's own library. The University of Georgia under Dr. John Gratzek has presented a short 3-day course for the last 8 or 9 years that has proven invaluable to everyone fortunate enough to participate. It is a fish disease workshop for fish keepers. Keeping fish healthy is hard work and often will be the difference between success and failure in ornamental fish farming. If one's business involves good numbers of fish kept inside for inventory or breeding purposes, one must have the fish cared for 7 days a week.

Crop Protection

Florida is subtropical. One must actually go south of Miami to have any real security against winter freezes. In the last 25 years many cold resistant strains have been developed and have also evolved (through natural selection). This has helped considerably. Large numbers of fish are protected through the winter by utilizing the greenhouse effect received by covering an outside dirt pond with polyfilm. Between 4° and 6°C (8° and 10°F) additional warmth is achieved by utilizing this proven method. In almost all cases this is sufficient to get even the most sensitive fish through the winter. The terrible cold winter of 1976–77 was an exception and this extraordinary prolonged cold killed a large number of the fish stored under polyfilm covered ponds. This I would have to call a natural disaster and not to be anticipated as happening with any regularity. One can always provide additional insurance by keeping some additional stock inside through the winter. If this is not practical, then (when a freeze is anticipated) bring a backup stock into warm buildings to be returned outside after the danger has passed.

During the winter, pumps should be used when needed, and the water levels in the ponds should be kept reasonably low. The reason for this is that when a cold freeze hits one must not turn the water off. Fish will congregate around the warmer water coming into the ponds. Turn the water off and scatter the fish, and the cold will damage or kill them. If the water is run for days on end, there must be as much room for the water coming in as possible. Overflow systems can be used and they work well for some. I do better without them. Pumping water out of the cold end of a pond will reduce the water level in all ponds in the area. The pumps can be moved to help prevent overflows. In a desperate situation it has proven less damaging to let the ponds flow together or into ditches, etc. When the water can be lowered, the fish are still alive and while the work of sorting is time consuming, hard, and expensive, it is much preferred to having nothing to sort to sell.

High water and flooding can and do come from extraordinary storms. If one is near the coast, the potential for tidal waves exists. This is a remote possibility. As much as 52.5 cm (21 in.) of rain in 2 days has fallen on my farms. This is very unusual, but when it happens it can cause severe problems. Most water problems can be avoided. If one is aware of the low areas, as time permits, one can continue to improve the diking around the ponds. Material can be obtained when ditches are deepened or pools redug.

A business cannot be run without an assured supply of water and the possibility of a reduced water table always exists. In past years, from time to time, and generally in the spring, Florida has suffered severe drought conditions. I doubt if any great percentage of ponds would ever dry out completely, but they do not have to dry out to kill fish. In warm weather, temperatures in the top 30 to 45 cm (12 to 18 in.) will be higher than fish can stand. The dissolved oxygen near the surface is also too low. This combined with the reduced amount of water available for each fish creates a deadly combination. The water supply must be adequate in drought situations. One can't wait for drought conditions to happen and then find the pumps cannot pick up the water from a reduced water table. Expert hydrologists are available for proper advice before the problem occurs. Eutrophication problems will occur from time to time, but vigilance is the key. Running water and moving fish are the best available methods to reduce or, hopefully, eliminate losses. Experience in handling foods, fertilizers, and chemicals will reduce this problem.

Indoor Facilities.—The size and scope will vary with the type of business and the type of fish being sold. The additional monies spent in constructing properly insulated buildings have two great advantages. The primary one is the fuel cost savings. The second is peculiar to this industry. Gradual temperature changes up or down will not affect or stress fish, but rapid temperature changes will. A tin roof and wooden building without insulation cool fast. In the event of power interruptions, a well insulated building can prevent bad losses simply by having the structural ability to cool slowly. Retaining sufficient heat for an additional one-half day can be of an incalculable financial asset.

When the buildings are constructed always keep in mind that large amounts of water are being dealt with. Level floors are a mistake in this business. Sloping floors are essential so the floor always drains dry. If constructing concrete vats, then have troughs constructed when the floors are poured so that the vats can be drained without wetting the floor.

Floors can be finished smoothly or slightly textured. The latter is needed to provide secure footing. Natural sunlight is very beneficial to the health of fish. Skylights are also expensive and difficult to fit into a tight budget. If possible, obtain as much natural sunlight as possible in the breeding area. I would recommend that the building be of concrete block. High ceilings are not needed and waste fuel. If one can heat the water and not the air, additional savings can be obtained. Most fish are held and raised inside in concrete vats or wooden vats. Aquariums work better, but they are more expensive initially and require more labor to maintain properly.

Breeding stock and purchased fish should be kept in glass aquariums. This is best for many reasons which include: easy visibility, ease of treating diseases, reducing the number of fish vulnerable to problems per holding container, etc. All fish being prepared for shipment should be kept in aquariums. The metal clad tank is virtually extinct today. Glass aquariums are about all that is available. They have great advantages over every other type of glass aquarium except one. This is the potential danger when they are handled. The practice of holding fish ready for shipment in aquariums, prior to shipping, and then dumping the tank and contents cannot be done safely with a glass aquarium.

The following system of shipping is fine and recommended. The tank is siphoned down before the fish are removed. The precounted fish are in the tank, and the tank is light enough (38 liter or 10 gal. size) to be easily handled. It is poured through a net (preferably one held in a water container to prevent fish bruising themselves) and then into the shipping bag. Swinging glass aquariums around just doesn't work out. The potential for breakage is high and the broken glass is extremely dangerous. A better system that is now evolving is to place a holding net in each 38 liter (10 gal.) aquarium. The fish to be shipped are put in the net. The net is removed and emptied into the bag and the tank is not moved.

Racks should be constructed out of rustproof material. Angle iron of 3 ¾ cm (1½ in.) was most popular for years. There are many more desirable materials available today. Rustproof slotted steel works well. It is adjustable and movable. Adequate and inexpensive racks can be constructed by using nothing but concrete blocks and treated 10 × 10 cm (4 × 4 in.) lumber. Wolmanized (creosote) wood will not deteriorate rapidly even in the high humidity. Racks can be easily moved. PVC is the only type of pipe that should be used for the water and air systems. PVC pipe is available with predrilled saddles for insertion of water and air valves. The new plastic faucets and larger valves work very well. If a rack is constructed to hold three tiers of aquariums, it is not necessary to run 3 separate lines for air. One line under the top tier can service all 3 levels. Plastic tubing is available in 152 m (500 ft) spools and should be bought this way.

Air stones provide a fine mist, but require more pressure than an open-ended piece of tubing. I recommend the latter. Lead should be bought in sheets and cut into thin strips which are wrapped about the bottom of the air lines to keep the end under water. This works well and is a fraction of the cost of air stones. Large pumps and air compressors are a thing of the past. Better and less expensive air blowers are available today manufactured by Conde, Gast, Rotron and others. Air compressors are not needed.

Water is often available under pressure. This usually occurs when city water is available. If not, then use a water pump (electric) to move water. The water can be fed into the aquariums through the same type of valve used for air. These are made of brass or plastic. The plastic valves have evolved to the point where they are better than the brass ones and they are less expensive. Test valves before committing yourself to one type as there are many cheap and inadequate plastic valves on the market. By simply drilling a small diameter hole in the PVC water pipe, adequate water can be received through this small hole. The flow can be stopped with a round wood toothpick. The water supply must be clean, however, or this system fails due to detritus clogging the small holes; but with good clean water it is a simple way to save money and can work very well.

Heating is usually done with oil or propane gas. If possible, design the heating system to have the units high off the ground to prevent rusting. I prefer hanging air blowers.

Light with the 2.4 m (8 ft) slimline fluorescent tubes. Be careful to place the lights in the aisles, not over a rack of tanks as this will block proper illumination. All electricity should be grounded, and only qualified electricians should be employed. Water and electricity are a bad combination. Check and recheck to be sure you are not exposing yourself or your employees to potential electrocution.

In general, keep the hatchery open. Water attracts vermin. The easier it is to hose down, and the fewer areas for vermin to hide, the better. Insecticides are dangerous around fish. This includes residual types as the affected vermin can and do drop into tanks to die and thus can poison fish.

Tropical fish do best at a water temperature of 23.3° to 24.4°C (74° to 76°F). If heating the air, the air temperature must be considerably higher than this to obtain the proper water temperature. Breeding fish require higher temperatures, and it makes sense to keep the breeding area separate, so it can be heated higher than the remainder of the building. A water temperature of 27.7° to 28.9°C (80° to 84°F) will give the optimum results in breeding. The ambient air should circulate. There can be a difference of as much as 3.3°C (6°F) between lowest and highest tanks on a rack if the air does not move.

Be sure you are adequately insured, particularly insofar as public liability is concerned. Business interruption insurance is worth investigating. Keep in mind, before building, the possibility of future expansion. It costs very little more to build, anticipating future growth. Build the hatchery as wide as possible. Use the longest span the building code allows (probably about 12.2 m or 40 ft) and thus eliminate columns.

Drains are easily built when constructing but if built later are a terrible problem. Therefore, plan carefully beforehand. Most hatcheries will use open water systems. This simply means that the water is used one time. A closed system (where the water is reused) can be made almost fully automatic. It can save tremendous amounts of labor and give continuous safe and clean water. This is accomplished with a combination of proper filtration and ultraviolet light. Ozone may be able to do the same, but it has not been used widely enough commercially to form any conclusions.

If at all possible, the holding and inventory aquariums should have holes drilled in the bottom about 10 cm (4 in.) from the middle of the front glass. Use a 2.5 cm (1 in.) hole. A predrilled neoprene stopper goes into the hole and a 1.25 cm (½ in.) rigid plastic standpipe is inserted in this stopper. The water that overflows through the standpipe is collected into PVC runoff pipes and discarded in an open system or recirculated through the processing equipment in a closed system. This is highly recommended.

Shipping Methods and Equipment

The vast majority of ornamental fish shipments travel by air freight. Some fish are sold on a pickup basis, not only to local dealers, but also to medium sized wholesalers who periodically travel by truck to Florida. They place orders on arrival with several breeders. They pick up the fish a day later and drive back to their holding depots in one of the southeastern states. The round trip will be 1610 km (1000 mi.) or more. It pays to do this as they can see what they buy and can pack the fish looser with more water (since they don't pay air freight). They will also save several hundred dollars in transportation charges. A few of the larger fish wholesalers own their own small cargo planes. They fly into Tampa and meet their suppliers on a prearranged schedule. There may be as many as 6 or 8 different suppliers all ready to immediately load the plane for a quick turnaround flight.

With few exceptions the industry has a standardized shipping method. Most shippers subscribe to the Airline Guide which is an invaluable tool to help route the shipments properly. The customer, when he places the order, can be given an airway bill number, and the exact flight and time he will receive his order of fish. This is easily done using an Airline Guide. Few shippers will accept orders with less than 48 hr advance notice. Most ship Tuesday through Fridays.

It is important when selling fish to let the customer know what he can expect to get. He should always know terms and prices. Prices are generally based on quantities. It costs about the same to gather 25 fish from a pond as 250. As a result, the price of a 250 lot will be lower per fish. The majority of business is conducted on either a cash on delivery (COD) (payment is collected by the carrier) or a charge through one of the major credit cards. Open accounts exist for some larger customers. The majority of business is done for cash. The industry practice is to charge for packing and packing materials on each completed box, and the cost of packing materials has risen tremendously in recent years. Fish are sold FOB Florida airport with freight charges collect. This plus the cost of the fish and the packing charge is the complete landed cost for the buyer.

The packing is done as follows: The fish go into a plastic bag containing a predetermined amount of water. The size of the bag is dependent on the number and size of the fish. Air is squeezed from the bag and replaced with oxygen. The bag is then sealed with a rubber band or by a heat sealing machine. Most polyfilm bags are 3 mil in thickness. One shipper uses a 6 mil tube of polyfilm. He cuts, makes, fills, and seals his own bags and this is all done in one operation. The bags are sealed by the same machine. This pillow shaped 6 mil bag costs less than the 3 mil commercially produced bag, and initial reports indicate that it is a superior packing method. The trend is toward heat sealing the bags. The completed bags are then put into a styrofoam box, and the box is put inside of a corrugated box, which is the final container. The package is well insulated for both heat and cold. Most fish are received within 15 hr of packing, but most fish can survive, with no more than minimal mortality, for 48 hr and more. Most fish losses are caused by carelessness, indifference, and delay. The arrival losses on normal shipments from average shippers should never exceed 1 or 2%. Industry practice is to provide sufficient overcount to more than compensate for normal shipping losses.

Potential customers are the large and small jobbers, and the large and small retail pet shops. There are today somewhere between 300 and 400 ornamental fish jobbers in the United States. There are somewhere between 10,000 and 15,000 retail outlets. These include extremely large stores vending live fish from as many as 500 aquariums down to the smallest of the variety stores with 4 or 6 small aquariums.

Some of the largest variety store chains operate their own distribution centers, but there are some specialized markets which can be very worthwhile. Public aquariums acquire most of their specimens from breeders and dealers. The best of the foreign markets is Canada for obvious reasons. The European market for ornamental fish is large and has good potential for Florida producers. Recent increases in living costs in the Far East are gradually making their product more and more competitive with the Florida produced fish, which are healthier and larger. In the next few years we can anticipate increased trade between Florida and

Europe. The larger of the trade magazines have lists of potential customers available at nominal cost. Advertising is done by direct mail, telephone solicitation, or with advertisement in the trade magazines.

POLYCULTURE

Bait minnow culture is readily adaptable to polyculture. For example, between 125 and 250 catfish fingerlings (*Ictalurus punctatus*) can be stocked per ha (50 and 100 per acre) with bait minnows in the spring, and can be harvested as food size fish in late fall or winter. The largemouth buffalo *(Ictiobus cyprinellus)* can be stocked with either catfish or minnows. Between 60 and 125 can be stocked per ha (25 and 50 per acre) in the spring and harvested as food size fish in the late fall or winter.

The effects of stocking various combinations of different species, such as catfish and buffalo, on the surviving population of bait minnows is not clearly known. The low stocking rates just described indicate that supplementary feeding for the food fish is not necessary since the catfish can feed primarily on chironomid larvae and mollusca while the buffalo fish will feed primarily on chironomids and entomostraca and debris. With more intensive stocking rates, all species of fish (catfish and buffalo) will utilize the supplementary feeds and may compete with the bait minnows.

While it is not known if catfish and buffalo will eat significant numbers of bait minnows, tests with stocking tilapia and catfish indicate that as much as 80% of the tilapia young are preyed upon by catfish.

If one is to practice polyculture it is suggested that only fingerlings of catfish and buffalo fish be raised together in order to avoid problems. Small numbers of buffalo fish are sometimes stocked with catfish, since they will eat smaller particles of food fragments which may be wasted by the catfish. Polyculture of catfish, bluegills, and bass is often used in farm ponds, and resulting production is probably higher. The bass will control the bluegills and utilize the top and middle waters while the catfish are bottom feeders.

Total production may be limited to 335 to 445 kg per ha (300 to 400 lb per acre) in well managed, fertilized ponds. Yields can be increased with supplementary feeding. The bass aid in controlling the number of bluegills or trash fish in more natural waters.

REFERENCES

ANON. (Undated.) Catfish Feeding and Growth. U.S. Dep. Interior Bur. Sport Fish. Wildl., Stuttgart, Ark.

ANON. 1972A. A statistical reporting system for the catfish farming industry, methodology and 1970 results. In cooperation with the Ind. Res. Exten. Cen. and Dep. Agric. Econ. Rural Sociology, Univ. Arkansas. U.S. Dep. Commer. Econ. Dev. Admin. Tech. Assist. Proj. *99-6-09044-2.*

ALTMAN, R.W. and IRWIN, W.H. (Undated.) Minnow Farming in the Southwest. Okla. Dep. Wildl. Conserv., Oklahoma City.

BARDACH, J.E., RYTHER, H., and McLARNEY, W.O. 1972. Aquaculture—The Farming and Husbandry of Fresh Water and Marine Organisms. Wiley-Interscience, New York.

BROWN, E.E., LaPLANTE, M.G., and COVEY, L.H. 1969. A synopsis of catfish farming. Ga. Agric. Exp. Stn. Res. Bull. *69.*

BROWN, E.E., HILL, T.K., and CHESNESS, J.L. 1974. Rainbow trout and channel catfish—a double-cropping system. Ga. Agric. Exp. Stn. Res. Rep. *196.*

BROWN, E.E. 1977. World Fish Farming: Cultivation and Economics. AVI Publishing Co., Westport, Conn.

DILLON, O.W., JR. 1976. Aquaculture in the southern United States. F.A.O. Tech. Conf. Aquaculture, Kyoto, Japan, May 26—June 2.

DEVARAJ, K.V.D. 1970. Food of channel catfish and white catfish in ponds that received supplemental feed. Unpublished Ph.D. dissertation. Auburn Univ., Auburn, Ala.

DOBIE, J.R., MEEHEAN, O.L., SNIESZKO, S.F., and WASHBORN, G.N. 1948. Propagation of minnows and other bait species. U.S. Fish Wildl. Serv. Circ. *12.*

DOBIE, J.R., MEEHEAN, O.L., SNIESZKO, S.F., and WASHBORN, G.N. 1956. Raising bait fishes. U.S. Fish Wildl. Serv. Circ. *35.*

EASLEY, J.E., JR. and FREUND, J.N. 1977. An economic analysis of eel farming in North Carolina. Univ. N.C. Sea Grant Publ. *S.G.-77-16.*

ELLIS, M.M., WESTFALL, B.A., and ELLIS, M.D. 1946. Determination of water quality. Fish Wildl. Serv. U.S. Res. Rep. *9,* 1-122.

FEAST, C.N. and HAGIE, C.E. 1948. Colorado's glass fish tank. Prog. Fish. Cult. *10,* 29-30.

FLICKINGER, S.A. 1971. Pond culture of bait fishes. Colo. State Univ. Coop. Ext. Serv. Bull. *478A.*

GERKING, S.D. 1949. Urethane (ethyl carbamate) in some fisheries procedures. Prog. Fish. Cult. *11,* 73-74.

GREENBERG, D.B. 1960. Trout Farming. Chilton Co., Philadelphia.

GUIDICE, J.J. 1968. The culture of bait fishes. Proc. Commer. Bait Fish Conf., Texas Agric. Ext. Serv. Dep. Wildl. Sci., Texas A & M Univ., College Station, Tex., March 19—20.

HUET, M. (Undated.) Textbook of Fish Culture—Breeding and Cultivation of Fish. Fishing News (Books), London.

KRAATZ, W.C. (Undated.) The resume of the food of the blunt-nosed minnow, a forage fish for fishponds. Ohio Bur. Sci. Res. Bull. *22.*

LEE, J.S. 1971. Catfish Farming. Miss. State Univ. Curriculum Coordinating Unit, State College, Miss.

LEITRITZ, E. and LEWIS, R.C. 1976. Trout and salmon culture—hatchery methods. Calif. Dep. Fish Game Fish Bull. *164*.

MALOY, C.R. 1966. Status of fish culture in the North American region. Proc. F.A.O. World Symp. on Warm-water Fish Culture, Rome, May 18—25. F.A.O. Fish Rep. *44*.

MARKUS, H.C. 1934. Life History of the Blackhead Minnow *(Pimephales promelas)*. Copeia, Washington, D.C.

MARTIN, M. 1955. Minnow Culture in Kentucky. Ky. Dep. Fish Wildl. Resour., Frankfort.

MEYER, F.P., SNEED, K.E., and ESCHMEYER, P.T. 1970. Report to the fish farmers. Bur. Sport Fish. Wildl. U.S. Res. Rep. *113*.

MEYER, F.P., SNEED, K.E., and ESCHMEYER, P.T. 1973. Second report to the fish farmers. U.S. Fish Wildl. Serv. Bur. Sport Fish. Wildl., Resour. Publ. *113*.

MITCHELL, T.E. and USAY, M.J. 1969. Catfish Farming Profit Opportunities. Miss. Res. Dev. Center, Jackson.

NORDEQUEST, O. 1893. Some notes about American fish-culture. U.S. Fish Comm. Bull. *13*, 197-200.

OSBORN, P.E. 1951. Some experiments on the use of thiouracil as an aid in holding and transporting fish. Prog. Fish. Cult. *13*, 75-78.

PRATHER, E.D., FIELDING, J.R., JOHNSON, M.C., and SWINGLE, H.S. 1953. Production of bait minnows in the southeast. Ala. Polytech Inst. Agric. Exp. Stn. Circ. *112*.

REESE, A. 1953. Use of hypnotic drugs in transporting trout. Calif. Fish Game Mimeo Paper.

RICKARDS, W.L., JONES, W.R., and FOSTER, J.E. 1979. Techniques for culturing the American eel. Proc. World Mariculture Soc. 9th Annu. Meet., Honolulu, Jan. 22—26, 1979, N.C. State Univ. Sea Grant Program, Raleigh. (in press)

SCHAEPERCLAUS, W. 1933. Textbook of pond culture. U.S. Dep. Interior Fish Wildl. Serv. Fish. Leaflet *311*.

SHIRA, A.F. 1918. Fish cultural activities of the Fairport Biological Station. Trans. Am. Fish. Soc. *47*, 39-44.

SMITH, P.L. 1973. Effects of *Tilapia aurea* (Steindachner), cage culture, and aeration on channel catfish *Ictalurus punctatus* (Rafinesque), production in ponds. Unpublished Ph.D. Dissertation. Auburn Univ., Auburn, Ala.

USUI, A. 1974. Eel Culture. Fishing News (Books), London.

WASCHO, H. and CLARK, C.F. 1951. Pond propagation of bluntnose and blackhead minnows. Ohio Dep. Nat. Resour. Bull. *4*.

WILLIAMSON, J. and SMITHERMAN, R.O. 1975. Food habits of hybrid buffalo fish, tilapia, Israeli carp and channel catfish in polyculture. Proc. 29th Annu. Conf. Southeast Assoc. Game Fish Commissioners, St. Louis, Oct. 12—15, 86-91.

6

Nutrition and Feeding

Richard T. Lovell

NUTRIENT REQUIREMENTS

The nutritional requirements of fish are similar to those of land animals. For growth, reproduction, and other normal physiological functions they need to consume protein, minerals, vitamins and growth factors, and energy sources. A deficiency of one or more of the essential nutrients results in a reduced rate of performance, disease, or even death. These nutrients may come from artificial or prepared diets, or from natural aquatic organisms. If fish are held in an artificial confinement where natural foods are absent, their diet must be nutritionally complete; however, where natural food is available and supplemental diets are fed for additional growth, the diets may not need to contain all of the essential nutrients.

In most cases, the fish culturist is interested in maximum rate of performance or health of the fish. To achieve this the fish must receive all of the essential nutrients in liberal and balanced quantities. There may be other objectives in a feeding program, such as holding fish at a certain desired size for a long period, pigment enhancement, reproductive purposes, and the like. Under these conditions a nutritionally balanced diet is still important.

The nutritional requirements of fish do not seem to vary greatly among species, especially within warmwater and within coldwater groups. The most notable exception is probably lipids, or essential fatty acids. The quantitative nutrient requirements that have been ascertained for several species of fish are probably an adequate basis for estimating the nutrient needs of others. As more information becomes available on nutrient requirements of various species, the recommended nutrient allowances of diets for specific needs of individual species will become more refined.

Protein

Protein is the main constituent of the fish body; thus, a generous dietary supply is needed for rapid growth. Protein is more expensive than carbohydrate or fat; therefore, the amount of protein in the diet should be limited to that which is needed for growth and tissue repair, and the energy should come from the cheaper sources.

Protein Level in Fish Diets.—Fish require a higher percentage of protein in their diet than do warm-blooded animals. For example, the optimum level of protein in practical diets for warmwater food fish is 30 to 36%, whereas in poultry diets the practical level is 16 to 22%.

The optimum percentage of protein in fish diets is influenced by several factors such as the following.

(1) Size of fish. Fish, like land animals, have higher protein requirements during early life than during later phases of growth.
(2) Physiological function. Less protein is needed in a maintenance diet than in one fed for a rapid growth rate.
(3) Protein quality. A protein that is deficient in one or more of the 10 essential amino acids will produce less growth than a protein that is balanced in the essential amino acids; or, more of a low quality protein is needed in the diet for maximum growth than of a high quality protein.
(4) Nonprotein energy in the diet. If the diet is deficient in energy, the fish will use part of the protein to meet energy needs, thus reducing the amount of dietary protein available for growth.
(5) Feeding rate. Fish fed to less than satiation, as frequently occurs in intensive pond culture of food fish, will benefit from diets containing higher percentages of protein than fish fed at or near the satiation rate.
(6) Natural foods. If natural aquatic organisms contribute significantly to the daily food intake of the fish, the protein level in the prepared diet may be reduced. For example, aquatic fauna that are consumed by various fish contain from 60 to 80% protein; thus, if these are in abundance the supplemental diet would need only a low percentage of protein.
(7) Economics. The cost and availability of protein sources is a major factor in determining how much protein to use in commercial diets.

Channel catfish of small size (10 to 20 g) require 9 to 10 g of high quality protein per kg of fish per day for maximum growth when fed a nutritionally balanced diet. As channel catfish approach harvestable size (0.5 kg), the daily protein requirement decreases to 7.5 to 8.0 g per kg. If channel catfish fingerlings will consume, or are fed, at the rate of 3% of their weight per day, a 33% protein ration will provide all of their protein requirement (10 g of protein per 30 g of feed per kg of fish). However, if fed at the rate of 2.5% of their weight per day, they will need a 40% protein diet. Most warmwater fish have protein needs similar to channel catfish; thus protein levels of 30 to 36% will probably be adequate for most warmwater fish diets.

Protein Quality.—Protein quality is influenced primarily by amino acid composition. Proteins are made up of 20 to 25 amino acids. Ten of these cannot be synthesized in the fish's body and must be provided in the diet. These are the same 10 amino acids that are required in the diet of the rat. The 10 essential amino acids and the quantitative requirement of each for several fish are presented in Table 6.1.

TABLE 6.1. ESSENTIAL AMINO ACID REQUIREMENTS OF FISH

| | (% of Protein) | | |
Amino Acid	Channel Catfish	Eel	Salmonoid
Arginine	4.3	3.9	6.0
Histidine	1.6	1.9	1.8
Isoleucine	2.3	4.1	2.2
Leucine	3.4	3.6	3.9
Lysine	5.1	4.8	5.0
Methionine + cystine[1]	2.3	4.5	4.0
Phenylalanine + tyrosine[2]	R[3]	R[3]	5.1
Threonine	2.2	3.6	2.2
Tryptophan	0.5	1.0	0.5
Valine	2.8	3.6	3.2

Sources: Channel catfish: Wilson and Robinson (1978); eel: Natl. Res. Counc. (1977); salmonoid: Natl. Res. Counc. (1973B).
[1]At least 33% of this requirement must be methionine.
[2]At least 50% of this requirement must be phenylalanine.
[3]R means required but quantity has not been determined.

Response of fish to protein quality was evidenced by results of a feeding experiment in which channel catfish were fed diets containing different percentages and sources of protein (Lovell *et al.* 1974). Higher levels of protein were required for maximum growth rate in all-plant diets than in diets containing some fish meal. Growth response improved as fish meal replaced part of the plant protein sources in constant-protein diets. The data show that adding fish meal, which is rich in methionine, to all-plant diets, which are low in methionine, increased gain per unit of protein fed and reduced the amount of protein required for maximum growth. The benefit of adding fish meal diminished as amount of fish meal or percentage of protein increased in the diet which indicated that the fish's requirement for methionine could be met by feeding either more protein or more fish meal.

Another factor affecting protein quality is digestibility. The percentage of protein in a feedstuff determined by chemical analysis is not synonymous with the amount of protein available from the feedstuff to the fish. Fish digest protein in natural aquatic fauna quite well and in phytoplankton relatively well. Protein in some commercial feedstuffs, such as fish meal and soybean meal, is highly available to fish as well as livestock and poultry. Protein in grains and fibrous feeds is less digestible to fish than to farm animals.

Comparisons of digestion coefficients for several foodstuffs for fish are presented in Table 6.2. Cooking improved the protein digestibility in corn and perhaps in other grains. Protein in fibrous materials, such as alfalfa meal, is very poorly utilized by catfish but is somewhat more available to herbivorous fish like tilapia. The level of starch in the diet has been found to have an inverse effect on protein digestibility with carp and catfish. As starch increases above 50% of the diet, protein digestibility is significantly suppressed.

TABLE 6.2. DIGESTIBLE PROTEIN AND ENERGY IN VARIOUS FOODSTUFFS FOR CHANNEL CATFISH

| Foodstuff | Protein (%) | | Energy (kcal/kg) | |
	Crude Protein	Digestible Protein	Gross Energy	Digestible Energy
Glucose	—	—	4000	3680
Corn, uncooked 30% of diet	9.0	5.4	4320	1103
Corn, extrusion cooked 30% of diet	9.0	7.7	4320	2528
Corn, extrusion cooked 60% of diet	9.0	7.0	4320	2020
Wheat	12.0	10.1	4230	2552
Soybean meal	49.0	41.1	4600	2560
Cottonseed meal	42.0	34.0	4530	2670
Fish meal (menhaden)	62.0	54.0	4770	4080
Alfalfa meal	17.0	2.2	4180	598

Source: Stickney and Lovell (1977).

Energy

Energy levels in fish feeds have been treated casually because a deficiency or excess of energy will not affect the health of fish appreciably, and practical feeds made with commonly available ingredients are not likely to be extremely high or low in energy. Nonetheless, levels and sources of energy in fish rations can significantly affect fish growth and are of definite economic importance in commercial feeds.

Livestock and poultry nutritionists have long recognized the importance of meeting energy requirements in formulating practical feeds. Feeding tables for various farm animals list energy allowances along with protein and other nutrient allowances for various productive purposes.

Some feeding tables present protein allowances for various dietary energy levels, i.e., as the energy plane of the ration increases, the percentage of protein increases proportionally. The ratio of energy to protein in commercial rations for farm animals increases as the animal gets larger.

Energy needs of fish are less than those of warmblooded animals because: (1) fish do not have to maintain a constant body temperature; (2) fish require less energy for muscle activity to maintain their position in water than do animals on land; and (3) fish require less energy to excrete nitrogen waste products than do warmblooded animals. According to recommendations on the nutrient requirements of farm animals by the Natl. Res. Counc. (1971, 1973A), the optimum amount of metabolizable energy for each gram of protein in the ration is 14 to 16 kcal for poultry and 15 to 24 kcal for swine. This compares with values of 6 to 10 kcal per gram of protein which have been found adequate for fish rations.

Many commercial fish feeds, because of high protein percentages, are probably deficient in energy in relation to protein. The ratio of digestible energy to protein in several commercial catfish feeds was calculated and found to range from 6.6 to 7.4 kcal for each gram of protein. The optimum ratio of available energy to protein in channel catfish ration has been established at 7.3 to 9.6 kcal per gram (Lovell 1977; Page and Andrews 1973; Garling and Wilson 1976). Variations in the energy protein ratios are due to variation in energy source or fish size.

Energy Sources.—Most cultured fish species digest protein and fats well but starch less efficiently than land animals. Warmwater fish, such as catfish, carp and tilapia, can utilize starch better than coldwater fish. Variation also exists amoung sources of starch, how much is in the diet, and whether or not the starch was heated prior to feeding. Table 6.2 compares fat protein and starch digestibility by channel catfish in several feedstuffs. Data in the table show that cooking corn, as in extrusion processing of fish feeds, improves its digestibility. Also, as the amount of corn in the diet increases, the digestibility of corn starch decreases.

Availability of the gross energy in feedstuffs for only a few materials and for a few fish has been determined. For warmwater fish, the values in Table 6.2 determined for channel catfish may be used as guides for estimating digestible energy for various classes of feed materials, i.e., fish meal, oilseed meals, grains and by-products, fibrous feedstuffs, oils and fats, and purified diet ingredients.

Effects of High and Low Energy Diets.—Fish eat to satisfy their metabolic energy requirement and, consequently, cease feeding when their calorie needs are satisfied. Because of this phenomenon, fish will eat less of a high energy diet than a low energy diet. Therefore, too much energy in relation to the percentage of protein in the ration can prevent fish from consuming enough protein to meet their daily need for optimum growth rate, even though the fish are allowed to eat as much as they will consume.

High energy diets, especially when a high percentage of the calories is from fat, will produce fatty fish which have a lower dressing percentage than fish fed a lower energy diet. (Dressing percentage is the percentage of the live weight that is available for resale after gutting, and, in some cases, removal of the head.) Experience with channel catfish has shown that when the nonprotein calories in the ration are primarily from carbohydrates, body fat content is affected in only extremely high energy diets. High carbohydrate diets when fed to channel catfish have not produced the adverse effects that have been found with salmonoids.

In an experiment at Auburn University (Lovell and Prather 1973), when diets fed to channel catfish contained high percentages of protein (42% and above) and very low amounts of nonprotein energy (less than 1.5 kcal per gram of diet), growth was suppressed. When the protein level was reduced to 36% and the nonprotein level remained the same, growth increased. When the nonprotein energy in either the 42 or the 36% protein diets was increased, growth improved. This experiment indicated that when too many of the calories in a fish diet come from protein the efficiency of utilization of the ration is suppressed.

Vitamins and Essential Growth Factors

Fish from the wild seldom show signs of nutritional diseases. This is because natural aquatic foods are fairly nutritious, especially in the essential growth factors like vitamins and minerals, and the fish's growth rate is limited to the amount of energy and protein in the natural foods. It is when fish are confined to unnatural conditions and fed supplemental feeds for fast growth that nutritional deficiency symptoms occur.

Traditionally, nutritional diseases are not a serious problem in pond culture of fish, primarily because with moderate stocking conditions the natural pond organisms satisfactorily supplement the artifical feeds with any deficient growth factor. However, as the culture environment deviates from the conventional pond system to more intensive or artificial conditions and the availability of natural food decreases, the nutritional adequacy of the prepared feed becomes more critical.

One of the most dramatic instances of nutritional disease in fish was the "broken back syndrome" in channel catfish from artificial culture systems or heavily stocked ponds (Fig. 6.1). The cause was unknown until a series of studies (Lovell 1973; Wilson 1973) revealed that channel catfish need vitamin C (ascorbic acid) in their diet for development of bone matrix, blood vessels, wound repairs, and other cartilagenous tissues in the body, and to minimize susceptibiity to pathogenic organisms.

FIG. 6.1.   THE BROKEN BACK SYNDROME WAS A COMMON ANOMALY IN COMMER-
CIAL CHANNEL CATFISH CULTURE UNTIL A DIETARY DEFICIENCY OF VITAMIN C
(ASCORBIC ACID) WAS FOUND TO BE THE CAUSE.
B—A vitamin C deficient catfish from a commercial culture system. C and D—Radio-
graphs showing deformed spinal columns in experimentally induced vitamin C
deficient catfish. C—The fish has lordosis (vertical curvature of the spinal column).
D—The fish has scoliosis (lateral curvature of the spinal column).

Studies were conducted at Auburn University to determine the value of ascorbic acid in feeds for pond-fed channel catfish (Lovell and Lim 1978). When the maximum standing crop of fish was 2245, 3368, or 4490 kg per ha (2000, 3000, or 4000 lb per acre), the fish did not need ascorbic acid in their feed. At densities of 11,227 kg per ha (10,000 lb per acre) the absence of supplemental ascorbic acid in their feed resulted in fish with deformed backs, reduced growth rate, and greater sensitivity to pathogenic bacteria. The stocking of tilapia, which competed for natural food, in ponds with 9884 channel catfish per ha (4000 per acre), resulted in ascorbic acid deficiency symptoms in the catfish when the vitamin was not added to the feed. The presence of sublethal levels of the chlorinated hydrocarbon insecticides, such as toxaphene, in the water increases the need for vitamin C and has caused spinal deformities in fish in diets normally adequate in vitamin C (Mayer *et al.* 1977).

Vitamins.—Vitamins are essential growth factors that are required in the diet in only very small quantities. One of the first symptoms of a deficiency of practically any of the 13 to 15 essential vitamins for warmwater fish is depressed appetite and reduced growth rate. Other common symptoms are abnormal color, lack of coordination, nervousness, hemorrhage, fatty livers, and increased susceptibility to bacterial infections. A summary of the vitamins required by catfish and salmonoids is presented in Table 6.3.

All 15 of the vitamins listed in Table 6.3 have been found to be

TABLE 6.3.   VITAMIN REQUIREMENTS OF CHANNEL CATFISH AND SALMONOIDS

Vitamin	Amount/kg of Diet[1]	
	Channel Catfish	Salmonoids
A	5500 IU	4000 IU
D	500 IU	R[2]
E	50 mg	400 mg
K	10 mg	40 mg
Thiamin	20 mg	10 mg
Riboflavin	20 mg	40 mg
Pyridoxine	20 mg	10 mg
D-Calcium pantothenate	50 mg	80 mg
Niacin	100 mg	150 mg
Folacin	5 mg	3 mg
B$_{12}$	R	0.02 mg
Inositol	NR[2]	400 mg
Biotin	R	1 mg
Choline	550 mg	800 mg
Ascorbic acid	50 mg	100 mg

Sources: Channel catfish: Stickney and Lovell (1977); salmonoids (coho salmon): Natl. Res. Counc. (1973B).
[1]Divide by 2.2 to get requirements per lb of diet.
[2]R = required; NR = not required.

essential in diets of fish; however, all fish do not require all 15 of the vitamins. Trout require all 15. Vitamin B_{12} is synthesized by intestinal bacteria in warmwater fish like channel catfish and carp, and dietary needs are marginal. Some vitamins, such as biotin, inositol, and folic acid, are widely found in feedstuffs and deficiency signs are difficult to produce in fish unless purified diets are used. Because of the variation in vitamin contents in diet ingredients and the relatively moderate cost of synthetic vitamins or vitamin-rich food sources, commercial fish diets are usually supplemented with all of the vitamins required by the species.

Recommended levels of vitamins for channel catfish and rainbow trout diets (Table 6.4) may serve as models for vitamin requirements for warmwater and coldwater fish diets.

Essential Fatty Acids.—Fish fed fat-free diets grow poorly. Rainbow trout require specific fatty acids, of the omega-3 or linolenic series, for normal growth and health. Essential fatty acids for warmwater fish

TABLE 6.4. GUIDELINES FOR FORMULATING PRACTICAL DIETS FOR CHANNEL CATFISH

Item	Fish Size	Quantity in Diet
Protein (%)	20 g	36
	100 g	30
Amino acids (% of protein)	All	See Table 6.1
Digestible energy (kcal/g)	All	8
Available phosphorus (%)	All	0.5
Calcium (%)	All	0.75
Trace minerals (mg/kg)[1]	All	
copper		4.3
cobalt		0.05
iron		44
iodine		2.8
zinc		88
manganese		115
Vitamins	All	See Table 6.3

[1]Use of trace mineral supplement when the diet contains less than 15% fish meal.

have not been specifically identified. A recent postulation is that both omega-3 and omega-6 (linoleic) fatty acids are essential for warmwater fish.

Channel catfish have been found to grow as well on diets containing tallow, which is very low in omega-3 fatty acids, as on diets containing fish oil, which is high in the omega-3 fatty acids. Diets containing only corn oil have produced lower growth rates in channel catfish than diets containing soybean oil or animal fats.

Warmwater fish feeds for practical feeding are formulated without regard to essential fatty acids. These feeds, many of which contain primarily plant products with very little fish meal, usually contain 3 to 4% total lipids and have produced favorable growth and feed conversion rates.

The minimum requirement of omega-3 fatty acids for salmonoids (trout and salmon) is 1% of the diet. Minimum levels of essential fatty acids for warmwater fish diets have not been established.

Minerals

Fish probably require the same minerals as warm-blooded animals for tissue formation and various metabolic processes. In addition, fish use inorganic elements to maintain osmotic balance between fluids in their body and the water. Minerals in the water can make significant contributions to the fish's requirements for some minerals, such as calcium.

Calcium and Phosphorus.—Fish, like mammals, require large amounts of calcium and phosphorus for growth and development. Most fish appear to be able to absorb enough calcium from the water, across the gills, for normal growth, except when the water is unusually low in calcium. Carp, rainbow trout, and red sea bream were found to absorb enough calcium from the environment to meet their needs for growth, provided their diets were adequate in phosphorus, whereas eel were found to benefit from a dietary source of calcium. Dietary phosphorus is essential for fish.

As indicated in Fig. 6.2, channel catfish benefited only slightly from a

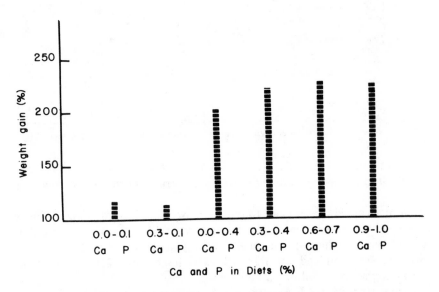

FIG. 6.2. WEIGHT GAINS OF CHANNEL CATFISH FED PURIFIED DIETS CONTAINING VARIOUS LEVELS OF CALCIUM (CA) AND PHOSPHORUS (P) IN A CONTROLLED ENVIRONMENT.
When Ca only was added to the basal diet (0% Ca and 0.1% P), there was no improvement in growth; when P only was added, there was a great improvement in growth; addition of both Ca and P caused very little growth increase.

dietary source of calcium (Lovell 1978). The data were obtained in aquarium studies with a level of 35 ppm of $CaCO_3$ in the water. The catfish showed a marked increase in growth response to the addition of phosphorus to the diet.

Levels of dissolved phosphorus are very low in natural waters in relation to calcium; consequently, the water in fish culture environments is not a significant source of phosphorus. Dietary deficiencies in phosphorus have caused reductions in growth rate, body content of calcium and phosphorus, and appetite in fish. Deformed backs (lordosis) and heads have been associated with phosphorus deficiency in carp.

Minimum requirements of "available" phosphorus in diets for channel catfish have been determined to be 0.4 to 0.5% using purified diets (Lovell 1978), and 0.8% using practical diet ingredients (Andrews *et al.* 1973). This difference is apparently associated with the availability of phosphorus to channel catfish from various dietary sources. If 0.5% is used as the minimum available phosphorus requirement for satisfactory growth of channel catfish the following availability percentages are suggested for phosphorus from various practical sources:

Source	Availability of Phosphorus (%)
Dicalcium phosphate	80
Fish meal, meat, and bone meal	50
Soybean meal	40
Grains	33

Other Essential Minerals.—Dietary requirements for most of the other minerals have not been established for fish. Differences in growth responses have been obtained by changing the dietary levels of magnesium, potassium, copper, iodine, and iron for several species of fish.

Natural feedstuffs are usually adequate in K, Mg, Na, and Cl for normal growth of animals unless there is a high rate of mineral loss. These elements are probably available in sufficient quantity in practical fish feeds without mineral supplementation. However, fish feeds low in animal products (fish meal, meat, and bone meal, etc.) may be deficient in trace minerals. When less than 15% of the ration is composed of animal products, a trace mineral supplement is recommended to provide the following amounts of minerals in the ration (mg/kg): Mn, 115; I, 2.8; Cu, 4.3; Zn, 88; Fe, 44; and Co, 0.05.

Guidelines for Meeting Nutrient Requirements

The most important limitations for formulating diets for finfish, with regard to nutritional adequacy for maximum growth and good health, are presented in Table 6.4. The nutrient requirements for warmwater fish are based on those determined with channel catfish and those for cold-water fish are based on determinations with trout and salmon. While the nutrient requirements vary slightly among species within warmwater and coldwater groups, the data in Table 6.4 will serve as a very good guide for general use in meeting nutrient requirements in experimental or practical diets for various fish where these values are unknown.

DIET FORMULATION AND PROCESSING

Practical Feeds for Food Fish

The livestock and poultry industries have been extremely successful in minimizing feed costs by computer calculation of least-cost production rations. To determine minimum-cost feed formulations by this method, the following information must be available:

(1) Costs of feed ingredients;
(2) Nutrient content of feed ingredients;
(3) Nutrient requirements of the animal (protein, energy, vitamins, minerals, amino acids);
(4) Availability of nutrients to the animal from various feed materials.

Items 1 (costs) and 2 (nutrient content) are readily available for most commercial feedstuffs. Table 6.5 contains the major nutrient contents (vitamins and trace minerals not included) of representative feedstuffs from various classes of feed materials. A respectable amount of information is available on item 3 (nutrient requirements) for several fish species, enough to formulate reasonably satisfactory production rations.

It is item 4 above in which there is a critical deficiency of information

TABLE 6.5. TEST DIETS FOR FISH

Ingredient	Channel Catfish Diet (%)	Test Diet H440 for Salmonoids (%)
Vitamin-free casein	29	38
Gelatin	6	12
Dextrin	30	28
Cellulose flour	20.25	8
Fish oil	3	3
Soybean oil	3	—
Corn oil	—	6
Carboxymethylcellulose	3	—
Mineral mixture[1]	4	4
Vitamin mixture[2]	1.5	1
Calcium propionate[3]	0.25	—

Sources: Channel catfish: Natl. Res. Counc. (1977); salmonoids: Natl. Res. Counc. (1973B).
[1]A salt mixture supplying all essential minerals for laboratory animals.
[2]Should meet the vitamin requirements for channel catfish and for coho salmon, Table 6.3.
[3]To retard mold growth.

for fish. The availability of nutrients from various types of feed materials and synthetic sources is quite well established for livestock and poultry. Feeding tables contain information for farm animals such as digestible protein, total digestible nutrients (TDN), digestible energy, metabolizable energy, and the like for most commercial feed ingredients. For fish, no such generally accepted measures of availability of nutrients in feeds have been established, and in many cases commercial fish feeds are formulated with the presumption that the availability of nutrients from various sources to fish is essentially the same as to simple-stomach farm animals, or that one feed will substitute for another strictly on the basis of nutrient content.

The feed analysis data in Table 6.5 include estimates of digestible energy values for fish which are based upon digestibility determinations made with channel catfish (Cruz 1975) for several materials from the various feedstuffs classes (oilseed meals, grains, animal by-products, etc.). Generally, the availability of protein and energy from oilseed sources and animal by-products is about the same for fish as for farm animals; the availability of energy from grains, especially corn, is less for fish than farm animals; as the percentage of starch (from grains) in the diet increases above approximately 30%, the availability of the dietary protein and energy to fish is decreased; and the availability of phosphorus in animal by-products, such as fish meal and meat and bone meal, is less for fish than for farm animals.

Complete Versus Supplemental Feeds

In many cases it is unnecessary and uneconomical to balance diets fed to fish in ponds according to the absolute nutrient requirements of the fish. In this case, *supplemental diets*, which may be deficient in some vitamins or minerals, are often used to feed fish at low densities in the pond on the premise that the diet deficiencies will be provided by natural food organisms in the pond.

In cases where the fish have limited or no access to other nutrient sources, such as in heavily stocked pond cultures or in artificial culture systems (cages, raceways and the like), all of the essential nutrients must be provided in the prepared diet in adequate quantities. Such *complete* diets are being used more frequently in catfish culture ponds because fish density has been steadily increasing in recent years and dietary deficiencies such as the broken back syndrome have been observed in pond culture.

Examples of practical diets for channel catfish, trout, carp and eel are presented in Table 6.5. These formulations have been used with good results in commercial culture.

Formulation Restrictions

Soybean meal and fish meal are used as major protein sources in all practical fish feeds in the United States. Soybean meal contains the highest quality (best balanced in essential amino acids) plant protein available. Fish meal has a higher quality protein but is much more expensive than plant protein sources; however, some fish meal has been found necessary in all practical fish feeds for maximum growth rate. A minimum of 7.5% fish meal has been recommended in catfish feeds. Salmonoid diets contain much higher percentages of fish meal. Trout diets contain 20% or more of fish meal and salmon diets contain about twice this amount. Apparently, salmon do not accept or assimilate plant protein sources well.

Cottonseed meal containing gossypol is toxic to some fish when fed in large amounts. Its toxicity has not been adequately evaluated with most fish; therefore, when gossypol-containing cottonseed meal is used in fish diets, the level should probably not exceed 15% of the formula.

Highly fibrous feedstuffs, like alfalfa meal, soybean millfeed and rice hulls, should be avoided since they are very poorly digested by fish. Fiber is beneficial for digestibility and absorption of nutrients in purified research diets, but studies with channel catfish (Leary and Lovell 1974) indicated no benefit of fiber in practical diets.

Some ingredients affect the binding or pelleting property of fish feeds. Fat reduces the binding quality in both pelleted and extruded diets; consequently, supplemental fat should be applied after the feed has been pelleted or extruded. Usually, 2% or less of fat can be added to the ingredient mixture prior to processing without seriously reducing the binding property of the processed feed particles. Starch is an important ingredient in fish feeds for binding ingredients together. Generally, a minimum of 20% grain is necessary in extruded fish feeds for proper expansion during processing. Organic binding agents such as lignin sulfonates and hemicelluloses (wood pulp by-products) are often used in pelleted feeds at a level of 2 to 2.5% to increase water stability.

Pelleting and Extruding

Early research demonstrated that practical feeds for carp and channel catfish should be in large particle form; that meal-type feed placed in the water was inefficiently consumed. Although some species, such as tilapias, can utilize meal-type foods well, most fish do not; thus, pelleting or extruding the feed in large particle form is necessary. *Pelleting*, through compression, produces a dense pellet that sinks rapidly in water. *Extrusion* is a process through which the feed material is moistened, precooked, expanded, extruded and dried, producing low density feed

FIG. 6.3. THIS MODERN PRODUCER-OWNED
FEED MILL WAS BUILT TO MANUFACTURE
CATFISH FEEDS FOR GROWERS IN THE
SOUTHEAST.
It processes primarily expanded (floating) fish
feeds but can make extruded pet foods during
the off-season.

particles which float in water. Pelleting is less expensive and generally
costs 10 to 12% less than extruded fish feeds. However, extruded, or
floating, feeds are very popular with catfish farmers; approximately
two-thirds of the commercially produced catfish in the United States are
fed extruded diets.

Pelleting involves the use of moisture, heat, and pressure to agglom-
erate ingredients into larger homogeneous particles. Steam or hot water
added to the ground feed mixture (mash) during pelleting gelatinizes
starch, which aids in binding ingredients. Generally, an amount of steam
is added to the mash to increase its moisture content to approximately
16% and temperature to 70° to 85°C (158° to 185°F) before passing
through the the pellet die; however, ingredient composition will influence
these conditions. The moisture must be removed by proper cooling and
ventilation immediately after the pellets leave the pelleting apparatus.

Additives that serve primarily as pelleting aids are frequently used in
fish feed formulas to reduce fines and increase water stability, although
research in fish feed technology has demonstrated that high quality fish

feeds can be made without binding materials by following good pelleting procedures. However, use of compounds such as hemicellulose and cellulose derivatives, lignosulfonates, bentonites, and others does allow the processor greater variation in ingredient selection and processing conditions to produce pellets of satisfactory quality.

The pelleted ration should be retained on a 0.32 cm (⅛ in.) mesh screen when immersed in water for 10 min, with no more than 10% of the original weight being lost. This specification may be met by grinding the formula through a 0.32 cm (⅛ in.) mesh screen after mixing, using high pressure, high quality (dry) steam to condition diet mixture before pelleting, cooling rapidly, and handling without undue breakage.

Extrusion requires higher levels of moisture, heat, and pressure than pelleting. Usually, the mixture of finely ground ingredients is conditioned with steam or water in "mash," and may be precooked before entering the extruder. The mash, which contains around 25% moisture, is compacted and heated to 135° to 175°C (275° to 347°F) under high pressure. As the material is squeezed through die holes at the end of the extruder barrel, part of the water in the superheated dough immediately vaporizes and causes expansion. The low density extruded particles contain more water than pellets and require more drying. Heat-sensitive vitamins are usually added on the surface after extrusion and drying. Extruded feeds are more firmly bound due to the almost complete gelatinization of the starch and result in less fines than pellets.

Extruded or expanded fish feeds have two definite advantages over pelleted feeds: (1) the particles float and are more resistant to disintegration in water; and (2) a floating feed allows the fish culturist to observe the condition of the fish and the amount of food consumed.

Storage of Fish Feeds

Fish feeds properly dried following pelleting or extruding and stored in cool and dry conditions will remain in good condition for relatively long periods. Generally, 90 days is the maximum storage time recommended for a complete fish feed stored at ambient temperature. High moisture conditions cause mold growth. Some molds produce toxins (mycotoxins) that are detrimental to fish. Mold inhibitors may be added to fish feeds that are prepared for use in warm, humid areas. Propionic acid (as sodium or calcium propionate) may be used for this purpose at a level of 0.25% of the ration.

Some nutrients are sensitive to oxidation and decrease in activity with storage time. Some ingredients are strongly pro-oxidative, such as fish oils, bloodmeal, or trace mineral additives. Fish feeds should contain antioxidants to protect the oxygen-sensitive nutrients from such agents.

Ascorbic acid is the most sensitive vitamin to deterioration during storage. The half-life for ascorbic acid in pelleted fish feeds is approximately 3 months at 26°C (79°F) and 50 to 90% relative humidity.

All fish feeds should be stored carefully, but storage time and conditions of storage are more important for complete than for supplemental feeds. Effects of nutrient deterioration, as with vitamins, will be more serious for fish without access to pond organisms.

Experimental Fish Diets

Whether designed for practical or basic research, all experimental diets should be designed with the following factors in perspective:

(1) Palatable;
(2) Desirable physical properties (size, texture, water stability);
(3) Nutritionally complete;
(4) All diets as near equal as possible in all respects except the variable under test.

Diets used to measure nutrient requirements or for other basic nutrition research should be prepared from highly purified ingredients. Examples of research diets for channel catfish and salmonoids are presented in Table 6.5. Diets similar to these have been used successfully in experiments with carp, tilapia, eel, and other fish. These diets may be prepared as moist or dry particles. Moist pellets are usually made by including a binding agent, such as gelatin, agar, or carboxymethylcellulose, in the ingredient formula, adding enough moisture after mixing to form a stiff, plastic consistency, extruding through a food grinder, and breaking the extrusions into short lengths for storage. Moist diets *must* be stored frozen and thawed only a day or so prior to feeding. This is to prevent loss of ascorbic acid (vitamin C) which is highly sensitive to oxidation in moist diets. Dry diets may be prepared by subsequently drying the moist diets described previously, or by adding about 10 to 12% water and pelleting in a laboratory model pellet mill. Binding agents should also be included in dry diet formulations. Propionic acid is advantageous in research diets, especially moist diets, to prevent mold growth.

Experimental diets for evaluation under practical conditions will usually be made from natural feedstuffs, but the same measures should be taken with practical types of diets as with purified diets to guard the palatability, physical, and nutritional properties. The diets should be designed so that the only difference in the diets to be compared is the variable in which the researcher is interested. This prevents the researcher from confusing the real cause of the fish response with that which he supposed when he designed the experiment.

Practical research diets should be pelleted or extruded under carefully controlled conditions. Errors in heating, drying or other factors can result in nutrient losses. If the diets are processed by extrusion or by drum drying, the researcher should be aware of the degree of nutrient loss to expect. Dry, experimental diets should be stored in a refrigerator (5° to 7°C or 41° to 45°F) to minimize losses or changes in nutrients, fats, or other oxidation-sensitive components, since research diets are often stored for several months with a long-term experiment.

Feeds for Various Species

Eel Diets.—Eel diets are fed in moist form. In Japan, eel feeds, containing as much as 25% precooked starch, are bagged as a dry meal. The farmer adds water and usually fish oil just before feeding. A cohesive, dough-like ball is formed and placed in wire feeding containers in the culture pond.

Aquarium Fish Diets.—The requirements of a satisfactory diet for aquarium fish are different and actually more demanding than those for commercial food fish. Generally, a diet for aquarium feeding should have the following properties: (1) nutritionally balanced; (2) palatable; (3) resistant to crumbling; (4) water stable; (5) buoyant; and (6) enhance pigmentation in ornamental fish.

Existing information on nutritional requirements of warmwater food fish (Table 6.4) is sufficient to formulate nutritionally balanced diets for aquarium species. The major challenge in preparing aquarium diets is to provide for desirable physical and palatability qualities. Experiences in the fishery research laboratory at Auburn University have indicated that diets with hard texture or which sink rapidly are poorly consumed by many aquarium species.

Flaked feeds processed on rotary drum dryers have met the criteria necessary for aquarium feeding. Processing recommendations developed by Boonyaratpalin and Lovell (1977) for preparing flaked fish feeds are presented in Table 6.6. Following these procedures the ingredient formula presented in Table 6.7 can be processed into flakes of desirable physical, nutritional, and pigmentation properties.

Flaked diets for aquarium fish, which are palatable, nutritious, water stable, buoyant, crumble resistant, and pigment enhancing, can be prepared from relatively inexpensive, simple ingredient combinations by carefully controlling processing parameters with a drum dryer. Functional ingredients are shrimp waste meal for hydrocolloidal properties; fish meal for nutritional value and palatability; soybean meal as a major protein source; grain by-product to supply starch and fiber; fish oil for

FIG. 6.4. COMMERCIAL DIETS FOR FISH.
A—Pelleted, for catfish and trout. B—Expanded (floating), for catfish and trout.
C— Extruded (nonfloating), for prawns and shrimps. D—Meal-type, for bait fish and
ornamental fish in ponds. E—Flaked, for aquarium fish. F—Flaked or freeze-dried
and reground, for larval fish in rearing tanks. G—Crumbled pellets, for larval fish.
H—Crumbled pellets, for catfish and trout fingerlings.

TABLE 6.6. PROCESSING RECOMMENDATIONS FOR PREPARING FLAKED FISH DIETS

Procedure	Specification
Mixing	Mix all dry ingredients, less fish oil and vitamin mix.
Grinding	Grind the mixture in a stud or grist mill so that 90% passes through a 60-mesh sieve and 67% passes through an 80-mesh sieve.
Wetting	Add oil and mix. Add hot water (90°C or 194°F) to adjust the solids content to 30% and mix in a high speed blender for 3 min. Add vitamin mix just before drum drying and blend 0.5 min.
Drum drying	Drum temperature, 160° to 165°C (320° to 329°F); rotation speed, 8 rpm; clearance between drums, 0.1 mm; exposure time on drum surface, 8 sec.
Packaging	Reduce flake particle size to approximately 0.5 to 1 cm in diameter and package in hermetically sealed bag within 30 min.

Source: Boonyaratpalin and Lovell (1977).

TABLE 6.7. INGREDIENT COMPOSITION FOR PRACTICAL DIETS FOR CHANNEL CATFISH AND RAINBOW TROUT

Ingredient	Channel Catfish (%)	Trout (%)
Fish meal: menhaden	10	
herring		31
Soybean meal: 44% protein	51.2	15
Yellow corn	22.7	
Distillers' dried solubles	7.5	
Wheat	5	
Wheat middlings		20
Dried whey		8
Brewers' yeast		5
Blood flour		10
Kelp meal		3
Animal fat	2	
Herring oil		2
Lignin sulfonate		2
Dicalcium phosphate	1	
Vitamin concentrate[1]	0.5	0.5
Iodized salt		3.5
Trace mineral mix[2]	0.08	

Sources: Channel catfish: Stickney and Lovell (1977); trout (Idaho Fish Game Dep.): Cuplin (1969).
[1]Vitamin requirements for channel catfish and salmonoids are presented in Table 6.3.
[2]Trace mineral mix should provide the following (mg/kg diet): Mn, 115; I, 2.8; Cu, 4.3; Zn, 88; Fe, 44; Co, 0.05.

palatability and water stability; and a concentrated pigment source. Important processing factors are fine grinding of ingredients, optimum solids content of the purée, and minimum heat exposure in drying.

Pond Diets for Bait and Ornamental Fish.—Meal-type feeds are satisfactory for feeding fish such as golden shiners, fathead minnows, and ornamental fish up to approximately 6 to 7 cm (2 ½ in.) in size, when reared in ponds. A research study demonstrated that golden shiners can use meal-type diets as well as pelleted, crumbled diets, thus indicating that there is no need to pellet these diets (Burtle and Lovell 1978).

Fine grinding of the ingredients, through a 0.16 cm (¼₆ in.) screen or smaller, is important. Use of ingredients which contain a high percentage of fat, or the addition of fat as an ingredient in the diet mixture, is necessary to provide a total fat content of at least 6% in the diet. This is important to prevent dustiness and to cause adhesion of smaller particles to larger ones and so minimize losses when the feed is put into the water. The fat also reduces the rate of water absorption by the particles and allows them to float longer in water.

Most bait and ornamental fish can obtain a large part of their nourishment from phytoplankton and zooplankton in the ponds; consequently, good pond fertilization is important. Where fish consume a large amount of natural food in the pond, the micronutrient composition of the supplemental diet would appear not highly important; however, the benefit of vitamin additives to these diets has not been evaluated. The more important ingredients in such diets are probably the sources of protein, energy, and phosphorus. A practical diet that has produced high yields in heavily stocked golden shiner ponds is presented in Table 6.8.

TABLE 6.8.   INGREDIENT COMPOSITION AND NUTRIENT ANALYSIS FOR PRACTICAL DIETS FOR GOLDEN SHINERS IN PONDS

Ingredient	Golden Shiners (%)
Fish meal (anchovy)	10
Soybean meal (49% protein)	57
Meat and bone meal	5
Corn meal	16
Distillers' dried solubles	7.5
Corn and wheat flakes	
Animal fat	3.5
Dicalcium phosphate	0.5
Dried whey	
Vitamin mix[1]	0.5

Source: Golden shiners: Felts (1979).
[1]Vitamin mix should meet the requirements for catfish (Table 6.3).

Foods for Larval Fish.—Some fish, such as channel catfish, trout, and salmon, have large yolk sacs when they hatch which nourish the fish until its digestive parts are relatively well developed. These fish can consume and assimilate a range of nutrient sources, including artificial diets, as their first food. Many other fish hatch with a small supply of endogenous nutrients and must begin feeding before their digestive system is well developed. The latter group of fish utilize artificial foods poorly, especially if they are the only foods available to them.

Much research is being directed toward development of artificial diets for larval fish, which traditionally require live organisms as their first food. The objective is to partially or completely replace live foods such as brine shrimp *(Artemia salina)*. Physical properties of these diets, such as particle size, density and water stability, as well as palatability and nutritional value, are important. Design of the culture container interacts with the diet characteristics; the diet particles must be kept in suspension in water currents to facilitate consumption by the fish.

Presently, there is no artificial food that will completely replace live foods, but progress has been made in partially replacing live foods with dry foods. Dry foods that show promise are diets that are flaked and subsequently ground to pass through a 30-mesh screen, freeze-dried mixed diets that are subsequently ground, freeze-dried and ground insects and crustaceans, and yeast cells. The technology for processing such diets is highly intricate inasmuch as the particle size and relative density are highly important.

Freeze-dried and flaked diets were prepared at Auburn University (Santiago 1978) for striped bass *(Morone saxatilis)* larvae. Both were consumed relatively well and partially replaced brine shrimp nauplii. Table 6.9 presents the ingredient mixture. Raw shrimp or fresh flesh was an essential binding agent. The mixture was made into a slurry containing approximately 80% moisture and either freeze-dried or flaked. The dried diets were subsequently ground. Particles passing through a 60-mesh screen were fed for the first week or so and those passing through a 30-mesh but not the 60-mesh screen were fed as the fish grew larger.

TABLE 6.9. INGREDIENT COMPOSITION AND NUTRIENT ANALYSIS OF DIETS FOR AQUARIUM FISH (FLAKED) AND LARVAL FISH (FLAKED OR FREEZE-DRIED)

	Aquarium Fish (%)	Larval Fish (%)
Ingredient		
fish flour (anchovy)	27	35
shrimp waste flour (with heads)	24	20
soybean flour (49% protein)	24	
rice polishings flour	10	
wheat bran flour	10.8	
fish oil	4	5
wet fish flesh		40
marigold petal flour	0.13	
vitamin mix	R[1]	R
Nutrient		
protein	40	65
lipids	6	8.4

Sources: Aquarium fish: Boonyaratpalin and Lovell (1977); larval fish: Santiago (1978).
[1]Vitamin supplement required; use the requirements for catfish (Table 6.3).

Larval fish requiring live foods grow best in waters containing natural, indigenous food organisms of plant and animal origin, provided the larvae are protected from predacious organisms. Several studies have been directed to culture and control growth of desirable natural planktonic organisms for larval fish food. Brine shrimp eggs (which are hatched into nauplii in the laboratory) are the major commercial source of live food for larval fish. With emphasis on artificial rearing of fry from a variety of fish species, the demand for brine shrimp eggs is great. The supply is relatively good but the cost is appreciable.

FEEDING PRACTICES

Feeding Food Fish

Feeding strategy can affect the profitability of a food fish production operation as much as any variable expense factor. Profitable fish farming requires the use of diets formulated to meet the nutritional requirements of the species and also the development of good feeding procedures. Fish culturists face a problem other animal husbandmen do not in that uneaten or unassimilated food contaminates the environment and may be hazardous to the health of the fish. Moreover, unconsumed fish food is soon dispersed in the water and makes relatively little contribution to fish production. Feeding procedures are affected by environmental factors such as temperature and water quality, physical factors such as rate of water exchange and type of rearing facility, management factors such as frequency and rate of feeding, and type and size of fish.

Feeding preferences and habits vary among species and sizes of fish and among culture environments. Method of food presentation is highly important in maximizing growth and efficiency of food utilization. Natural pond food is usually the most economical source of nutrients and should always be used to maximum advantage when possible.

All fish species have a temperature range in which the most rapid growth and best dietary efficiency can be obtained. Warmwater fish have an optimum growth temperature of near 30°C (86°F) and will make significant growth at temperatures between 21° and 31°C (70° and 88°F). Below this temperature, feeding is erratic and daily feeding is usually uneconomical. Coldwater fish such as trout grow economically at temperatures of 9° to 21°C (48° to 70°F) with the optimum for growth being near 15°C (60°F). In temperate areas where water temperature varies with season, it is important to manage the fish culture operation so that maximum feeding will be done during the warmer period of the year if maximum growth is desired.

Channel Catfish.—Most catfish produced in the United States are grown in ponds 4 to 16.1 ha (10 to 40 acres) in size. The fish are usually fed once daily, 6 or 7 days per week. Most catfish (approximately 67% or more) are fed extruded (floating) feeds. This allows the feeder to observe the feeding activity of the fish which prevents overfeeding and can serve as a check for disease or water quality problems. The fish are usually fed in the morning after dissolved oxygen (DO) level in the pond has begun to rise. Feeding should not be done late in the evening so that maximum oxygen consumption of the fish will not coincide with a decrease in DO level in the ponds, as occurs when photosynthesis stops.

Effect of Water Temperature on Feeding Activity of Catfish.—The optimum water temperature for food consumption and growth of channel catfish is near 30°C (86°F). Economical feeding can be done at temperatures above 21°C (70°F). An experiment was conducted at Auburn University to demonstrate the effect of water temperature on feeding activity of channel catfish in ponds during various parts of the season (Kubaryk 1978). When minimum (morning) water temperature was above 26°C (79°F), feeding two times daily resulted in maximum food consumption and growth; when morning temperature was 22° to 26°C (72° to 79°F), feeding once per day was optimal; and when morning water temperature was below 20°C (68°F), feeding on alternate days resulted in the highest food consumption. These data indicate that catfish should be fed twice daily as soon as water temperature warms up in the early summer and fed on this schedule until the water temperature begins to decrease in the fall.

Feeding Rate and Schedule for Catfish.—Early guidelines for feeding catfish in ponds recommended a daily allowance equal to 3% of the weight of the fish, not to exceed 39 kg per ha (35 lb per acre), 6 days per week. Research has shown that catfish in ponds will consume food at the rate of 3 to 4% of their weight only during the early part of the grow-out period, and that food consumption decreases to less than 1.5% of fish weight as the fish approach harvest size.

The values in Table 6.10 represent a guide for daily food allowance for channel catfish in ponds. The data come from an experiment in which channel catfish were fed to satiation and daily food consumption was carefully measured (Minton 1978). The experiment revealed that when channel catfish were fed a diet containing 35% protein and 2.8 kcal of digestible energy, a food allowance 12% less than satiation produced only 3% less weight gain. This indicates that the catfish farmer need not try to get the last bit of food into his fish; he can feed them just short of "all they will eat" and get almost as much production and greatly reduce the likelihood of overfeeding.

Winter Feeding of Catfish.—Winter feeding can effect a small weight

TABLE 6.10. FEEDING SCHEDULE FOR CHANNEL CATFISH[1]

Date	Water Temperature (°C)	(°F)	Fish Size (g)[2]	Feed Allowance (% of wt/day)
April 15	20	68	20	2
April 30	22.2	72	30	2.5
May 15	25.5	78	50	2.8
May 30	26.7	80	70	3
June 15	28.3	83	100	3
June 30	28.9	84	130	3
July 15	29.4	85	160	2.8
July 30	29.4	85	190	2.5
Aug. 15	30	86	270	2.2
Aug. 30	30	86	340	1.8
Sept. 15	28.3	83	400	1.6
Sept. 30	26.1	79	460	1.4
Oct. 15	22.8	73	500	1.1

[1]Food pond feeding of catfish stocked as 12.7 cm (5 in.) fingerlings in spring and harvested as 500 g fish in the fall in the southeastern United States (Stickney and Lovell 1977).
[2]28.4 g = 1 oz; 454 g = 1 lb.

gain and will increase disease resistance of overwintered fish. Results of several studies at Auburn University (Lovell and Sirikul 1974; Felts 1977) indicate that feeding 1% of fish weight on days that water temperature is above 12°C (54°F) is the best program for fish 114 to 454 g (0.4 to 1.0 lb) size. Fingerlings may be fed at a rate of 1% of body weight, 3 times weekly throughout the winter.

Trout and Salmon.—Trout and salmon are coldwater fish which are cultured as food fish. Trout culture is in fresh water, and salmon culture, from smolt to food fish size, is in a marine environment. This fact is alleged to be the reason that trout are grown very successfully on dry diets whereas salmon in a saline environment are fed wet diets. Salmon consume dry foods satisfactorily in fresh water but less well in salt water. Moist diets are more expensive to transport, store, and feed, which makes feeding costs greater for salmon than for trout.

Most trout cultured for food are grown in continuously flowing raceways. Although overfeeding will not endanger the life and health of the fish, as may occur in static pond feeding, it should be avoided for economic efficiency and to minimize the pollution load contained in the raceway effluent. A feeding schedule that has been used successfully with rainbow trout in raceways, which allows for fish size and water temperature variations, is presented in Table 6.11.

Trout are voracious feeders and consume feeds as soon as they hit the water; consequently, floating diets are not a mangement asset with trout as with the slower feeding channel catfish. Trout respond favorably to more frequent feeding than once daily, although they have a relatively large stomach and are "meal eaters." Twice daily feeding has been found economically feasible in practice.

TABLE 6.11. FEEDING SCHEDULE FOR RAINBOW TROUT

Water Temperature (°C)	(°F)	Fish Size (cm)[1]									
		2–5	5–7	7–10	10–12	12–15	15–18	18–20	20–23	23–25	25
		(% of Weight/Day)									
7	44.6	3.1	2.5	1.9	1.4	1.2	1.0	0.9	0.7	0.6	0.6
8	46.4	3.4	2.8	2.1	1.6	1.3	1.1	0.9	0.8	0.7	0.6
9	48.2	3.8	3.0	2.3	1.7	1.4	1.2	1.0	0.9	0.8	0.7
10	50.0	4.2	3.3	2.5	1.9	1.5	1.3	1.1	1.0	0.8	0.8
11	51.8	4.5	3.6	2.7	2.0	1.6	1.4	1.2	1.0	0.9	0.8
12	53.6	4.9	3.9	2.9	2.2	1.8	1.5	1.3	1.1	1.0	0.9
13	55.4	5.3	4.3	3.2	2.4	1.9	1.6	1.4	1.2	1.1	1.0
14	57.2	5.8	4.7	3.5	2.6	2.1	1.8	1.5	1.3	1.2	1.1
15	59.0	6.3	5.1	3.8	2.8	2.3	1.9	1.6	1.4	1.5	1.2
16	60.8	5.5	5.5	4.2	3.2	2.5	2.1	1.7	1.6	1.4	1.3

Source: Am. Fish. U.S. Trout News (1969).
[1] 2.54 cm = 1 in.

Salmon are grown satisfactorily to smolt size under managed conditions, with artificial diets, in fresh water. Relatively little information is available on feeding practices with salmon beyond this size when they become marine. Salmon rearing in pens in seawater is being practiced commercially in the northwest of the United States. The favorable market value of salmon should justify a great amount of effort in developing feeding technology for salmon.

Tilapias.—Tilapias are emerging as a valuable cultured food fish in the United States although they will not survive temperatures below 12°C (54°F) and, unless reproduction is controlled, they increase in number instead of size. However, tilapias are very efficient users of natural pond foods. They are omnivorous, continuous feeders with the ability to consume planktonic as well as benthic food sources.

Yields of 6736 to 8980 kg per ha (6000 to 8000 lb per acre) have been produced with only animal manure added to the pond (Lovshin 1974). The use of supplemental feed in addition to manuring has improved tilapia yields economically. Tilapias have responded to practical catfish diets with about the same conversion efficiency as catfish in both ponds and raceways. Thus a commercial catfish diet is probably adequate for tilapia.

Tilapia in ponds utilize meal diets almost as well as pelleted diets which is economically advantageous. Although large tilapia can consume particulate (pelleted) food more efficiently, the wasted food resulting from feeding a meal type diet benefits the fish by increasing production of pond organisms.

Another advantage of tilapia is that because they consume large amounts of natural pond organisms, they do not require a highly fastidious diet as do catfish. Essential growth factors, such as vitamins, trace minerals, and amino acids are not of great importance in supplemental pond diets for tilapia. Inorganic or organic fertilization of ponds to maximize yield of natural pond organisms is highly beneficial in tilapia culture.

Tilapia do not have as large a stomach as catfish or trout and cannot eat as much at one "meal"; thus, multiple daily feeding is beneficial for tilapia.

Bait Fish and Ornamental Fish in Ponds

Golden shiners (*Notemigonus crysoleucas*), other bait fish, and ornamental fish that are cultured in ponds make very efficient use of pond organisms. They obtain significant nourishment from phytoplankton, zooplankton, benthic insects and molluscs, and probably microorganisms growing on the surface of benthic debris. Therefore, maximum production of pond organisms is desirable. The benefit of supplemental feeding on yield of golden shiners in commercial types of ponds is evidenced from the following data (Felts 1979), which show that supplemental feeding increased production by at least 50% over fertilization only.

	Yield	
Treatment	lb/acre	kg/ha
Fertilization only	410	460
Fertilization, high fish meal diet	656	736
Fertilization, low fish meal diet	629	706
Fertilization, all plant diet	605	679

The preceding data indicate that a high percentage of fish meal is not necessary in golden shiner diets when pond food is abundant.

Golden shiners do not need pelleted, crumbled diets. A study at Auburn University (Burtle and Lovell 1978) demonstrated that golden shiners grew as well on non-pelleted (meal) diets as on pelleted, crumbled diets when the diet mixture was ground and subsequently supplemented with fat as described previously in this chapter. The meal diet remains afloat for a long time and wind will cause drifting; thus, it should be fed on the upwind side of the pond. A daily feeding rate of 3% of the estimated weight of golden shiners in the pond over a 90 day period resulted in a feed conversion ratio of 1.3 lb or kg of feed per lb or kg of gain (Felts 1979), which appears to be a satisfactory practical feeding rate.

Finely ground meal-type feeds with high fat contents, similar to those fed to bait fish, have been used successfully in pond rearing of ornamental fish. Meal-type pond diets should *not* be fed in aquariums; only water stable diets designed for aquarium feeding should be used.

In some situations a high concentration of zooplankton is desirable in ornamental fish ponds. This can be achieved by applying meat-and-bone meal or fish meal at the rate of 112 kg/ha (100 lb/acre) 2 or 3 times over a 10 day period prior to stocking the small fish in the ponds.

REFERENCES

AM. FISH U.S. TROUT NEWS. 1969. Revised New York State Fish Hatchery feeding charts. Dec., 14.

ANDREWS, J.W., MURAI, T., and CAMPBELL, C. 1973. Effects of dietary calcium and phosphorus on growth, food conversion, bone ash, and hematocrit levels in catfish. J. Nutr. *103*, 766-771.

BOONYARATPALIN, M. and LOVELL, R.T. 1977. Diet preparation for aquarium fishes. Aquaculture *12*, 53-62.

BURTLE, G.M. and LOVELL, R.T. 1978. Meal versus a crumbled pellet feed for golden shiners. Catfish Farmers Am. Res. Workshop. (abstract)

CHWANY, N. 1978. Development of practical diets for freshwater prawns. M.S. Thesis. Auburn Univ., Auburn, Ala.

CRUZ, E.M. 1973. Digestibility studies with channel catfish. Ph.D. Dissertation. Auburn Univ., Auburn, Ala.

CUPLIN, P. 1969. Performance evaluation of chelated minerals in Idaho open-formula diets. Trans. Am. Fish. Soc. 98, 772-776.

FELTS, S.K. 1977. Evaluation of winter feeding regimes for channel catfish. M.S. Thesis. Auburn Univ., Auburn, Ala.

FELTS, S.K. 1979. Evaluation of practical feeds for golden shiners. Catfish Farmers Am. Res. Workshop Annu. Meet., Jackson, Miss., Jan. 31.

GARLING, D.L., JR. and WILSON, R.P. 1976. The optimum dietary protein to energy requirement for channel catfish fingerlings. J. Nutr. 106, 1368-1375.

KUBARYK, J.M. 1978. Effects of frequency of feeding on growth and food consumption of channel catfish in ponds. M.S. Thesis. Auburn Univ., Auburn, Ala.

LEARY, D.F. and LOVELL, R.T. 1974. Value of fiber in production-type diets for channel catfish. Trans. Am. Fish. Soc. 104, 328-332.

LIM, C. and LOVELL, R.T. 1978. Pathology of the vitamin C deficiency syndrome in channel catfish. J. Nutr. 108, 1137-1146.

LOVELL, R.T. 1973. Essentiality of vitamin C in feeds for intensively fed caged catfish. J. Nutr. 103, 134-138.

LOVELL, R.T. 1977. Fish nutrition: energy. Comm. Fish Farmer 2, 29-30.

LOVELL, R.T. 1978. Phosphorus requirement of channel catfish. Trans. Am. Fish. Soc. 107, 617-621.

LOVELL, R.T. and LIM, C. 1978. Vitamin C in pond diets for channel catfish. Trans. Am. Fish. Soc. 107, 321-325.

LOVELL, R.T. and PRATHER, E.E. 1973. Response of intensively-fed channel catfish to diets containing various protein-energy ratios. Proc. Annu. Conf. Southeast. Assoc. Game Fish Comm. 27, 255-259.

LOVELL, R.T., PRATHER, E.E., TRES-DICK, G. and LIM, C. 1974. Interrelationships between quality and quantity of protein in feeds for channel catfish in intensive pond culture. Proc. Annu. Conf. Southeast. Assoc. Game Fish Comm. 28, 456-461.

LOVELL, R.T. and SIRIKUL, B. 1974. Winter feeding of channel catfish. Proc. 28th Annu. Conf. Southeast. Assoc. Game Fish Comm. 28, 208-216.

LOVSHIN, L.L., DA SILVA, A.B., and FERNANDES, G.A. 1974. Proc. FAO/CARPAS Symp. Aquacult. Latin Am. (Montevideo, Uruguay), May 28-31.

MAYER, F.L., MEHRLE, P.M., and CRUTCHER, P.L. 1977. Interactions of toxaphene and vitamin C in channel catfish. Trans. Am. Fish. Soc. 107, 326-333.

MINTON, R.V. 1978. Responses of channel catfish fed diets of two nutrient densities at levels of 75, 87 and 100 percent of satiation in ponds. M.S. Thesis. Auburn Univ., Auburn, Ala.

NATL. RES. COUNC. 1971. Nutrient requirements of poultry. Natl. Acad. Sci., Washington, D.C. Bull. *1*.

NATL. RES. COUNC. 1973A. Nutrient requirements of swine. Natl. Acad. Sci., Washington, D.C. Bull. *2*.

NATL. RES. COUNC. 1973B. Nutrient requirements of trout, salmon and catfish. Natl. Acad. Sci., Washington, D.C. Bull. *11*.

NATL. RES. COUNC. 1977. Nutrient requirements of warmwater fishes. Natl. Acad. Sci., Washington, D.C.

PAGE, J.W. and ANDREWS, J.W. 1973. Interactions of dietary levels of protein and energy on channel catfish. J. Nutr. *103*, 1339-1346.

SANTIAGO, C.B. 1978. Development and evaluation of dry diets for the first food of striped bass fry. M.S. Thesis. Auburn Univ., Auburn, Ala.

STICKNEY, R.R. and LOVELL, R.T. 1977. Nutrition and feeding of channel catfish. South. Coop. Ser. Bull. *218*.

WILSON, R.P. 1973. Absence of ascorbic acid synthesis in channel catfish, *Ictalurus punctatus,* and blue catfish, *Ictalurus.* Comp. Biochem. Physiol. B. *46*, 635-638.

WILSON, R.P. and ROBINSON, E.W. 1978. Amino acid requirements of channel catfish. Personal communication. Dep. Biochem., Miss. State Univ., Starkville.

7

Common Fish Diseases and Their Control

Disease is a departure from the typical normal state of health of a fish or any other animal. Diseases may be classified as *infectious* diseases when they are caused by microorganisms such as protozoa, bacteria, fungi, or viruses. Diseases not caused by microorganisms are *noninfectious* diseases which include nutritional deficiencies, intoxications, low oxygen levels, or gas bubble disease.

Parasitism is a term which means that a microorganism lives on or in the *host* fish and implies that the parasite does some damage to the host or may be capable of producing damage. The term *parasite* is a general term which can include any microorganism, but in fish pathology it is often restricted to protozoan (single cell) microorganisms, nematodes, monogenetic or digenetic trematodes, or parasitic copepods.

The visible effects which a parasite has on the host often depend on the number of parasites per fish and frequently on the environmental conditions which occur at the time of the parasitism. For example, a moderate gill infestation with the parasite *Ichthyophthirius multifiliis* may be tolerated well if the oxygen level is above 5 ppm, but infested fish may not tolerate oxygen levels below 5 ppm. Similarly, high levels of ammonia, extremes of pH, or low level toxicants may compound the damage already caused by the parasites. Frequently, these environmental stressors play a deciding role in the outcome of a particular disease.

Parasites may be facultative or obligate. Facultative parasites are free living in nature but may, under certain circumstances, begin to parasitize fish. Good examples of facultative parasites are the carnivorous protozoan *Tetrahymena pyriformis* and the stalked protozoan *Epistylis.* Both of these parasites may be free living in nature—usually in water high in organic pollution—and can transfer to fish where significant damage can result. Obligate parasites are those which require the presence of the host for reproduction. Good examples are *Ichthyophthirius multifiliis, Costia, Chilodonella,* and monogenetic trematodes.

TRANSMISSION OF DISEASE

The successful transmission of a disease requires the presence of a sufficient number of susceptible fish per unit volume of water and the presence of the organism in sufficient numbers. Frequently, if not in most cases, adverse environmental factors (stressors) are required to initiate a disease. Overcrowding of fish increases the likelihood of an initial contact between a parasite and a fish and also ensures that the parasite will spread rapidly among a group of fish. It does not take long before an introduced parasite can multiply to tremendous numbers on individual fish. During the initial stages of the parasitic infestation, the fish may appear to be normal. In this case, the infection is said to be *subclinical* or *inapparent.*

Parasites of fish can be transmitted in various ways. Contact transmission of parasites such as monogenetic trematodes can be encouraged by overcrowding. Obviously, most obligate parasites of fish are transmitted by the water. Parasitic copepods such as fish lice (*Argulus* sp.) have been implicated as vectors of bacterial diseases. It seems probable that monogenetic trematodes can also transmit bacterial diseases.

Stress plays an important role in the transmission of many bacterial diseases and possibly external protozoan diseases. Commonly, outbreaks of bacterial disease follow shipping, handling, grading, or environmental changes such as periods of suboptimal oxygen concentrations. The basic mechanism of how fish lose their natural resistance to bacteria after some stressful condition most likely is that hormonal imbalances result in a decreased efficiency of the natural ability of the fish to ward off the ever present bacteria in water.

The specific disease defense system of an animal is the immune system. The decreased efficiency of the immune system probably involves a combination of factors: a decrease in the effectiveness of the white blood cells of the fish to phagocytize bacteria (phago = to eat); and a decrease in the production of antibodies (specific proteins found in the blood which assist white blood cells in destroying bacteria). A commonly used diagram to emphasize the relationship among the fish, parasite, and stress is presented in Fig. 7.1 which suggests that many fish can function as inapparent carriers of potential disease inducing parasites as long as environmental stresses are minimized or not present.

NATURAL RESISTANCE AND IMMUNITY

All fish are not equally susceptible to all parasites. Scaleless fish such as channel catfish are extremely susceptible to severe outbreaks of *Ichthyophthirius multifiliis.* Presumably, the penetrating tomite, assisted

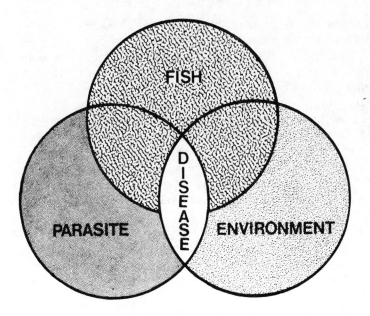

FIG. 7.1. THE FIGURE ILLUSTRATES THE INTERDEPENDENCE OF
HOST, PARASITE, AND ENVIRONMENTAL STRESS FACTORS IN A
DISEASE PROCESS

by the enzyme hyaluronidase, finds it relatively easy to penetrate the
epithelium of the skin. Many monogenetic trematodes will affect the
external surfaces of one species of fish or closely related species. Other
species of fish are not infested. Viruses of fish demonstrate a remarkable
species specificity in addition to infecting fry or fingerlings within a
narrow temperature range. Channel catfish virus disease, for example,
affects only channel catfish fingerlings during summer and early fall.

Immunity is the term used when an animal is rendered resistant to a
particular infectious disease by vaccination or previous exposure. A vaccine (the substance used to vaccinate fish) can be given by injection, by
immersing the fish in the vaccine, or by mixing with food. The vaccine
consists of a preparation of the disease organism which has been inactivated—usually with a chemical. Vaccines have been prepared and have
been shown to be effective against bacterial diseases of trout such as
vibriosis and "red mouth" disease. Fish have also been shown to develop
an immunity to the protozoan *Ichthyophthirius multifiliis.* The induc-
tion of the immune state following the administration of a vaccine is very
complex. Basically, antigens, which are specific bacterial, viral, or pro-
tozoal proteins in the vaccine, induce the production of substances in the
fish's blood called immunoglobulins which have the property of com-

bining with the invading disease organism. The antigens in the vaccine also stimulate specific cells within the fish's body to react specifically with the invading parasite. The latter type of induced immunity is referred to as *cell mediated immunity*; the former, where immuno-globulin proteins are produced, is called *humoral immunity* (humor = blood).

Vaccines for bacterial diseases of trout have been developed and marketed. The continued development of vacinnes will depend on the economic importance of a particular parasite and whether or not it can be controlled by less expensive medications or management practices. There are many technical problems hindering the development of vaccines: (1) One is the inability to grow the organisms in the laboratory (this includes most external parasites of fish). (2) The existence of many different strains of bacteria which produce an identical disease but which have sufficiently different surface protein antigens makes it necessary to include multiple serotypes within a vaccine. (3) A basic problem with virus vaccines is to develop stable avirulent strains. (4) A major problem is to develop easy methods of delivering sufficient amounts of vaccine to large groups of fish.

WATER QUALITY AND DISEASE

Good water quality is the key to successful fish production. An abundant water supply will solve many problems associated with intensive fish culture by diluting out accumulated wastes and toxic products as well as maintaining optimal water conditions.

Oxygen.—Oxygen is the limiting factor in fish production. A minimum acceptable level is considered to be 5 parts per million (ppm). At levels below 4 ppm fish may live but will not feed or grow.

Low oxygen levels are frequently a problem during the summer in fish ponds with a heavy algal bloom. Typically, the fish will be found dead or in severe stress at dawn when the oxygen levels may approach zero. Typically, larger fish die first, and water often changes in smell and color. Vegetation and algal blooms also die, which further complicates the problem by increasing the demand for oxygen in the decay process [the biological oxygen demand (BOD)]. Cold water contains more oxygen; this along with the fact that algae growth is restricted in the winter explains why low oxygen is not a problem during colder months. The basic problem during summer months is generally overcrowding with an abundance of water nutrients (nitrates, phosphates) resulting from fish wastes, artificial fertilizers, and decaying feed. The developing algal bloom produces oxygen during daylight hours, but consumes oxygen at night.

Observant fish farmers can anticipate low oxygen problems by odor and color of water and by observation of the fish at dawn. Low oxygen (2 ppm) measurements 2 ft below the surface are a sure sign of an impending oxygen problem. The oxygen can be elevated by use of commercial aerators or by pumping the water and spraying over the surface. Low levels of potassium permanganate (2 to 4 ppm) have been used as algicides as well as to quickly raise the oxygen levels; however, the increased BOD resulting from decaying algae will further complicate the problem. Surface agitation, increased inflow of aerated water, and thinning the population of fish are effective methods used singly or in combination to remedy the problem.

Oxygen levels in aquaria are rarely a problem because of the common use of mechanical aerators and bubblers, but problems can occur in heavily populated aquaria during a power failure. Emergency gas powered generators are used as a backup electrical energy source in large installations.

Gas Bubble Disease.—Gas bubble disease can occur in fish under conditions in which there is an excess of gases in the water. A gas saturation level of above 110% is usually considered to be a problem. Gases which can be involved include carbon dioxide, nitrogen, and oxygen. A supersaturation condition can arise from faulty water pumps with leaking gaskets; air is sucked into the pump, compressed, and distributed to aquaria. In one case involving a wholesaler of aquarium fish, dry activated carbon was placed in a cannister which was part of a water recirculation system. Shortly after each carbon change, many fish would begin to die within hours. Since the entire system was under pressure, gases within the activated carbon were released and supersaturation resulted. The problem was avoided by prewetting the carbon. An interesting observation was that many species were apparently more resistant than others. The signs of gas bubble disease can vary. Since gas emboli can reach the brain or heart, the fish can die suddenly with no other signs. Bubbles just under the surface of the skin may be seen. Many fish may develop a mottled finnage which under microscopic examination will be found to be gas bubbles. Hemorrhaging of the fins is common due to the occlusion of small blood vessels by gas bubbles.

Carbon Dioxide.—Carbon dioxide is a normal component gas of air and water. It is generated as a result of respiration as well as decay of organic substances. Water sources can contain 4 ppm. Levels in crowded home aquaria can reach 30 ppm without harm if enough aeration keeps the oxygen level maximal. The accumulation of carbon dioxide in sealed containers during fish transportation is a major problem in the tropcial fish industry. Research utilizing channel catfish in sealed shipping bags suggests that the carbon dioxide level must reach approximately 140 ppm (with an oxygen level at 10 ppm) before fish will die.

Hydrogen Ion Concentration (pH).—The quantity of hydrogen ions in water will determine if it is acidic or basic. Neutral pH is 7; values below 7 indicate increasing degrees of acidity while values above pH 7 indicate decreasing levels of acidity—often referred to as basic or alkaline. The pH of the blood of most animals is close to pH 7. Fish can tolerate wide ranges of pH, i.e., pH from 5 to 9. A desirable pH range for freshwater fish is from 6.5 to 7.5; the pH of marine culture systems is about 8.3.

Under pond culture conditions there may be a wide pH fluctuation due to the respiration process of algae. Algae utilize oxygen during dark periods and give off carbon dioxide as part of their normal respiration function. Carbon dioxide acidifies water; thus, in a very fertile pond pH values as low as 6.5 could be expected at dawn which could gradually shift to values as high as from 8 to 9 by late afternoon when carbon dioxide has been utilized by algae for growth. Thus, in nature, fish are subjected to normal swings of pH and tolerate them quite well. Water in recirculation systems as well as in ponds tends to become acidic since the nitrification of organic matter is both an oxygen consuming and a carbon dioxide and hydrogen ion producing function.

Alkalinity and Buffering Effect.—Fluctuations in pH in ponds or rapid downward pH swings occasionally seen in recirculation systems will be minimized if the water is buffered. If the water contains various dissolved minerals such as carbonates, bicarbonates, borates, or silicates, not only will the pH be naturally higher than neutral but these negatively charged ions will combine with hydrogen ions which essentially avoids a pH drop. Radical swings in pH are seen in soft water ponds with a low total alkalinity, that is, a low buffering capacity. In central recirculation systems in soft water areas of the country, waters can be buffered by addition of carbonates in the form of oyster shell or dolomite. Chemical buffers suitable for buffering in closed aquatic systems have been used successfully. Periodic addition of sodium bicarbonate can adjust pH as well as provide a temporary buffering effect.

Ammonia.—In aquatic systems ammonia accumulates as a result of the normal metabolism of the fish where it is excreted by kidneys as well as by the gill tissue. Ammonia is also formed by the normal decomposition processes of protein (uneaten foods) or dead phyto or zooplankton. Ammonia (NH_3) is toxic to fish while ammonium (NH_4) is relatively non-toxic. The proportion of NH_3 to NH_4 in an aquatic system is dependent on the pH and temperature of the water. As the pH and temperature increase there is proportionately more toxic ammonia. In ponds or raceways, high ammonia levels are a result of insufficient water flow for the amount of fish stocked. The amount of ammonia which is a lower end safe figure has been given as 0.02 ppm (Wedemeyer et al. 1976), but it is apparent that the toxicity varies with the particular species cultured, oxygen and pH levels, as well as water temperature. It has been demonstrated that bacterial gill disease of salmonoids is a syndrome requiring high temperatures and high ammonia levels, plus a bacterium in order to reproduce the disease. Bacterial gill disease is common in cultured warmwater species where poor water quality is a chronic problem.

In recirculating saltwater systems where dolomitic gravel is used as a biological filter, pH levels are usually at pH 8–8.4. Ammonia accumulations at these high pH levels are lethal to fish.

Ammonia accumulation can be a problem in transporting aquarium fish in sealed bags. Research done at the University of Georgia has shown that the addition of clay granules (zeolite) will reduce ammonia levels by a factor of eight. Additionally, such natural clay minerals act as ion exchangers and tend to buffer the water during the transport period.

Ammonia in water is a predisposing factor to bacterial gill disease. Low levels of ammonia over a long period of time may result in damage to gill tissue (Smith and Piper 1975). High levels of ammonia, for example over 0.3 ppm NH_3, probably interfere with respiration resulting in a physiological oxygen depletion.

In closed recirculation fish culture systems utilizing a gravel bed for both mechanical and biological filtration, ammonia and nitrites will accumulate during an initial period in which bacteria capable of oxidizing ammonia and nitrite to nitrates gradually colonize on the gravel and other surfaces (Fig. 7.2 and 7.3). During this initial "run in" period which may last from 30 to 60 days, total ammonia and nitrite levels are frequently high enough to cause extensive mortalities. In freshwater and saltwater systems, nitrites appear to be more toxic than ammonia.

The oxidation of ammonia (a breakdown product of proteins as well as an excretion product of fish) is facilitated by two types of bacteria: *Nitrosomonas* sp. convert ammonia to nitrites and *Nitrobacter* sp. convert nitrites to nitrates. Nitrates are utilized by plants and bacteria or denitrified to gaseous nitrogen and eventually "fixed" into plants by specific bacteria. Generally, closed systems do not have enough plant life to remove nitrates, leading to the necessity of ridding nitrates by water replacement. Nitrates are not toxic but they act as growth promoting

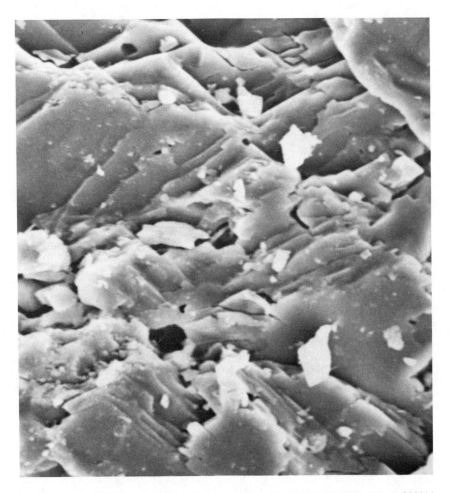

FIG. 7.2. SCANNING ELECTRON MICROGRAPH OF THE SURFACE OF A SMALL PARTICLE OF QUARRY GRAVEL
Note lack of bacteria on surface.

substances for many bacteria, which are undesirable in closed system fish culture.

Temperature Variations.—Temperature extremes compatible with fish life extend from just above freezing to 35° to 40°C (95° to 104°F); however, each species has an optimal growth temperature range. Fish respond to an increase in environmental temperature by increasing the respiration rate as well as cardiac rate and output. Fish eventually die

FIG. 7.3. SCANNING ELECTRON MICROGRAPH OF THE SURFACE OF A PIECE OF GRAVEL TAKEN FROM AN AQUARIUM
Note numerous bacteria on surface. The bacteria are necessary for the oxidation of ammonia to nitrates.

from abnormally high temperatures because of lack of oxygen and malfunction of enzyme systems.

Many fish can tolerate low environmental temperatures. These include trout, catfish and goldfish; however, most tropical adjusted species will die when temperatures drop below 10°C (50°F).

Slight changes in temperature are not deleterious to fish. Fish can tolerate upward changes in temperature better than downward changes. Importers of tropical fish have found that netting directly from transport bags into aquaria with slightly warmer waters rarely results in fish losses. It is always advisable to gradually condition fish to water temperature changes of 3°C (5°F) in either direction. Signs of temperature shock include excitability and rapid respiration, followed by collapse.

Intoxications.—Heavy metals such as zinc, copper, mercury, nickel, lead, or cadmium can enter into a fish culture system by industrial pollution. If the water is hard (300 ppm as $CaCO_3$) with a pH of 8, heavy metals will precipitate as carbonates or sulphates. In softer, low pH water (less than 100 ppm as $CaCO_3$), the metals are in their ionic form and are more toxic to fish.

Algal toxins have produced serious losses in pond culture. Generally, the appearance of algal toxins follows a period of elevated temperatures. Small fish die first after erratic swimming movements and convulsions followed by death. Since such toxins are produced during bright daylight hours, pH of the ponds will be high due to the use of carbon dioxide by the algae.

Hydrogen Sulfide (H_2S).—Hydrogen sulfide is the product of the anaerobic action of bacteria during protein degradation. Lethal levels can be reached during the harvest of fish from ponds with a thick layer of degenerating organic matter. Gas pockets are released leading to an abrupt increase of toxic levels of gas.

Other Chemical Toxicants.—Pollutants can come from various sources: heavy metals, cyanide, chlorinated diphenols, insecticides, herbicides, or sewage. In all cases of toxicant pollution, the course of the illness, as well as lesions, is related to the amount of toxicant found in the water. Consequently, deaths could be immediate (peracute) or occur over a longer period of time (chronic) depending on the level of pollutant. The effect of the pollutant on the fish is variable. There may be direct damage to gills or skin leading to metabolic asphyxia. The pollutant may be concentrated within the body causing liver and kidney damage.

Diagnosis of environmental intoxications requires a detailed knowledge of possible sources of toxicants and the assistance of a well equipped analytical laboratory.

NUTRITIONAL PROBLEMS

Nutritional problems may be encountered in fish fed "complete" artificial diets such as salmonoid fish, aquarium fish, or any other fish which do not have access to natural foods. Nutritional deficiencies are rare in channel catfish pond culture where, in addition to formulated pelletized food, ample natural food is available. When natural foods have not been available, as in floating cage culture of channel catfish, some pelletized fish foods have been found to be inadequate. Present day "cage chows" have improved nutritional value. Occasionally, fish farmers use artificial feeds which have been stored for too long a period, resulting in the gradual degradation of essential amino acids. In two cases examined at the University of Georgia, rainbow trout were submitted with cloudy lenses. A change in the feed remedied this problem.

Insufficient dietary proteins will eventually lead to reduced growth rate, poor reproduction, and increased susceptibility to infectious diseases. Deficiencies in fats and/or fatty acids leads to growth retardation, reproductive problems, and coloration abnormalities in skin. Deficiencies in carbohydrate levels and minerals are rare; however, an iodine deficiency can occur. Goiter in fish is characterized by a swelling behind the lower jaw of the fish. Dietary vitamins have been shown to be very importan in complete diets. Various signs have been attributed to specific vitamin deficiencies which include: convulsions, reduced growth, "blue slime," cloudy lens, apathy, inappetence, anemia, skin discolorations, and spine deformations (broken back or swayback).

Unbalanced diets with excesses of dietary ingredients can also lead to problems. Protein and fat excesses can lead to infiltration of livers and kidneys with fat (lipoid liver degeneration). Fish become sluggish, will not eat, and may develop ascites (swollen belly). As with other animals, excesses of carbohydrates lead to fat deposits in livers and other internal organs. Affected fish may have enlarged abdomens or pale gills. As with other imbalances, eggs may be retained and degenerate.

Dietary toxins can be found in fish foods. In general such toxins can come from microorganisms which contaminate feed, degradation of raw materials used to formulate diets, or breakdown of dietary ingredients during storage. These include rancid fats which impair liver function; mycotoxins from *Aspergillus flavus* causing liver tumors in trout; thiaminase in fish meal (processed herring) which destroys thiamin (vitamin B_1).

The avoidance of dietary problems in aquarium fish centers around the feeding of a diverse diet with the inclusion of live foods (brine shrimp) or frozen brine shrimp. Many breeders of aquarium fish prefer to feed live tubifex worms but suspect that such worms carry parasites. Another popular diet includes mixing beef liver and/or fresh shrimp with commercially formulated trout chows. Additions of vitamin premixes available from poultry supply outlets are sometimes used to supplement this mixture. Addition of the liver results in the food mass sticking together which facilitates addition of medicants as well as removing uneaten food from the aquaria. Unheated fish meal can be a source of mycobacteria and *Ichthyophonus hoferi*.

SEASONAL INCIDENCES OF DISEASE

Meyer (1970) has reviewed the seasonal occurrence of fish disease problems in warmwater fish farms in the southeast. His data suggest that most serious problems of fish occur during the warmer periods of the year—June, July, and August—periods which combine maximum water quality problems (oxygen depletions) with the greatest population of susceptible young fish. His data suggest that bacterial problems are maximal during the warmer months; that the monogenetic trematodes *Gyrodactylus* and *Dactylogyrus* are prevalent in late winter and early spring; and that the anchor worm is prevalent throughout the entire warm summer season.

MORTALITY PATTERNS AND BEHAVIORAL CHANGES DURING AN OUTBREAK OF DISEASE

Often the pattern of fish deaths will provide important evidence as to the basic cause of the problem. These relationships are presented in tabular form.

Mortality Pattern	Suspected Cause
Abrupt onset of deaths; large fish die first; fish at surface of water; water changes color and odor; fish die at dawn; phytoplankton die.	Oxygen depletion
Abrupt onset of deaths; not restricted to morning; young fish die first with convulsive swimming pattern; note death of frogs, turtles, and snails.	Intoxications
Continual mortality over extended period of time.	Parasites
Low mortality rate building up to high mortality rate with eventual decline in deaths.	Viruses Bacteria
Mortality rate initially low building up to a constant peak.	Nutritional problems

Observation of behavioral changes in fish is an indication of disease. The careful observation of the behavior of fish will often provide clues helpful in establishing the cause of the disease. The term "signs" of disease is used in animal medicine rather than "symptoms" since the word symptoms describes how a person feels. Listed are various signs of disease which will assist in arriving at the diagnosis.

Behavioral Signs	Suspected Causes
Many fish at surface of water "piping" for air; may congregate at water inlet.	Oxygen depletion Gill parasites
Fish at surface of water, listless, little movement.	Gill parasites Gill damage (pH extremes) or bacterial gill disease Anemia; nutritional or virus causes
Reduction of feeding activities.	Almost any disease; from severe water quality changes, parasites, bacteria, or poor quality of feed
"Flashing"—refers to fish rolling over with exposure of underbellies.	Internal or external protozoan or helminth (worm) infestation Virus diseases
Fish at bottom of aquarium, listless, off feed, fins may be clamped to body.	External parasites, internal parasites Poor water quality
Fish resting on bottom of pond at shoreline.	Gill parasites, expecially "ich" infestation
Scraping on rocks, sides of raceways, bottom of aquarium, etc.	External parasitic infestation; protozoans, trematodes, crustaceans
"Tail chasing" in trout.	Whirling disease
"Shimmies" expecially in aquarium fish. Descriptive of fish which appear to tremor.	Internal or external parasites

The preceding behavioral signs when initially observed serve notice that a serious problem is impending and immediate measures should be taken. These include an immediate check of water quality, coupled with an examination of the fish for disease producing organisms. Many fish farmers are adept at examining fish for external or internal parasites by using a microscope.

Sending Fish for Examination

When submitting fish to a diagnostic laboratory, it is most desirable to submit a live fish. Fish taken for examination should include apparent healthy specimens and those just showing initial signs of disease. Fish may be delivered and kept alive in pails equipped with portable aerators or with the occasional addition of hydrogen peroxide, or they may be packed in sealed bags containing oxygenated water. Bags with 3 liters of water will hold 1 large (15 to 20 cm long) fish, 5 to 7 fingerlings, or as many as 30 smaller aquarium species.

When sending fish include as much information as is available including the following.

Your name, address, and telephone number.

Species, number, and age of fish submitted for examination.

Water quality data should include: source of water, temperature, flow rate, dissolved oxygen, fertilization history, hardness, ammonia, nitrite, nitrate, and pH levels.

An immediate past history should include when the disease episode began and the death pattern noted. It is important to include all treatments administered including type of drug used and dosage.

Data regarding feeding should include type of feed, when purchased and time stored, frequency of feeding, size of pellets, and feeding responses.

Mention should be made of any possible stresses including handling, shipping, treatments, or abrupt water quality changes. Lastly, any behavioral signs should be mentioned.

CAUSATIVE AGENTS OF DISEASE

Introduction

As the disease causing organisms of fish become smaller in size, the difficulty and expense of making an exact etiologic diagnosis (etiology = cause) becomes increasingly difficult, time consuming, and expensive.

For example, the anchor worm can be seen with the naked eye, making the diagnosis easy. A bacterial disease diagnosis requires a great deal of technical background in initially culturing a diseased fish plus a laboratory equipped with incubators, autoclaves, and microscopes. A diagnosis of a virus (the smallest of pathogenic microorganisms) requires a laboratory equipped for cell culturing (viruses grow only in living cells), incubators, microscopes, plus a great deal of technical training in virology. Whereas a parasitological diagnosis may utilize 30 minutes of time, a virus diagnosis may, in the best equipped laboratories, take up to 14 days. In cases where no apparent etiological agent has been isolated, tissue may be subjected to histopathological examination where thin sections of fixed tissues are examined by a trained expert.

Examination of the Fish

In an outbreak of a disease, fish which are moribund should be examined along with a few apparently normal fish. Dead fish should not be used since after death many parasites leave the fish. The fish can be killed by a blow to the head if it is a large fish, or by pithing (severance of the spinal cord using a needle, scalpel, or scissors.) Overdose of common anesthetic agents can also be used. The appearance of the body can give valuable clues as to what caused the initial problem. In Table 7.1, changes in external features are listed with a selection of possible microorganisms. The table is presented as an overview of possible causes of a fish disease outbreak. It is important to note that many different organisms or conditions can cause identical signs of disease.

TABLE 7.1. EXTERNAL APPEARANCE RELATED TO FISH DISEASES

External Signs	Possible Disease Organism or Condition
Color: dark	(1) Whirling disease of trout, esp. posterior part of body (2) Broken back syndrone (avitaminosis C); may be unilateral (3) Virus disease of trout; infectious pancreatic necrosis (IPN) virus
red	(4) Circulatory collapse due to abrupt environmental change (shock)
Clubbed gills	(1) *Scyphidia* (channel catfish) (2) *Chilodonella* (3) Bacterial gill disease (4) Fungus infection of gills (5) Nutritional deficiencies

Ulcers	(1) Monogenetic trematodes
	 (2) Anchor worms, fish lice (*Argulus*)
	 (3) Bacteria
	     (a) *Aeromonas* sp.
	     (b) *Pseudomonas* sp.
	     (c) *Vibrio* sp.

TABLE 7.1. *(Continued)*

External Signs	Possible Disease Organism or Condition
	(d) *Mycobacterium* sp.
	(e) *Nocardia* sp.
	(f) *Flexibacter columnaris*
White cysts on skin, fins, and gills	(1) Metacercarial stage of digenetic trematode
	(2) Sporozoans
Black spots	(1) Metacercarial stage of digenetic trematode (black color is host reaction)
Localized hemorrhages	(1) *Argulus* (fish lice)
	(2) *Lernaea* (anchor worm)
	(3) Bacterial induced ulcers
	(4) Wounds
Swellings	(1) Tumors
	(2) Encysted parasites
	(3) Sporozoans
Deformities	(1) Genetic
	(2) Nutritional
	(3) Whirling disease (trout)
Skin changes:	
(1) Discrete white spots on skin to 1 mm diameter	"Ich" (*Ichthyophthirius*) or white spot disease
(2) Gray-white patches on skin in "saddle" position on on back or side of fish. May have yellow tinge	*Flexibacter columnaris* (columnaris disease)
(3) Frayed fins, tissues between fin rays necrotic (appear whitish)	Fin rot, caused by *Flexibacter columnaris*
(4) Fuzz ball appearance on localized areas of fish	*Saprolegnia* (fungus)
(5) Gray film on fish, fins opaque. Film may be localized toward upper half of body	*Costia* *Chilodonella* *Trichodina* *Scyphidia*
(6) Yellow speckled "gold dust" on body and fins. Best seen by reflected light	*Oodinium* (velvet disease)
(7) Mucus patches, some hemorrhage (with scraping movements)	Monogenetic trematodes
(8) Larger (1 to 4 mm) white spots on body, gills, or fins. Spots may appear black. Usually slightly elevated	Metacercariae of digenetic trematodes (grubs)
(9) White nodules over surface of body, may be elevated or smooth	Sporozoan disease (milk scale disease in shiners)
(10) Gray-blue areas on skin of larger fish (channel catfish, trout)	Nutritional

Microscopic Examination

A microscope is basically an optical instrument used for magnifying purposes. Very few external parasites can be seen with the naked eye. The advantages of microscopic examination of fish on a regular basis include:

(1) Keeping a record if regular and routine treatments are effective. Unfortunately, addition of various chemicals to ponds, tanks, or aquaria at a suggested dosage does not ensure effective results. A quick check with the microscope will confirm the effectiveness of the treatment.

(2) Keeping tabs on the parasite load in brood stock as well as in fingerlings. This will assist in planning preshipment treatments.

(3) Checking introduced stock during and after a quarantine period.

(4) Checking sick fish to determine the cause which can serve as the basis for selection of a course of treatment.

Dissection

Equipment required includes a microscope, glass slides, cover slips, two dissecting needles, heavy and light scissors, a scalpel with blade replacements, and a small bottle with water and eye dropper. If marine fish are to be examined, wet mount preparations of gills or skin will be prepared using 3% salt water.

Examination of Skin and Fins.—*Procedure.*—

Fig. 7.4. If fish is not dead, pith by inserting needle just posterior to head severing spinal cord.

Fig. 7.5. Scrape fish with scalpel blade from just behind opercula (gill covering) to tip of tail fin. With small fish, the tail may be placed on the end of the glass slide to facilitate transfer of the scales and mucus to the slide. Cut small pieces of various fins and transfer to slide.

Fig. 7.6. Immediately place drop of water (from tap or 3% saline in case of saltwater fish) directly on scales and pieces of fin, and cover with cover glass, which will prevent drying.

Note: It is advisable to make preparations of skin, fins, and gills as soon after death as possible and to examine the preparations immediately.

Fig. 7.7. Remove operculum with scissors, exposing gills. Use scissors to cut away attachment of gills to the body. Remove entire gill arch to slide and proceed to cut away the cartilaginous arch. Use needles to separate gill lamellae. This will facilitate observation of parasites. Only a small part of the gill lamellae is

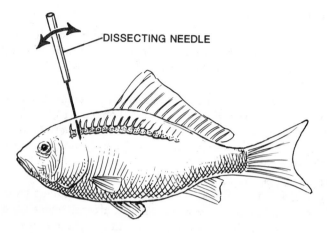

DISSECTING NEEDLE

FIG. 7.4. LIVE FISH MUST BE STUNNED BY A BLOW ON THE HEAD PRIOR TO EXAMINATION
Smaller fish are pithed by severing the spinal cord with a dissecting needle as shown. The needle is inserted into the spinal canal and moved laterally to sever the spinal cord.

needed. Preparations which are too thick will not permit light and nothing will be seen. Cover with enough drops of water to ensure a uniform distribution under the entire cover slip.
Fig. 7.8. Begin examination using lowest power objective—usually 5 or 10× magnification. Most protozoan parasites can be visualized with a total magnification of 100× (magnification is calculated by the product of the magnification of the eyepiece times the objective magnification). Refer to text for description of parasites.

Internal Examination.—The normal position of the internal structures of the fish is included in Fig. 7.9. The internal organs can be recognized by the following characteristics:

Intestine: Tubular structure leading from mouth to anal opening. Length will be variable depending on species of fish.

Liver: A discreet lobed organ found in the forepart of the internal body cavity. In goldfish it is closely intertwined with forepart of the intestine. Color should be dark red-brown.
Problems: If liver is white, gelatinous, or has white spots, may be indication of nutritional problem, or infection with acid fast-staining bacteria such as

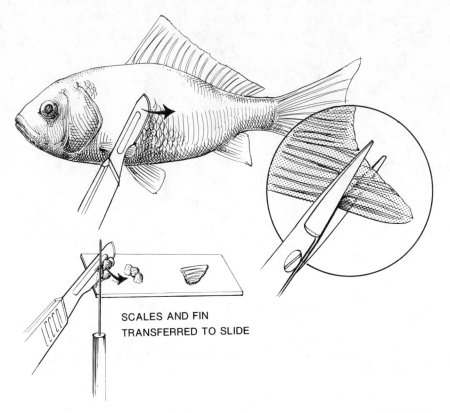

SCALES AND FIN
TRANSFERRED TO SLIDE

FIG. 7.5. PROCEDURE FOR MAKING SLIDE PREPARATION OF SCALES AND FINS
See text for detailed explanation.

	Mycobacterium, Nocardia, or a systemic fungus infection. May be site for encysted digenetic trematodes and pleurocercoid stages of tapeworms. *Hexamita* may invade the liver from the intestinal tract.
Heart:	Found as a small organ at very anterior part of body cavity. *Problems:* White spots may indicate systemic fungus infection *Ichthyophonus hoferi.*
Gall bladder:	Appears as green to yellow-green structure in midst of liver. *Problems:* May visualize *Hexamita.* Site for many sporozoan parasites in marine fish.
Head kidney:	Found just above gills as a homogeneous organ. It is the site of antibody production. Parasites are seldom if ever found.

FIG. 7.6. IMMEDIATELY AFTER PLACING TISSUE SPECIMENS ON SLIDE, DROP WA-
TER (OR 3% SALINE IN CASE OF MARINE FISH) ON SPECIMEN AND COVER WITH
GLASS COVER SLIP

Posterior kidney:Located just below the spinal column. Swim bladder
must be removed to expose. In catfish it is a Y-shape;
in many other species it appears as a reddish streak
along the length of the spinal column.
Problems: Most problems cannot be visualized with
a microscope. The kidney, which acts as a filtering
organ for the body, is one of the organs commonly
infected by viruses or bacteria. Bacteria which may
appear to infect the external surfaces of the fish
invariably kill the fish by spreading to vital internal
organs of which the kidney is the prime target. For
this reason, the kidney serves as a good site for
recovery of possible bacteria involved in a disease
episode. White nodules may indicate mycobacter-
iosis, nocardiosis, or systemic fungus infection.

Spleen: Usually appears as a bright red small organ found
adjacent to intestines. May be site of white nodules
indicating sporozoan cysts or acid fast bacterial
infections. The spleen functions as a reservoir for

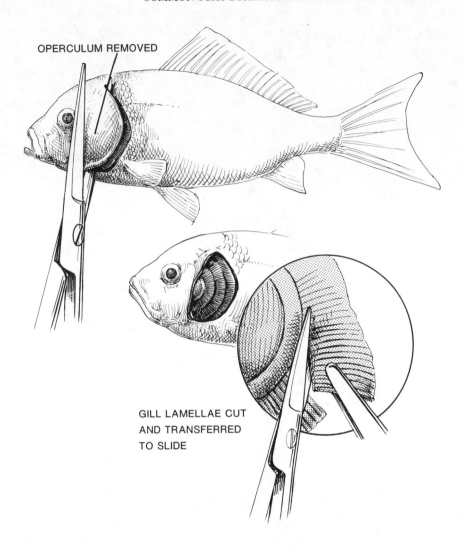

OPERCULUM REMOVED

GILL LAMELLAE CUT
AND TRANSFERRED
TO SLIDE

FIG. 7.7. GILLS ARE REMOVED BY FIRST REMOVING OPERCULUM FOLLOWED BY DISSECTION AND REMOVAL OF GILL TISSUE TO A GLASS SLIDE AS ILLUSTRATED IN FIG. 7.6.

                    red blood cells and may be the organ of choice to examine for blood parasites such as trypanosomes.

Swim bladder: Seen as clear or white dilated sacs found toward the top of the body cavity. In some species, a small tube connects with the inner ear; in others, with the forepart of the intestinal tract.

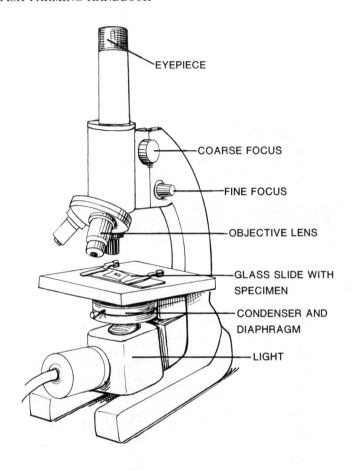

FIG. 7.8. ILLUSTRATION OF A MONOCULAR COMPOUND MICROSCOPE
Successful visualization of various protozoan parasites will require practice and familiarization with the normal microscopic appearance of tissues.

	Problems: May be infested with nematodes; hemorrhages are common in systemic bacterial diseases.
Gonads:	Seen as organs with tubular structures leading to common opening with end of intestinal tract. Testes of males appear to be shiny, white organs which on microscopic examination will reveal motile tadpole shaped sperm which are visible using a 40× objective. Female gonads are identified as egg filled organs.
	Problems: Sporozoan parasites frequently found in ovaries of shiners and many ornamental species.

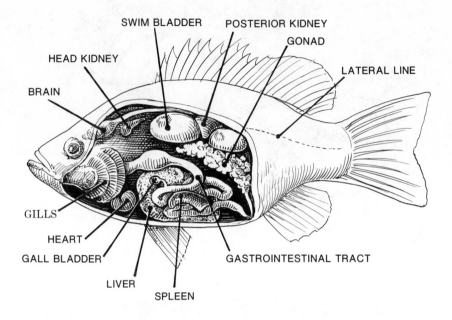

FIG. 7.9. RELATIVE POSITION OF INTERNAL ORGANS OF A FISH
See text for description.

Procedure for Examination of Internal Organs.—From a practical viewpoint, the two most important internal organs which should be examined in some detail are the intestine and liver. The fish should be dissected as shown in Fig. 7.10 to expose the internal organs. This is most easily done using scissors. If the brain is to be examined, a cranial incision will be made according to Fig. 7.11.

For very small fry, or small ornamental species, the entrails may have to be teased onto a slide and separated by using dissecting needles. Add enough water to cover the tissues, cover with glass cover slip, and examine using lowest power microscope objective. Examination of the intestinal tract (in small fish) may be facilitated by compressing a longer section of the intestine between two glass slides. This will facilitate the microscopic visualization of parasites. Larger parasites such as nematodes or tapeworms can be easily visualized by holding the slide up to the light and using the inverted eyepiece as a hand lens to carefully inspect the section of mounted intestine as shown in Fig. 7.12.

Intestines of larger fish are removed from the body cavity. The feces are removed after lengthwise slitting of the intestine, and sections of opened intestine mounted between glass slides. Feces should be examined for presence of parasites by preparing a wet mount between slide and cover slip.

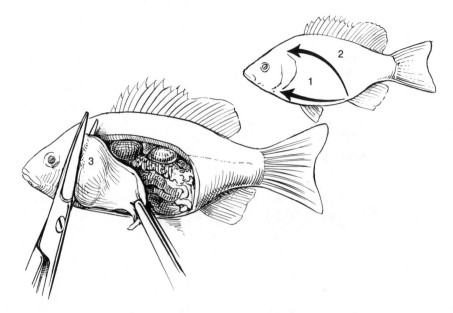

FIG. 7.10. TO EXPOSE INTERNAL ORGANS, THREE INCISIONS ARE REQUIRED
The end result is to remove the body wall exposing the internal organs.

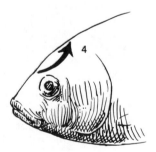

FIG. 7.11. BY USE OF A SHARP SCALPEL, KNIFE
OR SCISSORS, THE BRAIN CAN BE EXPOSED
AS SHOWN

Other organs can be examined by snipping small pieces of tissues and making wet mounts as previously described. It is important to use very small pieces of tissue so that they will flatten out when placed between the glass slide and cover slip.

Microscopic Examination.—After an initial scan of the intestines using the inverted ocular eyepiece, start a systematic observation of the tissues using the lowest power objective. A microscope must be adjusted to the

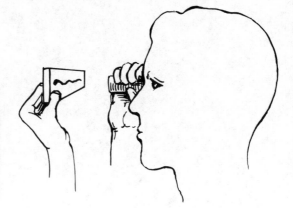

FIG. 7.12. INTESTINES OF SMALL FISH OR INTESTINAL CONTENTS OF LARGER FISH CAN BE PLACED BETWEEN TWO GLASS SLIDES AND EXAMINED FOR LARGER WORMS USING AN OCULAR FROM A MICROSCOPE
The ocular is used in the inverted position.

eyes of the user. Too much light is a common problem for those unfamiliar with microscopes. Light can be reduced depending on the particular microscope. This may include closing down the iris diaphragm, rheostat, or manipulation of the condenser (Fig. 7.8).

What to Look For.—Most parasites of fish either have a characteristic movement, form, or size. These principal characteristics are presented in Fig. 7.13.

External Protozoan Disease

Ichthyophthirius multifiliis.—*Common Name.*—"Ich," white spot disease.

General Description of the Organism.—The organism is a ciliated (cilia =hair) organism. The organism may be round, or near round. Occasionally, a U-shaped nucleus can be observed (Fig. 7.13L). Under 10× magnification, the white spots removed from the fish (up to 1 mm in diameter) will be seen to move very slowly. Higher magnification using a 40× objective will reveal that the cilia completely cover the organism. Figure 7.14 is a scanning electron micrograph of the organism. Microscopic examination of an "Ich" infested fish will reveal various sizes of the organism which are commonly called "trophs." Small, pear shaped forms, seen only with the microscope, are invasive forms called "tomites."

Host Distribution.—The parasite is found wherever fish are cultured.

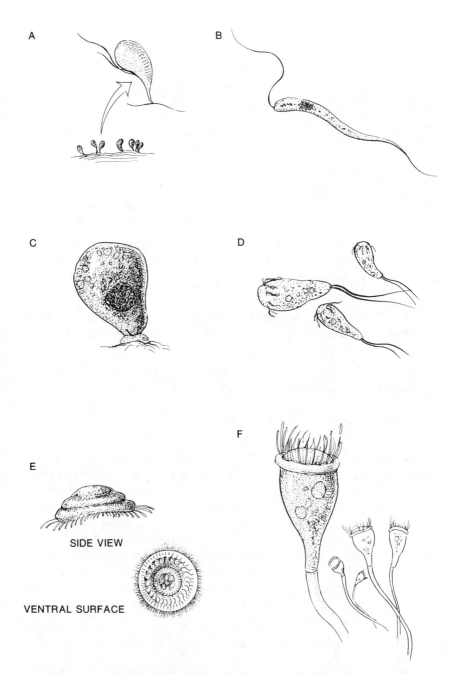

FIG. 7.13. A—*COSTIA*. B—TRYPANOSOME. C—*OODINIUM*. D—*HEXAMITA*. E—*TRICHODINA* (SIDE VIEW AND VENTRAL VIEW). F—*EPISTYLIS*.

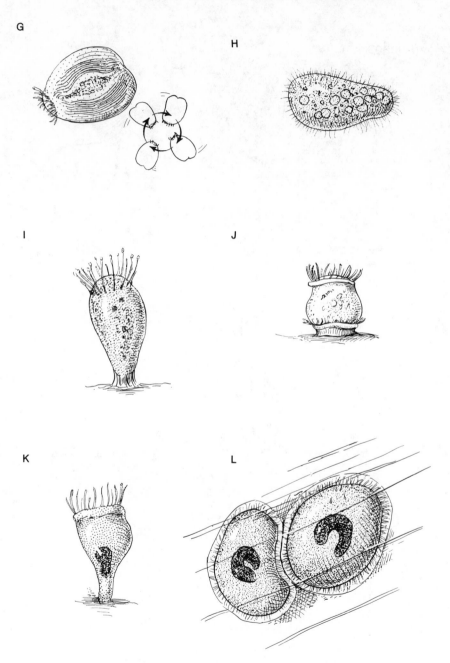

FIG. 7.13. G—*CHILODONELLA*. H—*TETRAHYMENA*. I—*TRICHOPHRYA*. J—*SCYPHIDIA*.
K—*GLOSSATELLA*. L—*ICHTHYOPHTHIRIUS*.

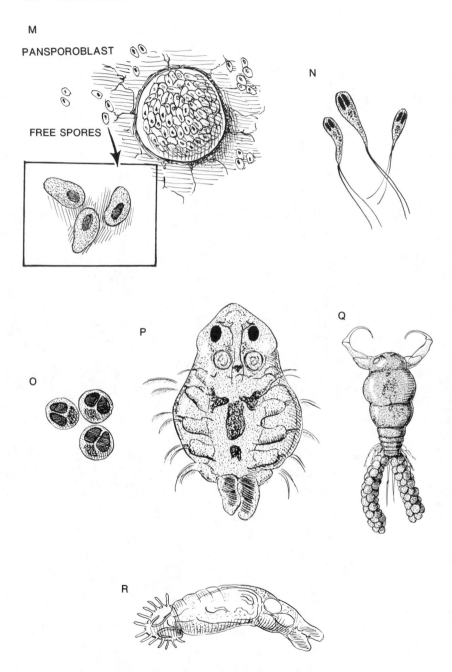

FIG. 7.13. M—*PLISTOPHORA.* N—SPORES OF *HENNEGUYA.* O—SPORES OF *MYXO-SOMA.* P—*ARGULUS.* Q—*ERGASILUS.* R—MONOGENETIC TREMATODE.

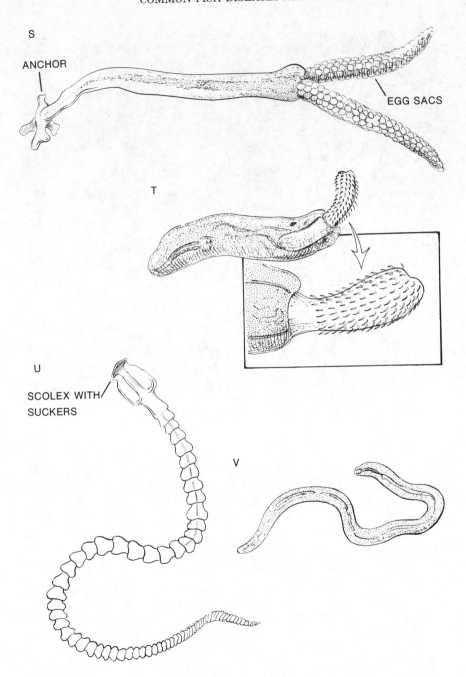

FIG. 7.13. S—*LERNAEA*. T—ACANTHOCEPHALAN (THORNYHEADED WORM). U—TAPEWORM. V—NEMATODE (ROUNDWORM).

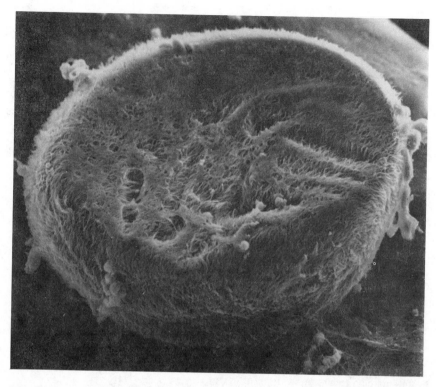

FIG. 7.14. SCANNING ELECTRON MICROGRAPH OF *ICHTHYOPHTHIRIUS MULTI-FILIIS*

Seasonal.—The optimal water temperature for "Ich" is 20° to 22°C (68° to 72°F). In warmwater fish culture the disease is considered to be a fall, winter, and spring occurring disease. At 29° to 30°C (82° to 85°F) the tomites will not survive, which possibly explains the seasonal nature of the parasite.

Life Cycle.—*Ichthyophthirius multifiliis* has a complex life cycle, which complicates treatment of the disease. The mature parasite, which is seen as a white spot, is encysted just under the skin of the fish. Eventually, the adult parasite leaves the fish and becomes a free swimming form for about 2 to 6 hr, after which time it attaches to any suitable substrate (rocks, plants, tubing, etc.). A membrane which assists in attachment is secreted over the organism, although attachment to a fixed substrate need not take place. The cyst then undergoes multiple fissions forming young forms called tomites. Figure 7.15 is a magnification of a cyst showing the outlines of tomites forming within. At the extreme right is a tomite which appears to be emerging from the cyst (Fig. 7.16). The

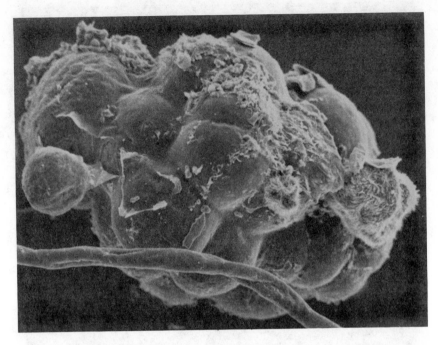

FIG. 7.15. SCANNING ELECTRON MICROGRAPH OF *ICHTHYOPHTHIRIUS MULTIFI-LIIS* IN CYST FORM JUST PRIOR TO RELEASE OF TOMITES
Note emerging tomite on far right of cyst.

number of tomites produced appears to be a function of the size of the encysted adult form. Large adult forms have produced up to 2000 tomites. The development of mature tomites can be completed within 12 hr at 25°C (77°F). After release, the tomites seek a host to penetrate, but die in 24 hr if a host is not present. The tomites penetrate fish by their ciliary action and apparently this is aided by an enzyme called hyaluronidase. Figure 7.17 is a scanning electron micrograph of the cilia of an adult organism. Once having penetrated, the tomites, now called trophozoites, feed on the cells and fluids of the fish. Optimal temperature levels for the development of trophozoites to adults ready to emerge from the host are 21° to 24°C (70° to 75°F) and at this temperature maturation takes 3 or 4 days. At 15.6°C (60°F) the entire cycle takes 10 to 14 days, and more than 5 weeks at 10°C (50°F). At lower temperatures the cycle may take over several months. It is for this reason that outbreaks can occur from groups of apparently normal fish which in fact may harbor an unseen cyst in gill tissue.

Signs of the Disease.—White spots are a common feature of this disease.

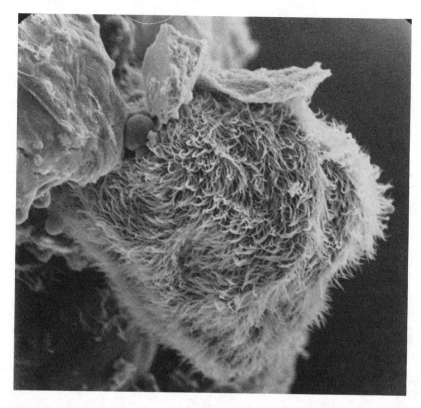

FIG. 7.16. HIGHER MAGNIFICATION OF TOMITE WHICH APPEARS TO BE EMERG-
ING FROM CYST

Fish may appear sluggish and lie at the bottom of the pond, raceway, or aquarium. In catfish ponds, fish frequently will rest at the bottom of the pond near the edge of the water. Flashing and rubbing may be a common early feature of the disease. In some cases, spots will not be seen on the body, but gills will be heavily infested; this phenomenon has been observed in channel catfish and various ornamental species. In advanced cases, bloody fins are common with a thick mucous layer covering the body.

Control.—Treatment of *Ichthyophthirius* is difficult because of the variability of the time of completion of the life cycle. Since the life cycle is not synchronized, and no drug has been found which will kill encysted forms, treatments must be prolonged and repeated frequently. Chemical treatment schedules for "Ich" and other parasites are included in the last section of this chapter.

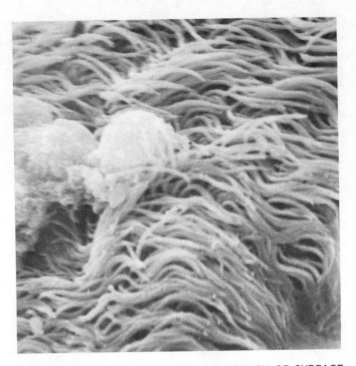

FIG. 7.17. SCANNING ELECTRON MICROGRAPH OF SURFACE
OF ADULT *ICHTHYOPHTHIRIUS* ORGANISM
Note cilia which probably contain protein antigens which can
immunize fish.

Control methods without the use of drugs have been used successfully.
These methods are based on getting rid of the cysts containing the
developing tomites and the tomites themselves.

One successful method is to place the fish in shallow, swiftly moving
water, removing dead fish promptly and sweeping the bottom of the
raceway or ponds daily. This method removes the nonencysted trophozoites as they leave the fish, as well as tomites.

Under aquarium conditions, elevating aquarium temperatures to 28° to
29°C (82° to 85°F) will eventually break the cycle due to the low tolerance
of the tomites for elevated temperatures. Transferring fish from one
aquarium to another every day for a week will eventually effect a cure
simply because the cysts and tomites can be continually circumvented.
Fish are transferred daily followed by cleansing and preparation of the
recently used aquaria for transfer the following day. The water temperature should be kept between 24° and 28°C (75° and 83°F). Some
aquarists effectively avoid reinfestation by tomites by wiping the sides of
aquaria followed by vigorous filtration with a diatomaceous filter. In this

method, the tomites are removed and the water may be sufficiently turbulent to minimize the initial penetration of a tomite.

The best control for *Ichthyophthirius* is prevention. Introduction of the disease into established fish culture systems can be avoided by quarantining "new" fish held at 24°C (75°F) or above for 2 to 3 weeks. During this time the fish can be treated for the disease. Water from quarantine ponds or tanks should not be circulated into other water containing fish. The parasite can also be introduced into home aquaria by plants.

If an outbreak of "Ich" has occurred in a pond, fish can be treated by addition of chemicals to pond water or transferred to tanks to facilitate isolation and treatment. Moving fish, however, is one additional source of stress. Prior to restocking ponds where "Ich" was a problem, the vacated ponds should be drained, disked and disinfected prior to the addition of other fish. Drying alone with a suitable time interval until fish are added should break the cycle. Disinfectants such as calcium hypochlorite added to the remaining water of ponds which cannot be drained will ensure that the remaining organisms are killed.

The marine counterpart of freshwater *Ichthyophthirius multifiliis* is *Cryptocaryon irritans*. A copper ion level of 0.15 ppm on a continual basis has been effective in controlling the parasite in marine systems.

Costia.—*Costia* are the smallest of protozoan parasites which can be found on the skin or gills of many species of cultured fish. The parasite has 4 flagella and is pear to oval shaped. When the organism is viewed on its side, it appears crescent shaped. A helpful characteristic in identifying the parasite is its characteristic flickering motion. Because of its extremely small size (10 to 20 microns by 3 to 8 microns) it may be overlooked unless tissue preparations are examined with 40X objectives. A few species have been described. In trout, it has been suggested that *Costia pyriformis* infests both gills and body, whereas *Costia necatrix* infests only the body surface. Figure 7.18 is a scanning electron micrograph of costia on the gills of an infested goldfish. Figure 7.19 is a scanning electron micrograph of one *Costia* organism.

Host Distribution.—The parasite is a problem in the culture of all freshwater fish, especially fry and fingerlings. Serious problems occur in channel catfish, goldfish, trout, baitfish, and tropical fish production. Recently, the parasite has been implicated as a serious problem in salmon smolts in sea cages (Wooten 1978).

Signs of Disease.—Fish may be seen at the surface of the water or accumulating at a point of water inflow. Flashing and scraping are common. In aquarium fish, fish are listless and remain on the bottom of

FIG. 7.18. SCANNING ELECTRON MICROGRAPH OF *COSTIA NECATRIX* ON THE GILLS OF AN INFESTED GOLDFISH

the aquarium. If infestation is chronic, fish will eat less and will lose condition. In all species a blue-gray sheen may be seen on the body. Diagnosis of the disease should be differentiated from an oxygen depletion problem. Wet mounts of skin and gill tissue will confirm the diagnosis. Wet mounts should be observed not longer than 5 min after the preparation is made, since the organisms will detach from the tissues.

Control.—Chemical treatments are presented later in this chapter. Experiments at the University of Georgia suggest that infested channel catfish fingerlings placed in flow-through holding tanks will spontaneously lose *Costia* as well as other external parasites. On the other hand, if infested fish are placed in closed aquatic systems, the infestation will worsen and mortalities will result. The adage of "changing water" has a twofold purpose of dilution of the parasites in the water as well as reducing stress due to possible poor water conditions.

Chilodonella.—This ciliated protozoan is approximately 50 to 70 microns in length and is easily observed with a 10× objective. The organism

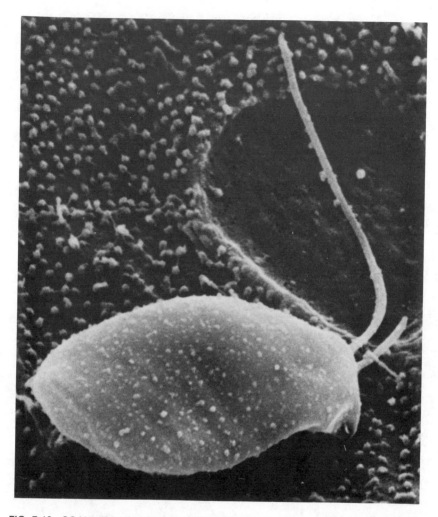

FIG. 7.19. SCANNING ELECTRON MICROGRAPH OF A SINGLE *COSTIA* ORGANISM
Note flagella.

is oval to heart shaped, flattened, with faint bands running the length of
the organism. Figure 7.20 is a scanning electron micrograph of a *Chilodonella*; note the distinct cilia on one end of the organism. Figure 7.21
illustrates many organisms on the gill of a channel catfish fingerling.

Host Distribution.—Chilodonella have been implicated in deaths of trout,
channel catfish, goldfish, bait fish, and a variety of ornamental fish. It is
most likely widespread in freshwater fish culture operations. A saltwater
species has been implicated as a problem in saltwater fish.

FIG. 7.20. SCANNING ELECTRON MICROGRAPH OF A SINGLE *CHILODONELLA* ORGANISM
Note presence of cilia at apical end of organism (bottom of picture).

Chilodonella is considered to be a highly destructive parasite. Normal appearing fish can be infested and signs of disease occur when the population of parasites increases. Under pond or raceway conditions the disease may progress from causing low mortalities initially to building up to a high mortality rate with time. Fish may die suddenly and some will eat just prior to the time of death. The latter phenomenon has been described for channel catfish and has been seen in a group of *Serpae* tetras in an aquarium. The fish may develop a blue-gray color and flared opercula, with gills appearing pale and covered with mucus. Infested fish held in aquaria (goldfish, channel catfish, and ornamental fish) may remain listless at the bottom prior to death.

Diagnosis is made by identification of the organism by microscopic examination of wet mount preparations of gills and skin. Wet mounts should be examined soon after preparation since the organisms tend to die soon after and lose their characteristic movement. Unlike *Ichthyophthirius multifiliis*, the parasite does not roll about but more often appears to move in a wide circular fashion (Fig. 7.13G), possibly due to the prominent anterior cilia as seen in Fig. 7.20.

Control and Treatment.—Avoid problem by quarantine and treatment of fish prior to stocking. Periodic checks of fingerlings will indicate if the parasite is present. Treatment should be initiated on a prophylactic basis. Thinning fish with addition of fresh water in conjunction with treatment is an excellent approach to control.

Trichodina.—*Trichodina* species are circular flattened organisms with a prominent denticular ring. The organism is ciliated and moves rapidly over the surface of gills, fins, and body of fish. Figure 7.22 shows the prominent denticular ring and cilia. The diameter of trichodinids can vary from 120 microns to that of smaller genera such as *Trichodonella*, which can be as small as 15 microns in diameter.

Host Distribution.—These parasites can be found in all species of cold and warmwater cultured fish. Trichodinids also occur on saltwater fish.

Signs of Disease.—White to whitish-gray blotches can occur on body and fins of fish. Heavily infested fish behave as if there is an oxygen depletion. They may show signs of lethargy and congregrate by incoming water or remain listless by the edge of the pond. Infestations of pond reared aquarium fish can be detected by placing small fry in a glass container and noting milky opaqueness on the tail fins. Treatment effectiveness has been judged by the loss of the opaqueness. Small hemorrhages may be observed on the skin. Many fish can carry this parasite without signs of disease, and many fish pathologists suggest that the problem is related to poor water quality conditions which permit the

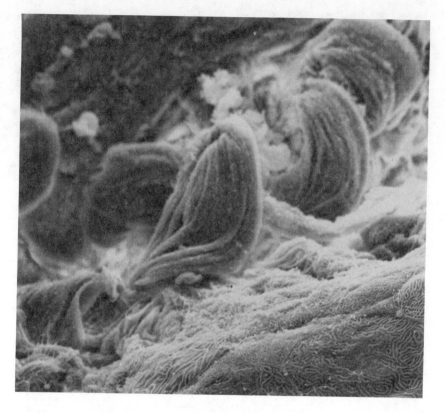

FIG. 7.21. SCANNING ELECTRON MICROGRAPH OF SURFACE OF CHANNEL CAT-FISH GILL INFESTED WITH *CHILODONELLA*

proliferation of the parasite. That poor water quality is associated with the problem is supported by the observation that parasites will spontaneously drop off fish placed into fresh flowing water.

Smaller forms of trichodinids have been observed in channel catfish fingerlings and various tropical species, as well as in marine species. It is generally believed that the smaller varieties such as *Trichodonella* are more pathogenic. Microscopic observation of the smaller types suggests that they are "cupped" over the secondary lamellae of the gills. While a denticular ring may be present, it is difficult to visualize.

Control and Treatment.—Trichodinids frequently affect fry in production ponds. Rapid inflow of fresh water along with the use of chemical treatments will solve the problem.

Scyphidia, Glossatella.—Vase-shaped organisms approximately 50 ×

FIG. 7.22. SCANNING ELECTRON MICROGRAPH OF VENTRAL ASPECT OF A SINGLE *TRICHODINA* SP.
Note the prominent denticular ring and two rows of cilia.

20 microns. *Scyphidia* has 2 rows of cilia, one being located around the mouth parts (peristome) and the other midway between the peristome and attachment organ (scopula). *Glossatella* has one row of cilia at the peristome. Figure 7.23 shows a colony of *Scyphidia* on the skin of a channel catfish fry. Figure 7.24 shows the double row of cilia and the scopula attached to the skin of a channel catfish fingerling. Parasites attach to gills, skin, and fins of fish. They are occasionally found in pond cultured tropical fish.

Host Distribution.—They are found in all cultured warmwater fish species with occasional infestation in salmonoids. They are mostly a problem of young fish.

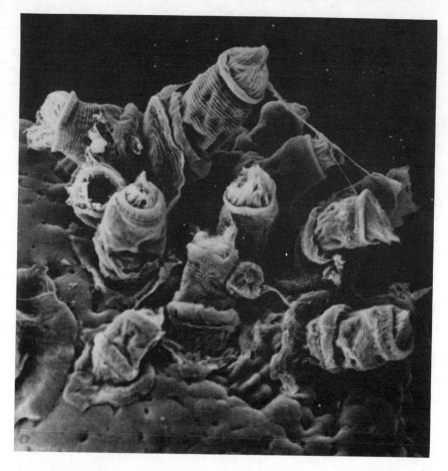

FIG. 7.23. SCANNING ELECTRON MICROGRAPH OF A COLONY OF *SCYPHIDIA* ON THE SKIN OF A CHANNEL CATFISH FRY

Signs of Disease.—Fry will not eat and may accumulate at the surface of the water or seek shallower water. Gills may appear enlarged, pale, and with pinpoint hemorrhages. Hemorrhages may be seen on the body.

Control.—The parasites can normally be found on the surfaces of plants, rocks, or debris in a pond. They attach to fish with their scopula; food such as bacteria and small pieces of organic matter is taken in through the apically located mouth parts. Overcrowding, leading to high organic content of the water, provides optimal conditions for the proliferation of the parasite. In a case presented to the University of Georgia, channel catfish swim-up fry were heavily infested in rearing troughs supplied by

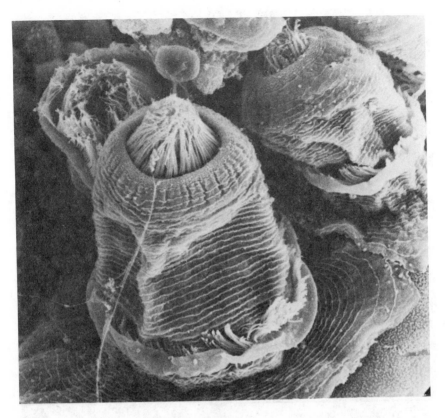

FIG. 7.24. SCANNING ELECTRON MICROGRAPH OF A SINGLE *SCYPHIDIA*
Note double row of cilia and scopula attached to skin.

recirculated water from a pond where adult fish were held. Treatment of fry with formalin baths prior to release in ponds resulted in significant increases of fingerling production. Obviously, control can be effected by ensuring that hatchery water is taken from ponds without "carrier" fish. If fry are infested, treatment with formalin baths will remove the parasite. In the case cited it was apparent that stocking fry with *Scyphidia* cuts their production significantly although no dead fish may be seen. Experience at the University of Georgia with treatment of fry prior to release suggests that production can be doubled. Pond treatments should be accompanied with thinning of fish and inflow of fresh water.

Epistylis.—These are stalked parasites generally found on the body of the fish. A row of cilia at the apical end of the organism is used to draw food into the main body of the parasite. In wet preparations made from skin scrapings, the main cell body is frequently separated from the stalk portion. Typically, the stalks will contract periodically, which is an aid in the identification of the organism.

Host Distribution.—They are frequently found on rainbow trout fingerlings in the western United States and have been seen in trout from north Georgia; found frequently on bass; less frequently on goldfish. They are frequently found in septic wounds on fish, possibly as secondary opportunistic parasites. Imported *Plecostomus* ornamental species have developed *Epistylis* colonies on the surfaces of their bodies. The parasite is frequently found to infest eggs. The organism is illustrated in Fig. 7.13F.

Signs of Disease.—When present in large numbers, they will cause fish to flash. In bass, hemorrhages and excessive localized mucus production are commonly seen. Infested eggs appear to be "fuzzy."

Control.—In bass in the southeastern United States, *Epistylis* infestations are thought to be associated with water highly polluted with organics. Increased temperature of the water favors reproduction of the parasite. Control of heat and organic pollution will minimize the problem. Treatments are covered in a later section.

Tetrahymena pyriformis.—*Tetrahymena pyriformis* is considered to be a free-living parasite most frequently found in stagnant waters. In closed aquatic systems (and home aquaria) these parasites can be observed by microscopic examination to be associated with decaying organic matter such as uneaten food or dead fish. The parasite is pear shaped, measuring approximately 20 microns in length, with longitudinal rows of cilia (Fig. 7.25). A consistent characteristic feature is the circular vacuoles found within the organism as illustrated in Fig. 7.13H. Unlike other ciliates infesting the skin of fish, *Tetrahymena* tends to move rapidly in a straight line with frequent abrupt changes in direction.

Host Distribution.—*Tetrahymena pyriformis* is a natural inhabitant of stagnant waters and can attack stressed fish. Case studies at the University of Georgia suggest that the organism is a frequent secondary complicating factor following bacterial skin infections. The parasite often infests guppies and other livebearing ornamental species. A counterpart has been observed to infest the skin of marine ornamental species.

Signs of Disease.—White areas may be seen on the fish. These areas may be intermixed with hemorrhagic areas. The lesion may be localized or may extend as a whitish band around the body of the fish. Frequently, eyes of infested fish will appear to have a white rim which, when examined microscopically, will contain hundreds of organisms. The organisms

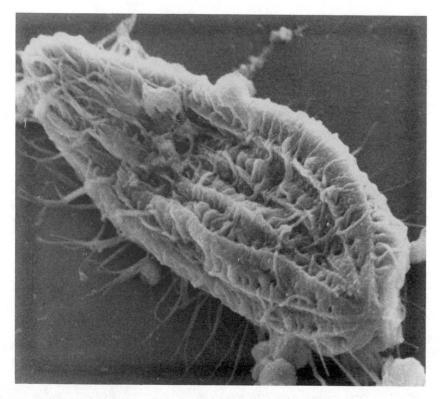

FIG. 7.25. SCANNING ELECTRON MICROGRAPH OF *TETRAHYMENA PYRIFORMIS*

can invade the tissue proper; cases examined at the University of Georgia have been seen where the parasite was found to invade the kidney. It has also been found in the brain. Fish appear listless at the bottom of the aquarium.

Control.—Control of the parasite is effected by reducing the conditions favoring the growth of the organism as well as by minimizing stress to the fish. Minimizing organic buildup in ponds or aquaria will minimize the natural growth of the parasite. Since the infestation is most often considered to be secondary to some stressful change, avoidance of any source of stress will be beneficial.

Treatment is difficult since organisms are able to penetrate the tissues of the fish which protect the parasites from various parasiticides. Formalin, malachite green, and copper sulfate have been used for treatment.

Trichophrya.—*Trichophrya* is a protozoan parasite belonging to the class Suctoria (see Fig. 7.13I). These parasites attack the gills of fish. On microscopic examination of gills, the organisms appear reddish-orange, nonmotile, with characteristic tentacles. The organisms are from 40 to 70 microns long (Fig. 7.26). Their role as a primary pathogen is conjectural.

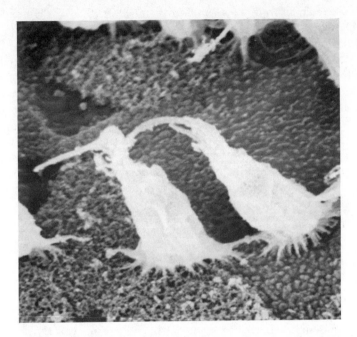

FIG. 7.26. SCANNING ELECTRON MICROGRAPH OF TRICHO-PHRYA ATTACHED TO GILL TISSUE OF CHANNEL CATFISH FINGERLING

Host Distribution.—This parasite is widespread in warmwater fish culture and has been found in catfish and bass fingerlings. This organism has not been recognized as a problem in ornamental fish farming operations.

Signs of Disease.—Where parasites are numerous, fish are off feed, lethargic, and may accumulate around inflowing water, suggesting anoxia. Infested gills may have a tendency to collect silt due to the presence of the organisms. Diagnosis is by microscopic examination and identification of the parasite.

Control.—Fish may be infested without showing signs of disease. Decision to treat will depend on severity of infestation and presence of clinical signs of disease. Copper sulfate, potassium permanganate, and formalin have been used to treat fish.

Oodinium.—*Oodinium* species have been described in freshwater, brackish, and marine fish culture (Fig. 7.13C). The parasites are classified as dinoflagellates. The disease is often referred to as "velvet" disease since the organisms may be so numerous on the surface of the fish that a fine yellowish sheen may be seen. The organism attacks gills (Fig. 7.27) and skin (Fig. 7.28) and may be found attached to the intestinal mucosa. It has a complex life cycle. The adult parasitic stage is pear shaped and varies in length from approximately 20 to 140 microns, depending on its stage of maturity. The organisms are attached to the tissues by root-like appendages. The adult stage is nonmotile, frequently having a yellow hue and frequently found in clusters. Young forms have an ovoid nucleus. A chitinous capsule surrounds the organism. After the organism reaches maturity, it drops off the fish and begins to multiply, forming motile dinospores which die if a host is not infested within 24 hr. On penetration, the flagella disappear and root-like appendages anchor the parasite to the tissue while maturation continues.

Host Distribution.—This is mostly a problem of fish in closed fish culture systems. Marine and freshwater ornamental species are susceptible.

Signs of Disease.—Afflicted fish may scrape themselves or, if the gills are heavily infested, show signs of suffocation including increased respiratory movements and gasping. The maturing organisms on the surface of the skin can be best visualized by use of a flashlight to inspect the fish. The light is directed to the back of the fish from directly above the aquarium. The *Oodinium* organisms are accentuated by the reflected light. This technique works best in darkened rooms. Diagnosis is confirmed by microscopic examination.

Control.—Dinospores can be spread from aquarium to aquarium by water on nets, plastic bags, or plants. Control should be directed to routine quarantine and treatment regimens, such as by avoidance of transfer of water between aquaria and by maintenance of low copper levels (0.11 to 0.18 ppm).

Hexamita, Spironucleus.—These organisms are small (10 microns) flagellated protozoans found in the intestinal tract of fish. The organisms are pear to oval shaped and have 8 flagella (Fig. 7.29). On microscopic examination of intestinal contents the organisms are seen to be actively motile. Flagella are best observed with phase contrast microscope.

Host Distribution.—Many fish pathologists believe that the organism

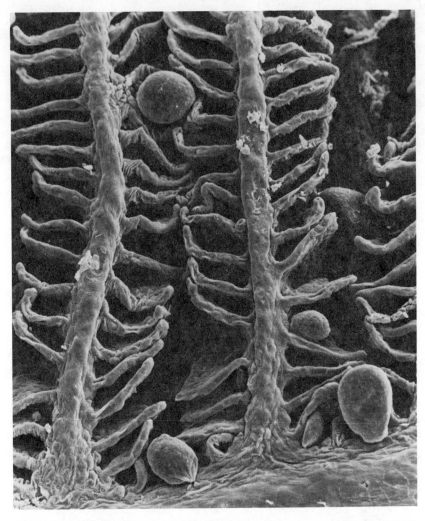

FIG. 7.27. SCANNING ELECTRON MICROGRAPH OF THE GILLS OF A GLASS CAT-
FISH (*KRYPTOPTERUS BICIRRIS*) INFESTED WITH *OODINIUM*
Note pear shape of organisms and physical interference with gill structure.

Hexamita salmonis does not cause disease in trout or salmon; however, in young salmonoids the condition known as "pinheads" has been attributed to the parasite. At the University of Georgia, flagellated intestinal protozoans have been found in goldfish (*Carassius auratus*), various types of gouramis (*Colisa* spp.), angelfish (*Pterophyllum scalare*) and discus fish (*Symphysodon discus*). The parasite commonly associated

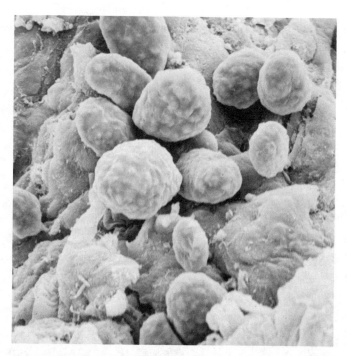

FIG. 7.28. SCANNING ELECTRON MICROGRAPH OF THE SKIN OF A GLASS CATFISH (*KRYPTOPTERUS BICIRRIS*) INFESTED WITH *OODINIUM* ORGANISMS

with angelfish and possibly discus fish has been called *Spironucleus elegans.*

Signs of Disease.—In young trout the organism has been found in the intestine, gall bladder, and blood, which suggests that it is pathogenic. Signs described in trout include pinheadedness, wasting, inappetence, and death.

Angelfish fry 10 days of age have been infested. The fish remain at the bottom of the aquarium, refuse to eat, and gradually waste away until death.

In young angelfish, as well as some gouramis and cichlids, the condition is characterized by a wasting disease accompanied by refusal of food and death. It is not clear whether or not *Hexamita* is responsible for the disease in discus fish and large red oscars referred to as "hole in the head." There is no doubt that both *Hexamita* and *Spironucleus* do at times enter into the bloodstream and, either alone or in combination with the bacteria, may contribute to the "hole in the head" syndrome. Treated angelfish have been shown to grow faster than untreated controls.

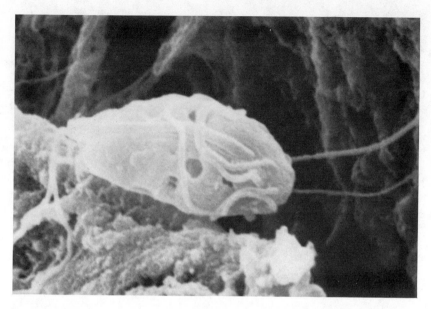

FIG. 7.29. SCANNING ELECTRON MICROGRAPH OF A SECTION OF SMALL INTES-
TINE OF A BLACK ANGELFISH (*PTEROPHYLLUM SCALARE*) INFESTED WITH *HEXA-
MITA*
Note flagella.

The parasite may be responsible for poor hatchability of angelfish eggs.

Diagnosis.—Diagnosis is made on the basis of the observation of wasting, inappetent fish coupled with microscopic examination of intestinal contents, gall bladders, liver, and heart blood wet mounts. Numbers of organisms per microscopic field of intestinal contents can be remarkably high.

Control.—Control can be effected by incorporation of drugs such as Metronidazole in food. Spironucleus infestations in infested pond-raised ornamental fish can be treated in the same manner. Metronidazole at a rate of 3 to 5 ppm in aquarium water will rid fish of the parasite. Since these parasites can live for extended periods of time in polluted water, keeping aquaria cleared of excess organic materials will tend to reduce the populations of the parasite. Breeders of angelfish, discus, and various species of ornamental cichlids and gouramis should check breeding stock for *Hexamita* on a regular basis.

Cryptobia, Trypanosoma.—*Cryptobia* and trypanosomes are flagellated protozoans found in the circulatory system of fish. Trypanosomes have one flagellum; cryptobia are biflagellate. Both are elongated with an undulating membrane present associated with one flagellum. Length may reach 30 to 50 microns. Both are spread by leeches.

Host Distribution.—The parasites have been reported in a variety of marine and freshwater fish including carp, eel, and rainbow trout. A trypanosome species has been consistently found in the blood of imported *Loricaria,* an ornamental catfish variety from South America.

Signs of Disease.—In California in rainbow trout brood fish, the disease is characterized by anemia, popeye and ascites (swollen abdomens). The ascites is presumably caused by upset of normal kidney function by the parasite. Other signs attributed to these blood parasites are general weakness or "sleeping sickness"—a common name used for the mammalian counterpart disease. Most infected fish do not show signs of disease.

Control.—Control of leeches will prevent spread of these parasites.

Sporozoan Diseases

Sporozoans are a class of protozoans which are the causative agents of some of the most serious diseases in fish culture. In general, the organisms are recognized by the morphology of their spores (Fig. 7.13M, N, and O) and by the number and location of polar filaments. The group includes the coccidia, which are of minor importance when compared to the myxosporidians, which have two polar capsules, and the microsporidians, which have one polar capsule.

The myxosporidians and microsporidians have some general features in common which include a tendency toward host and tissue specificity; both have a complex life cycle, and both are untreatable.

The signs of disease vary depending on the particular fish and species of parasite. Lesions may include deformed heads (*Myxosoma cerebralis*) of trout, blanched musculature in ornamental species, "milk scales" in golden shiners, and discrete white nodules. Microscopic examination of squashed wet mounts of such areas will reveal many spores frequently encased in packets called pansporoblasts.

The life cycles of the micro- and myxosporidians are complex. The cycle is initiated by the spore which on the death of an infected fish may remain viable in mud, gravel, or filters for long periods of time. Spores are probably eaten by fish, and the effect of the digestive juices may assist the spore in releasing its polar filament which serves to attach the parasite to the cell. The nuclear material (DNA) of microsporidia passes through the polar filament into cells at the site of the intestinal epithelium. Transfer of the parasite to the eventual site of the visible lesion may be by infected white blood cells.

Myxosporidian spores use their polar filaments for attachment to a cell, after which the internal part of the spore (sporoplasm) emerges as an ameboid form and and actively penetrates a cell of the intestinal lining, eventually making its way to its final destination directly or as a passenger in a white blood cell. At this site, the parasite, now called a trophozoite, continues to divide (shizogony) and fuse (sporogony), eventually resulting in masses of spores which are responsible for the pathology observed in the afflicted fish.

Control and Treatment.—There is no known effective treatment for the micro- and myxosporidian diseases. Control measures have been directed toward avoiding the infective spores by various methods. These will be discussed under each disease discussed below.

Myxosoma cerebralis.—This myxosporidian was not found in the United States until the mid-1950s when it was thought to be introduced by imported infected trout. The disease is restricted to various species of trout. The trophozoites have an affinity for cartilaginous areas of the head and spine which results in misshapen head and spine (scoliosis) and frequently a black tail due to nerve destruction. Fish will frequently swim in a tail-chasing fashion due to nervous system damage. This behavioral change suggested the common name, "whirling disease."

Deformed survivors are common but should not be sold, stocked or used as brood stock. Diagnosis is based on signs of the disease and positive identification of the spores by qualified experts. In Europe, where the disease is endemic, replacement of earthen raceways with concrete raceways has eliminated the disease as a problem, presumably by eliminating mud where spores either are trapped or pass through a necessary maturation period. The disease is endemic in certain eastern states of the USA. The disease can be avoided by purchasing fingerlings or brood stock from hatcheries which are certified free of the disease. In the face of an outbreak of the disease, the Fish Health Section of the American Fisheries Society recommends disposal of all fish followed by disinfection of premises.

Ceratomyxa shasta.—This is a myxosporidian of trout and salmon from western states. The parasite can be found in most tissues of infected fish. Apparently, the parasite in California is found in specific drainages having a lake or impounded water as part of the system. It has been demonstrated that certain strains of trout show an increased resistance to the parasite. No treatment exists, but use of resistant strains of trout in infested drainage areas may circumvent the problem.

Myxobolus notemigoni.—This myxosporidian causes "milk scale" disease of golden shiners. The fish appear to have white elevated areas, often covering the entire body. The parasite causes minimal mortality but damages the appearance of the fish. Diagnosis is made on the basis of the appearance of the fish and confirmed by a microscopic search for typical spores. Control of the disease can be effected by killing spores which by definition are resistant to many chemicals. According to Hoffman and O'Grodnick (1977), thorough air drying of contaminated mud will kill spores of *Myxosoma cerebralis*; and this practice may work for other myxo- and microsporidians. They were able to show that 10 ppm of chlorine would kill spores in water but levels of chlorine as high as 1200 ppm did not kill spores in mud. It would seem that partial drainage of ponds followed by addition of chlorine and eventual drying would radically reduce spores.

Henneguya.—The spores of this myxosporidian are characterized by having two elongated tail-like appendages (Fig. 7.13N). They are found in many species of freshwater fish. Signs of the disease are generally noted as white cyst formations in skin, muscles and gills.

In channel catfish, *Henneguya* causes white cysts in the skin and also can be found as cysts in the gills. *Henneguya* infestations have been observed in cardinal tetras (*Cheirodon axelrodi*) imported from South America where the principal signs of disease were discrete white circumscribed areas distributed throughout the skin of the fish. The fish were useless for commercial purposes. Control measures should include draining, drying, and disinfecting ponds prior to restocking.

Plistophora ovariae.—These microsporidia are a problem in golden shiners. The spores can be found in the liver and kidneys, but the principal problem is growth in the ovarian tissues which reduces the spawn. Infected fish frequently show reduced weight and growth. Serious consequences from the parasite infestation have been avoided by management practices. Breeder fish from high production ponds are carried over to the next breeding season while breeders in low production ponds are not used, with the assumption that fish in low production ponds are diseased. Use of younger stock as breeders has minimized the problem since it appears that older fish are more likely to be infested.

Plistophora hyphessobryconis.—This microsporidian is the causative agent of "neon tetra disease" which is an unfortunate name since other species of fish are susceptible to the parasite. Typical signs of the disease in neon tetras and other ornamental fish is a whitening of the musculature with a loss of color in the affected areas. Under pond conditions, afflicted fish will be noted by their whitish color when viewed from above against the dark background of pond bottom. Attempts to net a school of fish will invariably result in capturing a greater percentage of infested fish, probably because of reduced swimming efficiency due to muscle destruction.

The diagnosis is confirmed by microscopic examination of wet mounts of muscle. Spores with dark-appearing pansporoblasts are easily identified using 40 magnification objectives (Fig. 7.13M).

A microsporidian closely resembling *P. hyphessobryconis* has been observed in Florida raised tiger barbs (*Puntius tetrazona*), as well as other varieties of ornamental species by University of Georgia researchers. While this parasite has been found in musculature, it has also been seen in ovarian tissues, suggesting that breeding efficiencies may wane if the parasite is established in breeding stock. In actual fact, evidence suggests that the lack of production in certain Florida ornamental fish ponds may be associated with *Plistophora* infestations or early mortalities in fry.

Plistophora infestations of ornamental fish are frequently complicated by bacterial infections.

Control and Treatment.—No treatment for the disease is available. Claims are frequently made within the ornamental fish industry that the disease can be treated. In all cases of reputed treatment, the initial dose of drug (Formalin) will kill the most seriously ill fish, leaving the subclinical fish to live on and die at a later date. Our experience suggests that at least 25% of normal appearing fish in an infested group of fish will be harboring the parasite as subclinical carriers. These fish are probably destined to die as the disease progresses.

Control of the disease in a fish pond should be possible by adhering to the good management practices of draining and drying; however, many Florida ponds cannot be effectively drained or dried. An alternative method would be to line smaller ponds with plastic sheeting to avoid spores. This could be done on selected "low production" ponds, on ponds known to be infested, or on ponds used to grow breeding stock.

Coccidiosis.—In carp farms in Europe, *Eimeria cyprini* has been found to affect the intestinal mucosa of young carp, resulting in emaciation, sunken eyes, and general debility. The diagnosis is made by microscopic examination for intestinal scrapings and by identification of typical sporozoites. Control of the disease in Europe is by draining ponds and liming.

Other *Eimeria* spp. have been found in gouramis and in marine fish, where they have been found in kidneys, gonads, and livers.

Members of the genus *Hemogregarina* are intracellular blood parasites of various fish and diagnosed by microscopic examination of blood smears.

Monogenetic Trematodes

Monogenetic trematodes are worm-like parasites, most frequently found infesting the gills or skin of fish. The term monogenetic refers to the fact that these parasites have no intermediate hosts and, with the exception of egg laying types, spend their entire life on the host. The parasites range in length from 50 microns to 3 mm. The parasites attach themselves to the skin or gills by specialized organs of attachment consisting of hooks, spine-like anchors, or clamps (Fig. 7.30). There are

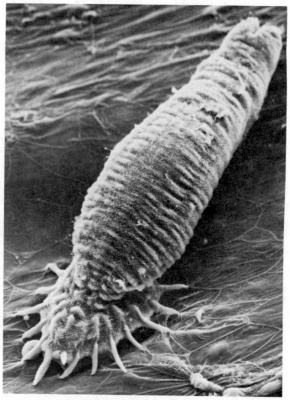

FIG. 7.30. SCANNING ELECTRON MICROGRAPH OF MONOGENETIC TREMATODE ON TAIL FIN OF A GOLDFISH

hundreds of species of monogenetic trematodes of which the most common belong to the following genera: *Gyrodactylus, Dactylogyrus, Cleidodiscus,* and *Benedinia.* Genera are differentiated by morphology and reproduction. The many species of these genera are classified by morphologic characteristics such as sucker arrangement or number of hooklets. Some types of monogenetic trematodes may be egg layers (Dactylogyridae) while others are live bearers, giving birth to a single individual (Gyrodactylidae).

In general, monogenetic trematodes are not problematic in wild fish. Under relatively crowded fish culture conditions the parasites can attain high enough populations to cause pathologic changes and signs of disease.

Behavioral changes in fish suffering heavy infestations may include flashing or gill scraping. Signs of infestation may include localized hemorrhages, which may appear to be fringed with whitened areas. Heavy infestations may be followed by secondary bacterial invasion leading to systemic bacterial disease or enlarged localized ulcers.

Host Distribution.—*Salmonoids.*—*Gyrodactylus* spp. are found in all species of salmonoids under culture. Fingerling and catchable size rainbow trout appear to be the most severely infested with most problems occurring during winter and spring periods. One or more fins of infested trout may have whited edges which on microscopic examination will reveal many parasites. Gyrodactyli will have internal developing embryos.

Goldfish.—*Gyrodactylus* infestations may cause serious problems in goldfish, especially during winter months. Most parasites are found on the skin and fins. Concurrent infestations of *Dactylogyrus* on gills are common.

Research at the University of Georgia suggests that *Gyrodactylus* may harbor the bacterium *Aeromonas salmonicida*, which has been implicated as an initiating cause of ulcer disease of goldfish.

Channel Catfish.—*Cleidodiscus* is associated with the gills of channel catfish. Severely affected fish may have pale gills with increased mucus production. Fish may appear weak and may scrape gills. More seriously afflicted fish may show signs of anoxia. Lesions induced by worm infestations frequently are invaded by *Flexibacter columnaris*. *Cleidodiscus* can be recognized by the 4 eye spots at the anterior end of the parasite, presence of eggs, 4 anchors, and 14 hooklets on the attachment organ.

Golden Shiners.—These fish, like goldfish, can be infested with *Gyrodactylus* and *Dactylogyrus* with similar pathological effects as in infestations in other fish.

Cultured Ornamental Fish.—A variety of species of monogenetic trematodes have been observed on imported ornamental species of fish, as well as many fish species cultured in Florida. Most damaging are the egg laying species found on gills; however, heavy infestations of *Gyrodactylus* spp. have been noted in livebearers. Signs of disease include localized hemorrhages, ulcerations, and complications with secondary bacterial infection.

Marine Fish.—Several species of monogenetic parasites affect fish in marine aquaria. Whereas the parasites rarely cause problems in the native marine environment, confinement in a marine aquarium intensifies the parasite load. There are numerous genera of parasites affecting marine fish and over 100 species. Members of the genera *Benedinia* and *Microcotyle* are common. Marine fish have been presented to the University of Georgia with hundreds of trematodes on the gills. The fish died with typical signs of anoxia—open mouths and flared opercula. Fish with species infesting the skin will exhibit flashing and scraping motions.

Control and Treatment.—Avoidance of monogenetic trematodes can be done by good management practices. Known infested ponds should be drained, limed, and dried prior to restocking with treated fish. Infested wild fish can act as a reservoir of infestation for cultured fish if the water source is from a lake or stream. In such cases, routine treatment may be required. In all cases where new fish are introduced into a hatchery, they should be examined prior to stocking. Such fish should be held and quarantined, during which time treatments can be administered. Formalin baths or treatment with organophosphates have been shown to be effective (see section under treatment).

Digenetic Trematodes

Digenetic trematodes are hermaphroditic, flat worms with two suckers—one located at the "head" of the parasite and the other roughly at mid-body. The term "digenetic" refers to the complex life cycle which usually involves snails and frequently a bird. Fish may serve as the final host for digenetic trematodes, in which case they will be found in the lumen of the gastrointestinal tract or, in the case of blood flukes, in the arteries leading from the heart to the gills. In the case of the intestinal inhabiting forms, little, if any, damage is done to the host even though eggs are shed which perpetuate the infestation. Eggs develop into a ciliated larval form called miracidia which penetrate a snail, eventually finding their way to the liver, where the miracidium develops into a second larval stage called a sporocyst. Within the sporocysts are produced numerous *rediae* which in turn develop into many *cercariae,* which is the larval form of the adult worm. The cercariae, which have a tail-like structure, leave the snail and penetrate a fish, during which process the tail is lost. In cases where the fish is the final host, the miracidium is swallowed, develops into the adult fluke in the intestine, and the cycle is completed. In the case of blood flukes (*Sanguinicola* spp.), the miracid-ium penetrates the gills and makes its way to the heart and blood vessels, where it develops into a mature worm.

In the majority of digenetic trematode infestations, however, the fish serves as the secondary intermediate host, with the first intermediate host being a snail and the final intermediate host being a bird or mam-mal. In these cases, the adult forms are found in the intestinal tract of birds or mammals where eggs are released which develop into miracidia, which attack snails. Released cercariae from snails attack the fish by penetrating the skin and encyst in the skin and internal organs as metacercariae.

The cycle is completed when the final host eats the fish. The meta-cercariae are released in the digestive tract and develop into the adult form.

Host Distribution.—Trout can serve as the final host of *Sanguinicola davisi*. The parasite locates in the arteries between the heart and gills. The snail *Oxytrema circumlineata* serves as the intermediate host.

Golden shiners, bluegills, and green sunfish serve as the second in-termediate host for *Uvulifer* sp. where the final hosts are kingfishers and the first intermediate hosts are *Helisoma* sp. snails. The encysted meta-cercariae are encysted in fins, skin, and muscles where they appear as small black nodules, hence the common name "black spot" disease. Meta-cercariae located in internal organs appear as discreet white spots.

A wide variety of freshwater fish, including cultured tropical varieties, serve as second intermediate hosts for *Clinostomum* spp. where the metacercariae form cysts in the skin and internal organs. The disease is called "yellow grub"; herons serve as the final host.

Digenetic trematodes have not been a problem in channel catfish cul-ture.

Ornamental fish cultured in Florida ponds frequently serve as inter-mediate hosts for digenetic trematodes. Adult forms are most likely parasites of various predatory water birds which are a major menace to Florida tropical fish producers. Infested fish have 1 to 3 mm white spots on the skin, gills, and internal organs.

Wild caught ornamental fish imported from South America or Asia frequently will be infested with internally located metacercariae. In some cases the infestation is very heavy and is responsible for the destruction of tissues of internal organs. In South American imported tropical fish, this problem appears to be seasonal, with the greatest amount of infes-tation being toward the end of the dry season. It also appears that in South America, fish taken from small rivers are less likely to be infested than those captured in larger ponds—possibly due to the fact that aquatic birds are less likely to feed in rivers canopied by thick forests.

Signs of Disease.—White, yellow or black distinct nodules are located in the skin (grubs). In some cases involving ornamental fish, metacercariae may be found only in gill tissues or only in interior organs, in which case they are most frequently found in mesentery tissues adjacent to the intestine or liver. In some cases, heavy infestations are seen in the liver and kidneys.

Control and Treatment.—Snail control is the best method of controlling the infestation. Prior to stocking, ponds should be dried and limed. Various methods for controlling aquatic birds have been used with variable results. Since herons do not land directly in ponds, a low wire fence near the edge of the water or a few wire strings around the pond will serve to discourage the birds. Noise-making machines resembling guns have discouraged birds for a period of time but eventually lose their effectiveness. Migratory aquatic birds are protected by law.

Cestodes (Tapeworms)

Tapeworms, like digenetic trematodes, have a complex life cycle and may be found within the intestine of a fish as a sexually mature adult in which the fish is the final host. Larval forms may be found encysted in various internal organs or free in the visceral cavity, in which case the fish is an intermediate host, where the final host may be other fish, mammals, or birds.

Tapeworms typically are hermaphroditic, flat, segmented, and with a head (scolex) equipped with four suckers and occasionally hooks. Tapeworms are classified by whether or not they are segmented, by the form of the suckers, or by the arrangement of hooks on the scolex. Each segment (proglottid) contains a full component of male and female sexual organs; eggs develop within the segments and are shed periodically in the feces. When eggs reach the water, a ciliated larva called a coracidium is released and eventually eaten by a copepod inside which an unsegmented larval form called a procercoid develops. Eventually, the copepod is eaten by a fish where the procercoid is released, growing into an advanced larval form called the pleurocercoid. The pleurocercoid may remain within the intestine but more frequently migrates to the visceral cavity, liver, or muscles. Certain classes of pleurocercoids become very long with a segmented appearance and literally pack the abdominal cavity. If the final host is a larger mammal or bird, the pleurocercoids are taken in with ingested fish and develop into adults in the intestinal tract.

Host Distribution.—Adult tapeworms and larval forms are widespread in wild fish and common in cultured fish. A few examples are presented in the following paragraphs.

Channel Catfish.—Corallobothrium is frequently found in the posterior intestine of channel catfish fingerlings. Infestations can become very severe but there is some question whether the parasite causes a problem. Claims have been made that heavily infested fry are stunted. Fry become infested by eating copepods harboring the larval stages of the para-

FIG. 7.31.   SCANNING ELECTRON MICROGRAPH OF A TAPEWORM REMOVED FROM THE LARGE INTESTINE OF A CHANNEL CATFISH. ACTUAL SIZE IS LESS THAN 12 MM

site. Figure 7.31 is a scanning electron micrograph of *Corallobothrium* from a channel catfish fingerling.

Golden Shiners.—*Bothriocephalus achaeilognathi,* frequently called the Asian tapeworm, can kill golden shiner fingerlings by occlusion of the intestine. In some cases, the forepart of the intestine is ruptured by the occluding tapeworms. The tapeworm is also a problem in fathead minnow culture.

Ornamental and Marine Fish.—Marine ornamental fish may be the final host for tapeworms or may serve as intermediate hosts. Marine ornamental fish submitted for necropsy examination at the University of Georgia have been found to be heavily parasited both with adult tapeworms and in many instances with pleurocercoids. Whether or not these infestations harm the host is an open question.

Freshwater ornamental fish from Florida as well as from South American exporting countries will occasionally harbor pleurocercoids in small numbers. Adult forms are infrequently found in South American imported ornamental species.

Control and Treatment.—Control of adult tapeworms in fish can be accomplished by treating held-over brood stock with di-n-butyl tin oxide and draining and liming ponds. In catfish breeding stock, a bolling gun (used for administration of capsules to small domestic animals) can be used to treat the fish. Treatment of smaller fish must be done by incorporating the drug in the food. Since free-living copepods such as *Cyclops* or *Diaptomus* or tubifex worms can act as the intermediate host, aquarists are adivsed to be wary of feeding this type of live food.

Nematodes

Nematodes, often referred to as roundworms, are thin elongated worms which, when viewed through a microscope, are round with a smooth surface (Fig. 7.13V). Many genera and species of roundworms infest fish in the wild and under culture conditions; however, most infestations do not result in visible disease signs.

Nematodes are usually found within the intestine of the host fish. Larval forms as well as adult forms can be found in the liver, visceral cavity, swim bladder, muscles, and other internal organs. Many species have one intermediate host which may be an invertebrate, such as *Daphnia* or *Cyclops.*

Host Distribution.—Many genera and species of nematodes parasitize a wide variety of wild and cultured fish.

Signs of Disease.—Members of the family Dracunculoidea are frequently blood parasites or move about freely in the peritoneal cavity of fish. In bluegills, *Philometra* sp. can be found in the eye and brain with resulting convulsions and death. These roundworms appear blood red in color. Species of *Philometra* occur sporadically in various tropical varieties and are found as red worms in the abdominal cavity. These fish may appear bloated. Species of the genus *Camallanus* are frequently found in a variety of fish. The worms frequently extrude from the anus of the fish and of livebearing ornamental varieties. Heavy infestations may lead to emaciation. Various types of ornamental fish have been shown to be heavily infested with members of the genus *Capillaria* and heavy infestations of this genus also lead to emaciation. Many types of nematodes are found as larval forms and may appear as tumor-like swellings in the musculature.

Control and Treatment.—Larval forms are untreatable. Intestinal infestations can be treated by incorporation of specific medicants in food. Since various invertebrates (*Cyclops,* etc.) can harbor intermediate stages, feeding of live foods from pond waters can introduce the worms into freshwater aquaria.

Acanthocephalans

Acanthocephalans, also called thornyheaded worms, are characterized by a proboscis with hooks (Fig. 7.13T) at the front end of a bulb-shaped body. Adult forms are found in the intestine of fish where the "nose" is embedded in the intestinal lining of the fish. Heavy infestations can cause intestinal blockage.

The life cycle of many acanthocephalans is completed when larvae of water insects or copepods ingest the eggs. The cycle is completed when a fish eats the insect. Fish may be the intermediate host for a larger vertebrate such as birds or mammals. In this case, cysts containing larval stages of acanthocephalans may be found in the abdominal cavity of the fish.

Host Distribution.—Acanthocephalans are usually not a problem in cultured freshwater fish. Likewise, the worms are of minor importance in wild caught freshwater ornamental species; however, marine fish are common hosts for acanthocephalans.

Signs of Disease.—Emaciation is common in heavily infested fish. Diagnosis is by examination of intestines and identification of the characteristic spiny proboscis.

Control and Treatment.—Since insect larval forms are the intermediate forms, attention can be directed to the avoidance of this type of feed by aquarists. Suggested chemical treatment is presented in a later section.

Parasitic Copepods

These parasites are small parasitic crustaceans which attach to the exterior surfaces of the body of the fish, including gills. Parasitic copepods are common and can be difficult to control. Many genera of freshwater and marine parasitic copepods exist. Nonparasitic copepods act as intermediate hosts for various roundworms, tapeworms, and acanthocephalans. Many of the parasitic copepods burrow into the flesh and cannot be dislodged by chemical treatment. Parasitic copepods have a complex life cycle which involves mating of the parasites in the water and attachment of the female to the fish with subsequent production of eggs which pass through several distinct larval stages. Overwintering may occur as the adult female or in the copepodid (larval) stage.

Host Distribution.—Various species of fish are subject to infestation with parasitic copepods. Salmonoids, channel catfish, bait fish, goldfish, and pond-raised ornamental fish are susceptible.

Signs of Disease.—Parasitic copepods are visible with the naked eye or use of a hand lens. Fish may exhibit flashing; signs of disease include hemorrhaging from bites (*Argulus*) and ulcerations from infection with secondary bacterial invaders.

Treatment and Control.—Attached forms of parasitic copepods are not easily killed. Treatment is directed toward killing larval forms using organophosphates. Avoidance of the parasites by careful selection of noninfested fish used for stocking or breeding is the best method of control. Individual cases of copepod infestation in aquaria can be managed by careful extraction with forceps. Effective control of larval stages in aquaria is possible by using organophosphates.

Parasitic copepods do not often develop epizootic populations in natural waters but they may cause serious problems to fish culturists. Body conformations vary greatly in this group of parasites and appendages may have been greatly modified or even lost. As a consequence, each form which attacks various fish will be illustrated and described separately.

Argulus spp. are commonly referred to as "fish lice." They have a flattened, saucer-like shape (Fig. 7.13P) and can be observed creeping rapidly about over the body of a fish. If motionless, the parasite resembles a scale. Close examination of individuals will reveal the presence of jointed legs and two large sucking discs for attachment which may give the organism the appearance of having large eyes. When attached to fish, the parasites feed on blood and body fluids. Even large fish may be killed by this organism if it becomes abundant.

Lernaea cyprinacea, the anchor parasite, is found on non-scaled and scaled fishes. Inflamed areas accompany the site of attachment and a secondary bacterial or fungal infection frequently develops. *Lernaea* are firmly attached to their hosts. After the juvenile forms penetrate the host, the appendages of the head become so modified that the parasite resembles an anchor at its anterior end (Fig. 7.13S). These branching protrusions prevent release of the parasite. The parasite frequently is a primary factor in lethal bacterial infections.

Ergasilus sp. is a parasitic copepod which attacks the gills of various fish. This form grossly resembles the free-living copepod, *Cyclops,* but the second antennae are enlarged, terminating in large claws which serve as a means of attachment (Fig. 7.13Q). The parasites feed on blood and body fluids. Secondary infections in the wounds thus caused are common. The parasite is brightly colored. Close relatives which affect gills of salmonoids are *Salminicola edwardsii.*

Bacterial Infections

Bacteria are single celled organisms which have the following characteristics. They are a very small size (0.3 to 0.5 microns), are found everywhere in nature, and fulfill a multitude of natural functions such as decomposition and nitrogen fixation. Many nonpathogenic bacteria are useful in fermentation processes and in other industrial uses. Most pathogenic bacteria of fish can be cultivated on artificial media such as blood agar or trypticase soy agar where visible colonies can be seen with the naked eye. Some bacteria are motile, others are not.

Two large groups of bacteria exist, depending on chemical reactions of the bacterial cell with the Gram stain. Gram-negative bacteria appear pink or red when examined with a microscope. Most fish pathogens are included in the Gram-negative group. These include *Aeromonas, Pseudomonas, Flexibacter,* and *Vibrio.* Gram-positive bacteria appear blue by microscopic examination. *Corynebacterium* spp. are one of the Gram-negative genera which affect fish. Figure 7.32 is a scanning electron micrograph of individual rod-shaped *Flexibacter columnaris* bacteria. The tendency of these bacteria to aggregate into "haystack" formations is shown in Fig. 7.33.

Another large group exists which is called "acid fast" bacteria because, when stained with carbol fuchsin, acid alcohol will not remove the stain. These include mycobacteria which are common infections in fish and which are chronic in nature. Another group of acid fast bacteria is

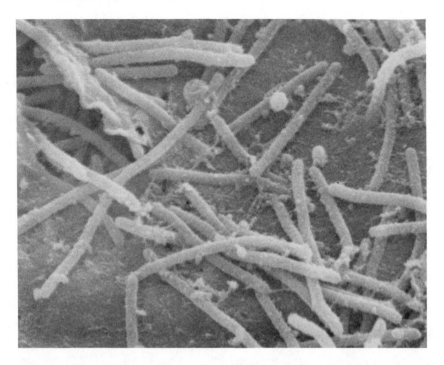

FIG. 7.32.   SCANNING ELECTRON MICROGRAPH OF *FLEXIBACTER COLUMNARIS* ON SURFACE OF SKIN OF INFECTED CHANNEL CATFISH

Nocardia. Like mycobacteria, *Nocardia* produce chronic disease in freshwater and saltwater fish.

Bacteria are classified according to size, motility, colony characteristics, Gram reaction, carbohydrate fermentation reactions, growth on selective media and by specific serological reactions with specific antisera.

Bacteria which are capable of producing disease in fish are almost always present either in the surrounding water, on the surface of the fish, or within the fish where they may be present in the intestine or other internal organs. In the case where bacterial pathogens are associated with a fish and not found free in water, they are called *obligate bacteria.* *Aeromonas salmonicida,* which causes furunculosis in trout, is thought to be an obligate bacterium.

The approach to the diagnosis of a bacterial infection should be to first *exclude* water quality or environmental problems (toxicities, oxygen depletion) which could be the cause of the mortality. Secondly, exclude the presence of external or internal parasites. If water quality is good, with the absence of heavy parasitism, then suspect a bacterial problem.

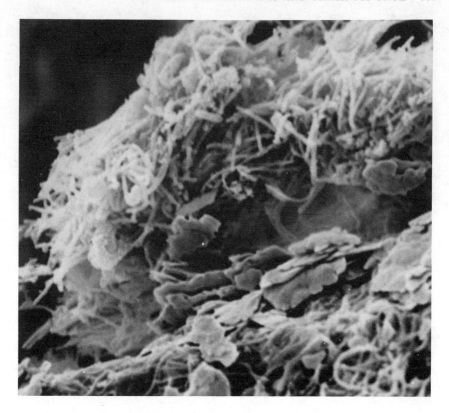

FIG. 7.33. SCANNING ELECTRON MICROGRAPH OF A MASS OF *FLEXIBACTER COLUMNARIS* SHOWING TENDENCY OF THESE ORGANISMS TO FORM "HAY-STACKS"

Bacterial infections do not arise spontaneously but are the result of some stress or series of stresses to the fish which make them more susceptible to bacterial infection.

Stress.—Stress factors of almost any type will reduce the innate resistance of a fish. The basic mechanism by which stress decreases resistance to infection is probably hormonal, resulting in decreased humoral (antibody) response and cellular (phagocytic) response. The weakened fish cannot maintain its normal fight against the ever present bacteria.

Stresses are many and varied. In general, a strss is any change from what is an optimal condition for the fish. Stresses can be arbitrarily divided into chemical, environmental, and biological. Chemical stresses could include low oxygen levels, high carbon dioxide, ammonia, and nitrite levels. Sublethal levels of insecticides and heavy metals could also be stressors. Environmental stresses could include extremes in temperatures, excessive saturation of water with gases (gas bubble disease), or excessive sunlight. Biological stresses could include infestation with external or internal parasites or a lack of a balanced diet.

Most stress conditions are multiple rather than being single. For example, the stresses associated with shipping fisinclude netting, ammonia accumulation, carbon dioxide accumulation, and finally the chance of being subjected to a radically different water quality and a strange environment on arrival.

Diagnosis.—A fish kill after a severe stress or series of stresses is often caused by a bacterial infection. Since many fish with a variety of bacterial infections may exhibit identical signs and lesions, a diagnosis of a *specific* bacterial involvement is likely to be inaccurate. The signs and lesions often seen in bacterial infections include inappetence, lethargy, hemorrhages, fin rot, "mouth fungus," "popeye," dropsy (bloating), blanched areas of the skin, and color changes.

The isolation of the bacteria, while requiring the assistance of a trained expert, is very important in that the bacteriologist, by isolating and identifying the bacteria, can easily predict if the antibiotics used are in fact able to inhibit or kill the bacteria. The test for antibiotic sensitivity will assist in selecting the antibiotic to be used and provides a permanent record which is useful in selecting antibiotics for future use. Continued use of antibiotics can lead to the emergence of antibiotic resistant strains.

Isolation of Bacteria.—Occasionally, the fish culturist may require that diseased fish be cultured for bacteria. It is important to take a kidney culture shortly prior to the death of a fish rather than to submit to a laboratory a dead fish which has been invaded by normal bacteria from the intestine. A bacterial isolation from the kidney of a recently killed fish may be very significant whereas an isolation of a bacterium from a fish which has been dead for an undetermined time is useless.

The equipment required is minimal. A blowtorch is required for sterilization of a standard bacteriological needle or a regular dissecting needle. Sterile bacteriological media can be obtained through a scientific supply house or from local medical pathology laboratories. Plates or tubes of bacteriological media will develop visible colonies on media such as trypticase soy agar or blood agar if bacteria are present. If the sample was taken without contaminating the needle by touching unsterilized skin or other surfaces, development of numerous white colonies after 18 to 24 hr of incubation is significant. Under practical conditions, the inoculated media can be incubated at room temperature or on surfaces which may be a few degrees warmer than room temperature, such as under lamps or near heating registers. A step-by-step method for bacterial isolation is presented below.

Specification of the bacteria is best left to experts who may wish to repeat the primary isolation from a moribund fish. Shotts and Bullock (1975) present detailed laboratory procedures for the identification of bacteria isolated from fish. Figure 7.34 presents an isolation schema for the identification of various Gram-negative bacteria commonly associated with fish diseases.

Necropsy techniques for large fish include:

(1) Wrap fish in paper towel for ease in handling.
(2) Examine for gross lesions (discoloration of skin, hemorrhage, etc.). Smears can be made of skin lesions and Gram-stained.
(3) Strike the fish a stunning blow to the top of the head or kill by an overdose of anesthesia.
(4) Sear area immediately behind dorsal fin with a glowing red spatula, using a rolling motion so sides as well as dorsal surface of the fish are well seared (Fig. 7.35).
(5) Dip scissors in 70% alcohol and then flame.
(6) Insert cooled scissors into the middle of the seared area, directly over the spinal column, push downward until the spinal column is reached, then sever the spinal column (do not cut deeper) (Fig. 7.36).
(7) Bend head and tail together, keeping dorsal fin uppermost.
(8) With the spinal column severed, the surrounding tissue will tear. The dark red mass exposed just ventral to the spinal column is the kidney.
(9) Insert a loop *down into* the kidney tissue (Fig. 7.36) and then streak the culture plates (Fig. 7.37).

Kidney Culture of Aquarium Fish.—Frequently, aquarium fish are so small as to make surface sterilization impossible. This difficulty can be circumvented by the following procedure which should sterilize the external portions of the fish.

(1) Kill the fish by pithing or anesthesia.
(2) Place the fish in a beaker or jar containing 1:5000 Roccal or other suitable surface disinfectant (chlorine, etc.).
(3) Transfer the fish to a small beaker of 70% alcohol. Use forceps previously sterilized by flaming to transfer the fish.
(4) Cut the fish behind the dorsal fin and proceed as described in (6) for larger fish.
(5) *Remember*: The point of external sterilization is to avoid bacteria which are always present on the surface of the fish. Without surface sterilization and care to keep sterility by flaming instruments, results will be misleading.

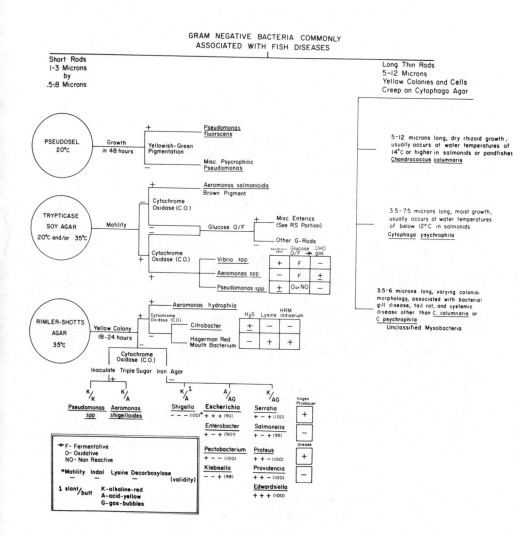

From Shotts and Bullock (1975)

FIG. 7.34. SCHEMA FOR IDENTIFICATION OF GRAM-NEGATIVE BACTERIA ASSO-
CIATED WITH FISH DISEASE

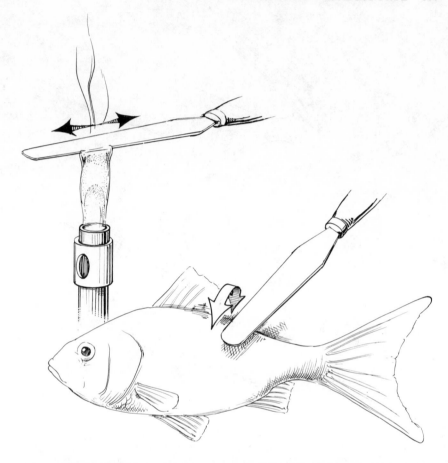

FIG. 7.35. THE INITIAL PROCEDURE IN CULTURING THE KIDNEY IS TO STERILIZE
THE SKIN JUST BEHIND THE DORSAL FIN USING A RED HOT SPATULA
A rolling motion will ensure that an adequate area is sterilized.

(6) Sterilize the bacteriological needle or loop by flaming until it is
cherry red. Hold near vertical.

(7) Place sterilized loop or needle into the kidney-blood area.

Plate Streaking Method.—After a sample of kidney tissue has been
recovered by using a sterile needle, the "sample" must be transferred to
suitable bacteriological medium. The purpose of "streaking" (Fig. 7.37) is
to isolate colonies of bacteria.

(1) Streak surface of the plate without digging into the agar.

(2) Push the loop into the agar at the edge of the plate twice before
continuing to streak at right angles to the previous streak.

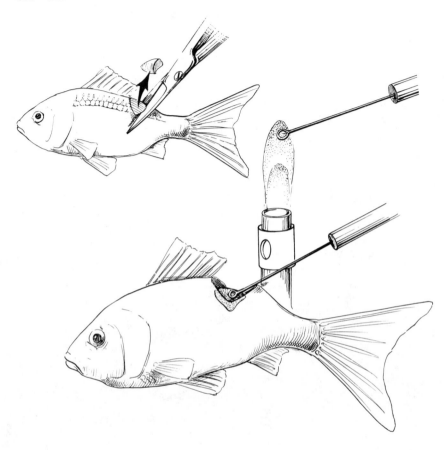

FIG. 7.36. A TRIANGULAR PORTION OF THE MUSCULATURE IS EXCISED OVER THE SEARED AREA
The cut must include the spinal column. Exposure of the kidney may require a further dissection if the cut is not sufficiently deep. A sterile bacteriological loop is placed into the kidney, recognizable as a red pulp.

(3) Cross the previously streaked area once as the new streak is initiated.

(4) Incubate the plates inverted at 26° to 30°C for 24 to 48 hr, or tape the plate and submit it to the diagnostic laboratory for identification and antibiotic sensitivity testing.

Common Bacterial Diseases.—It has already been stated that bacterial diseases of fish frequently follow stress conditions. It is also true that fish can carry potentially pathogenic bacteria on the surface of the

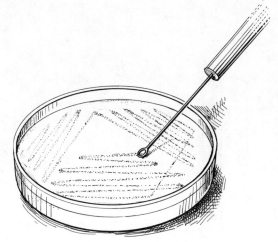

FIG. 7.37. AFTER THE BAC-
TERIOLOGICAL NEEDLE HAS
BEEN INOCULATED WITH
KIDNEY TISSUE, THE NEEDLE
IS STREAKED ACROSS A
PLATE OF SUITABLE AGAR
MEDIUM
Cross streaking is important
to facilitate isolation of indi-
vidual colonies.

skin or within the intestine (and possibly other organs) without showing
signs of disease. The listing of a few selected bacterial diseases presupposes that the reader recognizes that potentially pathogenic bacteria are
common in nature and that stress or a series of stresses is the determining factor of whether or not the fish will become clinically sick.
Since the visible signs of bacterial diseases are very similar, principal
bacterial diseases will be listed in tabular form (see Table 7.2).

Fungal Diseases

Fungi are microorganisms which frequently appear as filamentous
growth internally or externally on fish. The tangled white mass frequently seen on the surface of the skin of fish is called a *mycelium* and is
made up of single strands of the fungus which are called *hyphae* (Fig.
7.38). Microscopic observation of a mycelial mass will reveal terminal
distended areas of the hyphae which are called *sporangia.* These sporangia are filled with *spores* which are released into the water.

There are many different species of fungi which can infect both freshwater and saltwater fish. Most, if not all, external fungi are thought to
be opportunistic secondary invaders following a primary injury, ulcera-
tion, or parasitic infestation. These external fungi, typified by *Saproleg-
nia* spp., appear as puff balls or white filaments on the surface of the fish.
Death is caused by invasion and destruction of internal organs. *Saproleg-
nia* spp. as well as other fungi can attack fish eggs, resulting in widespread infection and death. Fungal infections of eggs can be minimized

TABLE 7.2. LIST OF FISH DISEASES CAUSED BY BACTERIA

Common Name	Bacteria and Characteristics	Signs and Pathology	Prevention and Treatment
Bacterial gill disease. Mainly in hatchery salmonoids	Flavobacteria. Associated with crowding, possibly high ammonia. Gram-negative	Asphyxia; gill destruction	(1) Avoidance of crowding (2) Quaternary ammonium disinfectants
Bacterial hemorrhagic septicemia. World-wide occurrence in most species of freshwater fish	*Aeromonas hydrophila.* Gram-negative	Hemorrhages on skin; exophthalmos; dropsy; ulcers on skin; death	(1) Avoidance of infection (2) Oxytetracycline or chloramphenicol orally or intraperitoneally
Coldwater disease of salmonoids	*Cytophaga psychrophila.* Gram-negative. Use Ordall's medium for isolation	Skin ulcerations; kidney involvement	(1) Oxytetracycline orally (2) Sulfamethazine or sulfisoxasole orally (3) Quaternary ammonium or diquat externally
Columnaris disease. Worldwide occurrence. Cottonmouth disease, tail rot, mouth fungus—in all species of freshwater fish; common in aquarium fish	*Flexibacter columnaris*	Cottonmouth, skin lesions, tail rot	(1) Oxytetracycline orally (2) Sulfamethazine or sulfisoxasole orally
Hagerman redmouth disease. Trout	RM bacterium (enterobacteria)	Small pinpoint hemorrhages	(1) Oxytetracycline orally. Vaccine available
Fin and body rot. Common disease of freshwater fish	Multiple causes, primarily *Flexibacter columnaris*	Fin necrosis	(1) Avoidance of crowding and water pollution (2) Antibiotics in food or in water (tropical fish)

Fish furunculosis. Worldwide except Australia and New Zealand. Chiefly disease of salmonoids	*Aeromonas salmonicida*	Boils (furuncles) on skin; ulcer disease of goldfish	(1) Sulfamerazine in food (2) Oxytetracycline in food (3) Furoxone (4) Oral immunization (only under experimental conditions)
Kidney disease (see disease). In salmonoids in North America and Europe	*Corynebacterium.* Inadequately described	Exophthalmia; skin hemorrhages; enlarged kidneys with white lesions	(1) Avoidance of exposure (2) Prophylaxis with sulfamethazine in feed
Mycobacteriosis. Universal—common in aquarium fish	*Mycobacterium piscium; M. fortuitum*	Skin lesions; debility; exophthalmos; wasting; color changes	(1) Avoidance of infection. Do not feed untested fish meal protein
Nocardiosis. Same as mycobacteriosis, but much less common	*Nocardia asteroides*	Same as mycobacteria	(1) Avoidance of infection
Pseudomonas diseases. Very similar to bacterial hemorrhagic septicemia	*Pseudomonas fluorescens.* Capsulated form very virulent	Hemorrhagic septicemia; dropsy; exophthalmos; hemorrhages and ulcers	(1) Antibiotics in food or water
Ulcer disease. Chiefly brook trout in the United States	*Hemophilus piscium*	External ulcers; septicemia	(1) Oxytetracycline and chloramphenicol orally
Vibriosis. Worldwide; mostly in marine and estuarine environment; also common in aquarium fish	*Vibrio anguillarum*	Hemorrhages; septicemia; dropsy; exophthalmos; ulcers	(1) Oxytetracycline, nitrofurazone and sulfamerazine orally (2) Furanace in bath externally (3) Oral immunization (4) Avoidance of infection

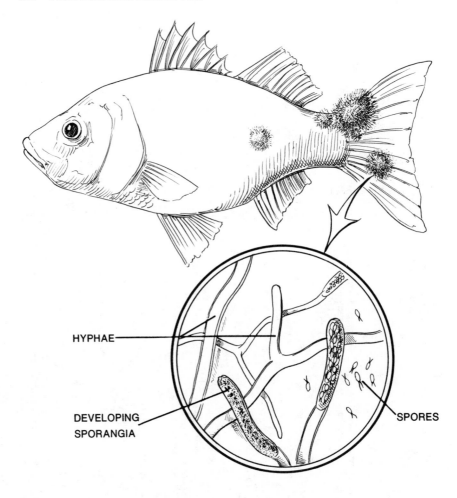

FIG. 7.38. FISH WITH SAPROLEGNIASIS
Microscopic examination of "puff balls" on surface of skin reveal hyphae, developing
sporangia, and sporangia filled with spores.

HYPHAE

DEVELOPING
SPORANGIA

SPORES

by reduction of organics in the water and by dips in antifungal solutions.

A fungus which is normally considered to be an internal and systemic
problem is *Ichthyosporidium hoferi*. Principally a problem of marine
fish, it has been readily transmitted to freshwater species where fish
meal contaminated with the organism has been used as a dietary com-

ponent. Signs of the disease include invasion of the internal organs of the body, resulting in whitish appearing nodules in the heart, liver, kidneys, spleen, and gonads. Skin infections result in roughened or "sandpaper" skin. Occasionally the infestation can cause holes in the head of discus fish. Microscopic examination of smears reveals cyst-like growths which may appear to be budding. This disease can mimic mycobacterial infections and should be differentiated by staining smears with acid fast stains. The disease should be suspected if fish are fed a fish protein source which has not been heated. There is no treatment available for this disease. Avoid the fungus by feeding properly processed fish food.

Gill rot, associated with fungi of the genus *Branchiomyces,* can be found in a variety of cultured fish associated with poor water quality conditions with special regard to increased organic matter. The fungus probably acts as an opportunistic secondary invader following gill changes associated with poor water. Fish appear oxygen starved, weak, lethargic, and lag behind the school. On gross examination the gills appear gray; on microscopic examination hyphae are visible in the gill tissue. Prevention and control should be directed toward thinning out fish, increasing water flow, and cleaning ponds between stocking in order to reduce organics.

Virus Diseases

Viruses are the smallest of microorganisms and are measured in nanometers (25,000,000 nm per in.) (Fig. 7.39). Viruses range in size from 25 nm to approximately 300 nm. At this size range, they can be visualized only through an electron microscope. Viruses are unique microorganisms in that they do not have an independent metabolism as protozoans or bacteria but are completely dependent on the metabolic machinery of a living cell of a host animal for their multiplication. The multiplication cycle is initiated by the virus's entering into a cell of the host where the nucleic acid of the virus [either ribonucleic acid (RNA) or deoxyribonucleic acid (DNA)] redirects the metabolic machinery of the cells to produce more virus RNA or DNA plus an additional component of new proteins which may have specific lethal properties or which may be assembled into a protective coating for the virus nucleic acid. The outer protein protective coating for the viral nucleic acid is called the *capsid,* which varies in morphology between various groups of viruses. Viruses of all species of warm- or cold-blooded animals fall into distinct groups depending on the nucleic acid type (either RNA or DNA), size and surface configuration, and resistance to lipid solvents. Differences within

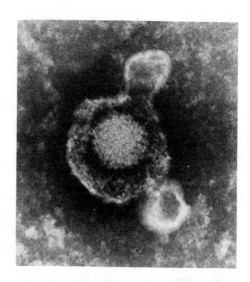

FIG. 7.39. ELECTRON MICRO-
GRAPH OF A HERPESVIRUS PARTI-
CLE
Note outer envelope and aggregate
of capsomeres which form the cap-
sid. See text for description.

main groups of viruses are usually detected by small differences in the protein characteristics of the capsid as measured by virus neutralization tests. For example, channel catfish virus is a member of the herpesvirus group of viruses because it has 162 protein structures (capsomeres) on the capsid. Morphologically, it is identical to other herpesviruses of man or domestic animals; however, it produces disease only in catfish. Generally, viruses show a high degree of species specificity, which means that channel catfish virus will affect only channel catfish or possibly closely related species of catfish.

Virus diseases of fish are generally limited to young stock during periods of optimal temperatures which closely correspond to the optimal temperatures for the growth of the particular fish species.

Viruses of fish are similar to most mammalian viruses in that they have a tendency toward affecting particular tissues within the body. Thus, infectious pancreatic virus of salmonoids affects pancreatic tissue severely, although it is found in other internal organs including the intestinal lining. Poxviruses of fish in Europe, like mammalian poxviruses, produce lesions of the skin. Channel catfish virus attacks many internal organs and is readily recovered from kidney and liver tissues.

The effect of viruses on fish may be *acute,* resulting in extensive tissue damage and subsequent rapid death as a direct result. Virus infections of both mammals and fish are frequently complicated with a secondary bacterial invasion of the weakened host—a fact which frequently complicates an accurate diagnosis. In any virus infection of fish, a certain percentage of the infected fish will not demonstrate any observable signs of illness, yet may be infected and remain as carriers of the virus. Frequently, fish which have recovered from a virus infection may remain as carriers of the virus. Once a fish is exposed to a virus infection, the immune system of the fish responds by the production of specific blood proteins (called antibodies) which can react with a virus and neutralize its activity. These antibodies can protect a fish from a second attack of the virus as well as serve to determine if a fish has ever been exposed to the virus. Specific antibodies against a variety of fish viruses can be prepared by injecting the virus into rabbits (or large fish) and recovering the antibody-containing blood serum after 3 to 4 weeks. The resulting blood serum containing specific antibodies for a particular virus can then be used in specific laboratory tests to identify a virus isolate.

Virus infections of fish are detected and accurately diagnosed by the isolation and identification of a specific virus. Since viruses can propagate only within the living cell, fish diagnostic laboratories carry test tube cultures of fish cells which are known to be susceptible to the virus (Fig. 7.40). Cell cultures can be purchased from biological supply companies and can be maintained in the laboratory. In practice, viruses are isolated in the following way: tissues from a suspected case are completely pulverized, treated with antibiotics, or passed through a bacterial retaining filter prior to inoculating a few drops of the tissue juice on a layer of fish cells growing on the inside surface of the test tube. The cell cultures are continually bathed in a sterile medium and can be easily observed with the use of a microscope. The presence of a virus is suspected when the cell cultures begin to die and detach from the glass surface. The killing of cell cultures by a virus is called the *cytopathic effect* (Fig. 7.41). If the inoculated cultures are covered with a semi-solid overlay of agar, the cytopathic effect is localized. These localized areas are called *plaques* and are frequently used for presumptive diagnosis of a virus because of their size and form (Fig. 7.42).

Diagnosis of virus infections in infected tissues of a fish or in suspected cell cultures can be done by fluorescent antibody staining. This technique is based on the fact that fluorescein tagged (antivirus) antibody will combine with virus in infected tissues or cell cultures. The localization of the virus is done by use of a microscope supplied with an ultraviolet light source. Figure 7.43 is a microphotograph of two fluorescing cells of a cell culture of brown bullhead cells infected with channel catfish virus.

An isolation of a virus is confirmed when in subsequent tests the cytopathogenic effect is neutralized by mixing the virus with antiviral antibodies previously made by injecting rabbits. This type of test is called a virus neutralization test (Fig. 7.44).

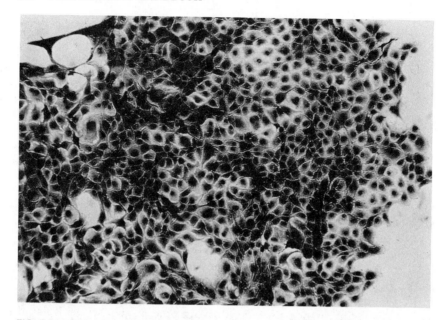

FIG. 7.40. CELL CULTURES ARE USED IN VIRUS DIAGNOSTIC PROCEDURES
Cultures of fish cells are propagated on the surface of glass or plastic flasks. The presence of a virus is indicated by the destruction of the cells (cytopathic effect) as visualized through a microscope.

There is no direct treatment for virus infections of fish. Experimental virus vaccines have been shown to be effective in protecting fish from virus infections but, at this writing, it is doubtful whether virus vaccination programs for commercially raised fish would be commercially feasible or ecologically acceptable.

The term "avoidance" of virus infestations suggests that careful husbandry practices will tend to keep the virus out of a hatchery. For instance, new brood stock of trout or channel catfish can be tested for specific virus antibodies prior to purchase of replacement stock. A positive virus neutralization test indicates that the adults have been exposed sometime in the past and may be carriers of the virus.

Use of water from streams which may contain infected carrier fish is obviously dangerous from the viewpoint of spread of virus as well as other microorganisms. Overcrowding of fish in hatcheries will aid in the spread of a virus simply by increasing the probability of contact with a susceptible fish. Disinfection of nets by suitable chemical dips or by drying will prevent virus transport from infected farm to non-infected farm.

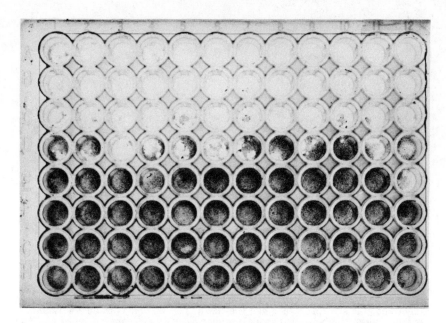

FIG. 7.41. THE CYTOPATHIC EFFECT OF CHANNEL CATFISH VIRUS CAN BE
DEMONSTRATED BY INOCULATION OF ALIQUOTS OF SERIAL 10-FOLD DILUTIONS
OF VIRUS INTO CULTURES OF BROWN BULLHEAD CELLS
The top three rows of cell cultures have been destroyed; the fourth row shows
partial destruction, while the fifth row shows no virus activity. In this manner virus
can be quantitated by calculating the dilution which causes destruction of 50% of the
cell cultures in any one dilution series.

Discussion of Selected Virus Diseases

Trout Diseases.—*Infectious Pancreatic Necrosis Virus (IPN).*—IPN is a
virus which affects salmonoid fry or fingerlings. Rarely are adult fish
infected. The disease was first reported in the United States and later in
Europe and Japan. The virus has been characterized as similar to the
reovirus group. Signs of the disease include a dark color, abdominal
swelling, "popeye," and abnormal swimming movements including
whirling. On post-mortem examination, the gut is normally without food
but the intestine frequently is filled with mucus. The liver and spleen
may appear anemic and hemorrhages may appear in the pyloric region.
The virus is easily transmitted by water and probably on the surface of
eggs, although it is difficult if not impossible to disinfect eggs. The virus
has been reported to be maintained in the intestines of fish-eating birds
for as long as 8 days after experimental exposure. Birds, therefore, could

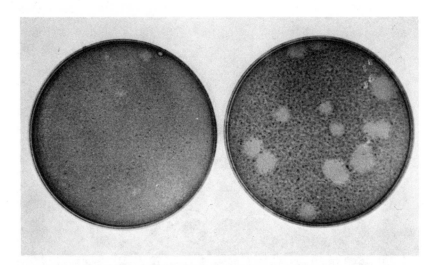

FIG. 7.42. LOCALIZED AREAS OF CYTOPATHIC EFFECTS ARE CALLED PLAQUES
The inoculated cell culture is overlaid with an agar nutrient mixture, incubated for 3 to 5 days, and stained with a vital stain. The size and shape of the plaque are frequently used to differentiate between two viruses such as infectious pancreatic necrosis and infectious hematopoietic necrosis virus of trout. The culture on the left was uninoculated.

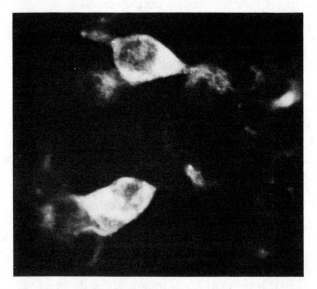

FIG. 7.43. POSITIVE FLUORESCENT ANTIBODY TEST FOR CHANNEL CATFISH VIRUS-INFECTED BROWN BULLHEAD CELLS
See text for explanation.

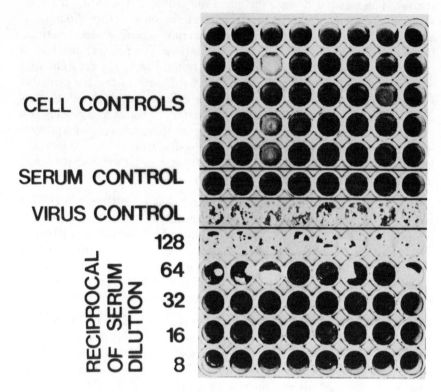

CELL CONTROLS

SERUM CONTROL

VIRUS CONTROL

RECIPROCAL OF SERUM DILUTION

128

64

32

16

8

FIG. 7.44. ILLUSTRATION OF A VIRUS NEUTRALIZATION TEST
The first five rows (from top) are uninoculated cell controls which confirm that the
cell cultures are living. The serum control indicates that the test serum is not toxic to
the cells. The virus control indicates that the virus is active and is killing the cells
(note white areas where cells are detached from the glass). The results indicate (bottom five rows) that the test serum from one channel catfish completely neutralized
the virus at a dilution of 1/32. This result suggests that channel catfish had been
exposed to the virus sometime in the past.

pose a problem in the control of the infection. The disease can be avoided
by stocking fish hatched from eggs derived from certified disease-free
hatcheries. Fry should be reared separately from older fish, which may
be carriers of the disease.

Infectious Hematopoietic Necrosis Virus.—This virus affects young sock-
eye salmon, chinook salmon, and rainbow trout. Infections are most common in western North America in water below 15°C. Fish less than 2
months of age are most susceptible, and the susceptibility decreases with
age. The causative agent is a member of the Rhabdovirus group of

viruses. The signs of disease include a darkening of the skin, "popeye," abdominal distention, and hemorrhages at the base of fins. Anemia, as evidenced by paleness of the gills and internal organs, is characteristic. Definitive diagnosis is done by virus isolation. The primary method of transmission of the virus is from carrier brood fish to the fry. The eggs are probably infected but the signs of the disease in fry are probably delayed until protective antibodies in the yolk sac are depleted. Control of IHN can be approached by adjusting hatchery water and rearing fish above 15°C, by isolation of suspected infected fish and avoidance of virus spread by use of disinfectants for nets, pails, and other implements. Dipping eggs with organic iodophors at 100 ppm of iodine activity for 10 min has been shown to reduce infections.

Viral Hemorrhagic Septicemia.—Viral hemorrhagic septicemia (VHS) is a virus of trout which is found only in Europe. It affects fingerling trout at temperatures lower than 8°C. The virus can induce mortalities up to 70% with signs of anemia, "popeye," and hemorrhages of skin and fins. Nervous disorders which are detected by erratic swimming patterns are common toward the end of the epizootic. The occurrence of VHS in European trout led to a ban on importation of frozen trout from Europe.

Channel Catfish Virus Disease.—Channel catfish virus is a member of the herpesvirus group which affects catfish fingerlings during the summer months when water temperature is between 24° and 27°C (75° and 80°F). The mortality rate can reach 80% of fingerlings in some epizootics. Signs of disease include hemorrhages of the fins, "popeye," abdominal distention, and anemia. Fish occasionally hang in a vertical position just prior to death. This disease is found in channel catfish culture areas in the south and midsouth states. Although the disease has been reported from a wide geographical area, the incidence of the disease is relatively infrequent. The disease can be avoided by careful selection of breeding stock which has tested negative for antivirus antibody.

Lymphocystis Disease.—*Lymphocystis* virus propagates in cells of fish resulting in the increase of the size of individual cells up to 50,000 times their normal volume. The disease has been demonstrated in marine and freshwater species. Signs of the disease vary from slight opalescence of the fins to raspberry-like growths on the fins. Examination of suspect lesions with a 10X hand lens will reveal that small white spots are actually enlarged cells. The tumor-like growths can outgrow the blood supply, resulting in frayed fins. Occasionally, the enlarged cells will develop under scales, resulting in protruding scales. Diagnosis is confirmed by microscopic observation of many enlarged cells. Since there is no treatment, control should be directed toward getting rid of affected fish. Ultraviolet irradiation units adapted for closed water systems will effectively kill the virus in water. Presumably the virus is spread from fish to fish in the water. Fish rarely die from the disease, but the resulting disfigurement in tropical ornamental varieties is of economic importance.

Other Viruses

Other viruses have been described; however, their importance may be more academic than economic. Viruses of bluegills and shiners have been isolated but apparently are of minor importance as disease-causing entities. *Herpesvirus salmonis* has been isolated from trout but apparently is not recognized as a serious problem when compared to IPN or IHN viruses. European carp farming has fought the disease known as "spring viremia" for years. Much controversy has surrounded the cause of the disease. Present day dogma suggests that a primary viral infection is followed by secondary invasion with a bacterium, usually of the *Aeromonas* group. The virus is in the *Rhabdovirus* group.

There are undoubtedly many viruses which are present as disease entities in populations of fish but which have not been isolated. Many epizootics of disease in ornamental fish may be virus related, but further study is hampered by the lack of available cell cultures.

PREVENTION AND TREATMENT

Stocking Healthy Fish.—Disease-free fish should be stocked regardless of whether the fish are to be put into a home aquarium, trout raceways, or catfish production ponds. Obtaining "specific pathogen free" fish for stocking purposes is possible if the primary producers are on a rigid disease control program. Specific pathogen free status could imply that the stock has been certified to be free of a specific disease by a qualified expert. Purchasing fish from establishments which are not regularly monitored for diseases and not under a treatment regimen is risky. In such cases, the seller may insist that the fish have never been sick. While the fish may never have shown signs of disease, they may, nevertheless, be parasitized.

Quarantine of newly purchased fish is a good practice, especially if the disease status of the fish is not known. Quarantine simply means that replacement stock or any newly introduced fish will be segregated from all other fish for a period of time, usually from two to three weeks. It is important to hold fish in a water containing system which does not flow into or recirculate with water used in the main culture system. In closed recirculating systems outfitted with diatomaceous filters and ultraviolet sterilizing lights, disease agents should not spread throughout the system. In pond or raceway systems, quarantined fish should be placed into a pond or segment which receives water from springs or wells (parasite free) and which is not discharged directly into other fish containing systems.

Treatment during the quarantine period should be based on an accurate diagnosis of which disease agents the fish may be carrying. In a commercial production unit, as well as individual aquaria, recently quarantined fish may need to be fed antibiotic containing feed (commercially available) to prevent bacterial disease resulting from shipping stresses. Specific chemical treatments should be based on the findings of the necropsy examination. In the case of individual aquarium fish, quarantine and treatment are an excellent way to avoid the introduction of various parasites; however, a "shotgun" approach to treatment is necessarily used. In such cases, treatment should be directed toward elimination of external protozoans and monogenetic trematodes, as well as internal parasites.

Water Source.—Well water or spring water should be free of parasites. Water from streams or lakes containing fish will be a source of parasites. If the water source is likely to contain parasites from free-living fish or if the water is to be recirculated, filtration of the water may be necessary but expensive. Highly sophisticated recirculation systems have been developed and are being used both experimentally and under practical conditions. Such systems utilize a sand or gravel filter which acts to both mechanically remove debris as well as oxidize ammonia to nitrates. Frequently, a diatomaceous earth filter will be used to further filter the water. Lastly, ultraviolet lamps have been used to effectively sterilize the water. Such systems represent a high initial investment, but, in specific cases such as public aquaria or wholesale tropical fish establishments, they will contain disease problems to individual tanks.

Natural water sources contain various free-living microorganisms which can become problems as the organic content of the water increases. These include *Epistylis, Trichophrya, Tetrahymena, Scyphidia,* and *Glossatella.* Increases of the organic content of water also increase the total bacterial counts of the water. Many bacteria which can infect fish can be found in organically polluted waters.

Sanitation.—Stocking of fish should be done with particular attention to cleaning and disinfection of the culture system. In pond culture, draining and removal of the organic "muck" followed by liming, disking, and drying of ponds will remove all parasites with the possible exception of sporozoans. In aquarium systems, new batches of fish should be placed into aquaria which have been vacated of fish for 4 to 5 days or disinfected. Disinfection of "old" water followed by complete water change is the best way to prevent carryover of disease agents from one batch of fish to the next. If aquaria or holding tanks contain sick fish, sterilization of the water with 25 ppm of chlorine prior to draining will disinfect the aquarium or vat as well as be a good ecological practice. Killing by chlorine is influenced by the quantity of organics, and a second disinfection with chlorine may be advisable after the insides of the container are scrubbed clean with brushes.

The numbers of fish which are stocked have a direct correlation to the ease of parasite transfer from fish to fish. Moreover, overcrowding leads to organic pollution, increased bacterial population, and favors the proliferation of external protozoans in addition to creating several stress conditions for the fish. Stocking practices depend on the amount of inflow water or, in closed systems, the amount of replaced water on a daily basis. It has been noted that external protozoan parasites will decrease when parasitized fish are placed in flow-through tanks supplied with dechlorinated municipal water.

Pollution Control.—Increases in organic load from overfeeding can lead to an increase in potentially pathogenic protozoans such as *Epistylis* as well as providing optimal growth conditions for bacteria. Under pond conditions, excessive feeding results in an increased demand for oxygen by the decomposition process as well as increases in ammonia levels. These accumulative factors stress the fish, resulting in reduced resistance to external parasitism and bacterial infections.

Monitoring Fish Health.—When fish begin to show signs of parasitism, treatment in many cases will give disappointing results since most of the damage inflicted by the microorganism has already been done. Treatment with various chemicals and drugs is frequently a stress in itself, and additional fish may die because of the treatment. Fish producers should be aware of potential problems in hatcheries. This can be done by arranging for periodic monitoring of stock for parasites. Monitoring for specific bacteria or virus diseases by specific blood serum checks may be advisable in salmonoid or channel catfish culture. A monitoring schedule will provide an early alert to potential problems and will allow preventive treatment to be carried out before serious problems arise.

Prevention by Chemical Prophylaxis.—Treatments of fish should be based on an accurate diagnosis of the problem along with an evaluation if the cost of the treatment is justified in light of the problem and expected results. Decisions as to when to treat fish prophylactically can be based on microscopic checks for parasite loads or based on past experiences. Assistance of a trained disease specialist is recommended on the basis that an accurate diagnosis is needed for selection of the specific treatment and dosage. As important is the fact that a trained disease specialist can assess whether or not a specific treatment is effective.

Specific Treatments

A few important considerations should be considered prior to any treatment.

Diagnosis.—The choice of a chemical to be added to the water or a drug to be incorporated into the food is based on an accurate diagnosis. A "shotgun" approach may work a certain percentage of the time; however, so-called "problem cases" are usually based on an erroneous evaluation of the cause of the disease and use of the wrong chemical or drug.

Water Quality.—A successful treatment presupposes a knowledge of existing water quality conditions including temperature, total hardness, oxygen levels, pH, and organic load. Dosages of chemicals frequently are dependent on water hardness, as in the case of copper sulfate; pH, as in the case of organic phosphate; or relative organic load, as in the case of potassium permanganate.

Species of Fish.—Some species of ornamental fish are sensitive to dosages of various chemicals easily tolerated by trout or channel catfish. Younger fish may be more sensitive to a particular chemical than young of other species. The young of many species are more sensitive to malachite green than adults; also, skinned fish such as eels have an increased sensitivity to malachite green.

Laws and Regulations.—Chemicals and drugs must be used in accordance with current regulations as promulgated by the Food and Drug Administration and by the Environmental Protection Agency.

Preliminary Biological Tests.—Whenever possible, pre-tests of chemicals to be added to the water or of drugs to be incorporated into food should be conducted on a small batch basis. Such tests provide toxicity as well as efficacy data.

Dosages.—Most chemicals used in treatment of fish diseases will be toxic to the fish if correct dosages are not used or if dosages are not adjusted according to varying water quality conditions. Dosages of chemicals in water are always accompanied by information on the length of treatment, which will vary depending on the concentration of chemical.

Methods of Treatment

Dip Method.—This method is used when very strong concentrations of chemicals are used. The chemical is added to water in a concrete vat or glass aquarium. Fish or eggs are dipped in the solution for a brief period— usually a matter of seconds. Nets or wooden boxes with net bottoms can be used to contain the fish. This is the preferred method of treating eggs.

Flush.—This method is used as a once-through treatment where water is continually running through an egg incubator, vat, or rearing trough. A relatively concentrated amount of chemical is added at the inflow point and is allowed to pass through the culture system. This method is well suited for use in egg incubators. The flush method is not as controllable as other methods since it is impossible to accurately estimate time-dose relationships.

Baths.—Treatments are carried out in vats, or aquaria, or any other suitable container. Metal containers should be avoided. The amount of chemical used is usually high enough to result in severely stressed fish if the treatment is not terminated when fish first begin to show signs of stress. Treatments are generally gauged to last between 15 min and 1 hr with ample aeration. Constant attention to the reactions of the fish is necessary to avoid over-treatment. The treatment should be terminated at the first signs of stress, which may include attempts to jump out of the treatment tank, lack of reaction to stimuli, or schooling behavior. Rapid termination of a bath treatment can be facilitated by reducing the treatment volume of the bath followed by quick drainage and the immediate addition of fresh water. An alternate method of termination of the treatment can be effected by rapid transfer of fish into fresh water using large nets.

Long-term Treatment or Indefinite Treatment.—This method consists of addition of chemicals to a culture system at dosages which are non-toxic to fish. This method is frequently used in pond culture, large recirculating systems, and hauling vats, as well as in transportation of aquarium fish in closed bags.

Calculating Chemical Treatments

The following discussion of the calculation of chemical treatments is taken from the booklet, *Second Report to the Fish Farmers,* published by the Bureau of Sport Fisheries and Wildlife and edited by Fred P. Meyer, Kermit E. Sneed, and Paul T. Eschmeyer (U.S. Dep. Interior 1973).

The calculation of the amounts of chemicals to be used in fish farming may involve English and metric units or their combinations, and may be confusing to some workers.

Units of Measure.—Most people are familiar with pounds, acres, and gallons. Fish culture uses additional standard units, discussed here individually.

A *cubic foot* is a unit of volume measuring 1 ft square and 1 ft high. One cubic foot of water weighs 62.4 lb and contains approximately 7.5 gal. The number of cubic feet in a pond equals the length (in ft) × the width (in ft) × the depth (in ft).

An *acre-foot* is a unit of volume having an area of 1 surface acre and a depth of 1 ft. One acre-ft of water contains 43,560 ft³, 326,000 gal., or 2,718,000 lb. Acre-feet are computed by multiplying the area (in acres) by the average depth (in ft).

One *gallon* contains 4 qt, 8 pt, or 3800 ml. One gallon of water weighs approximately 8.34 lb, or 3800 g.

A *pound* contains 16 oz, or 453.6 (454) g. An *ounce* contains 28.35 g.

Weights are expressed in the metric system as grams and kilograms. A *kilogram* equals 1000 g, or 2.2 lb. Volumes are expressed in *cubic centimeters* (cc) or *milliliters.* One *liter* holds 1000 cc or ml. A gallon contains 3.8 liters, or 3800 ml.

Metric units are frequently used when working with very small amounts of chemical. Thus, we may apply grams per gal., grams per ft³, ml per gal., or ml per ft³. Large units of volume will require large units of weight, so it is most convenient to use lb per acre-ft for such treatments.

Units of Treatment.—In the treatment of fish, it is a common practice to add chemicals to the water to produce a desired concentration. Concentrations are generally expressed as parts of chemical per million parts of water, usually written as ppm. One part per million refers to the addition of 1 lb of chemical to 999,999 lb of water, to give a total weight of 1,000,000 lb of solution. The amounts of chemical needed to produce 1 ppm in each of the standard units of water volume are:

2.7 lb per acre-ft

0.0038 g per gal.

0.0283 g per ft³

Another method of treatment is the incorporation of a chemical into fish feed. Such treatment is based on the weight of the fish. Standard units of treatment are given in grams of chemical per 100 lb of fish per day. If 100 lb of fish are to be treated with terramycin at the rate of 2.5 g per 100 lb of fish per day, the amount of feed fed each day must contain 2.5 g of the drug. Generally speaking, fish are fed at the rate of

3% of their body weight per day. At this level, 3 lb of feed are fed each day to each 100 lb of fish. For treatment, the feed requires 2.5 g of terramycin in every 3 lb of ration. The feed should contain $100 \div 3 \times 2.5$, or 83.3 g, of terramycin activity per 100 lb.

Formulations.—So far, we have assumed that all of the chemicals we will be using are pure compounds, or 100% active ingredient. However, few of the compounds used in fish culture are pure chemicals. Most of them are mixtures, and the level of active ingredient is stated on the label *unless* the recommended use specifically indicates otherwise.

As an example, let us assume that the chemical we are going to use contains 25% active ingredient. To find how much of this formulation is required, divide 100 by the percentage of active ingredient in the formulation. In this case, it is $100 \div 25$; it takes 4 times as much of this formulation as it would of a pure chemical.

Formulations of antibiotics generally have the drug in a premix form which can be incorporated into feeds. Usually such formulations contain 25 g of activity per lb of material. If we wish to incorporate 2.5 g of active ingredient into a quantity of feed using the above material, we need $2.5 \div 25$, or 0.1 lb of the formulation. Another way to express this quantity is $0.1 \times 16 = 1.6$ oz, or $0.1 \times 454 = 45.4$ g.

Some chemicals are provided as liquids containing a stated number of lb per gal. Typical products of this type contain 4 lb of active ingredient per gal. In this case, it is easier to work with the cubic centimeter instead of the gallon. Since 1 gal. contains 3800 ml, a 4 lb per gal. formulation will contain 1816 g (4×454), or $1816 \div 3800 = 0.48$ g per ml. *Always remember to use the weight of the active chemical when computing parts per million to be used.*

Other chemicals may be liquids in pure form. A typical one of this type is formaldehyde solution (formalin). This solution contains only 38% formaldehyde gas, but the liquid is a pure compound for fish cultural purposes. In using a chemical such as this, it is necessary to know how its weight compares to that of water. If it is heavier than water, fewer cc are needed to deliver the desired weight of chemical.

Calculating Treatments.—Consider some of the typical situations in which treatments are made. For a tank treatment, unless the volume is known in gallons, it is best to measure the tank to determine its length, width, and water depth in feet. Multiplying these gives the volume of the tank in cubic feet.

A formula for determining the amount of a chemical to use is the *volume × amount of chemical needed to produce 1 ppm in each unit of volume × ppm desired.*

Assume that we wish to treat a tank measuring 12 ft × 2.5 ft × 2 ft with 0.25 ppm of malachite green. Our calculation is 12 × 2.5 × 2 = 60 ft^3. It takes 0.0283 g to yield 1 ppm in 1 ft^3, so we multiply 60 × 0.0283 × 0.25 = 0.42 g of chemical.

If the area of a pond is known in acres, the acre-ft can be determined by measuring the pond depth along several transects with a weighted line to find the average depth. The average depth can be computed by adding all of the measurements and then dividing by the number of readings. Be sure to make measurements in both the deep and the shallow areas on the pond. If the pond is uniform in shape, one transect through the center along the long axis and another along the short axis is adequate. Do not rely on an estimate or a guess, because many fish have been killed by overdoses resulting from such procedures. For large reservoirs, an engineer can help determine the area. Every fish farmer should know the volume of his ponds *before* a treatment is needed, to avoid loss of valuable time.

The formula for deciding on how much chemical is needed to treat a 28,000 ft^3 pond with 2 ppm of a 25% active formulation is as follows: volume in ft^3 × 0.0283 × 2 × 100 ÷ 25 = 7924 g, or 17.35 lb.

Another way is to convert the number of cubic feet to acre-ft by dividing by 43,560 (the number of cubic feet in 1 acre-ft). Thus, the pond has a volume of 0.8 acre-ft. The calculation now becomes: acre-ft × 2.7 (lb needed to give 1 ppm in 1 acre-ft) × 2 ppm × 100 ÷ 25 = 17.45 lb of the 25% formulation. The slight difference between amounts of chemical as determined by the two methods is due to rounding off and will not affect the treatment.

If we wish to treat a tank which holds 500 gal. of water, the calculation is gal. × 0.0038 (conversion factor to give 1 ppm in 1 gal. of water) × ppm desired × 100, divided by the percentage of active ingredient in the formulation. A treatment with pure malachite green at the rate of 0.25 ppm in the same tank is computed as 500 × 0.0038 × 0.25 × 100/100, or 0.475 g. Treating the tank with 25 ppm of formaldehyde is calculated in the same way, but involves an extra factor because formalin is heavier than water. The calculation is: 500 × 0.0028 × 25 × 100/100 × 1/1.08 (specific gravity of formalin) = 43.98 ml of formalin.

If a tank is round and its volume in gallons is not known, we must follow a slightly different procedure. Volume is computed as area × depth; to find the area of the circular bottom, we determine the radius of the bottom by measuring the diameter of the tank and dividing by two. The formula for the bottom area of a round tank is 3.14 × the radius multiplied by itself. The volume is computed by multiplying the area by the depth.

If we have a round tank measuring 5 ft in diameter and 3 ft in depth, we determine the area as follows: 3.14 × 2.5 (½ the diameter) × 2.5 (½ the diameter) = 19.625 ft². The *volume* is calculated as 19.625 × 3 (depth in ft) or 58.875 ft³.

Chemical treatments for round tanks are computed as for other tanks; volume × conversion factor × ppm desired × 100 ÷ % of active ingredient in formulation to be used = the amount of chemical needed to treat at the desired level.

In computing the amount of a chemical to be used, one can begin with any unit of volume. In the calculation, you must be sure to use the correct conversion factor which will yield 1 ppm in the unit of volume with which you are working (see Table 7.3).

Chemicals and Drugs

The following procedures have in part been condensed from materials supplied by Dr. S.F. Snieszko of the Eastern Fish Disease Laboratory, Kearneysville, West Virginia, and Dr. G.L. Hoffman of the Fish Farming Experimental Station, Stuttgart, Arkansas.

Disinfection of Pails, Vats, and Miscellaneous Equipment.—*Chlorine.*—Used at 200 ppm, chlorine will disinfect in 30 to 60 min; but it is corrosive, irritating to mucous membranes and will destroy netting. Chlorine is available as Chlorox® containing 5.25% hypochlorite. HTH

TABLE 7.3. TABLE OF EQUIVALENTS

1 acre-ft	= 1 acre of surface area covered by 1 ft of water
	= 43,560 ft³
	= 2,718,144 lb of water
	= 326,000 gal. of water
1 ft³	= 7.5 gal.
	= 62.4 lb of water
	= 28,354.6 g of water
1 gal.	= 8.34 lb of water
	= 3800 ml
	= 3800 g of water
1 qt	= 950 ml
	= 950 g of water
1 lb	= 453.6 g (454)
	= 16 oz
1 oz	= 28.35 g
1 ppm requires:	2.7 lb per acre-ft
	0.0038 g per gal.
	0.0283 g per ft³
	0.0000623 lb per ft³

stands for hi-test hypochlorite and contains a 65% chlorine concentration. Chlorine in water will be inactivated by sunlight or will be driven off in gaseous form by agitation. Sodium thiosulfate will neutralize chlorine immediately at the rate of 1 part of sodium thiosulfate to 8 parts of chlorine. Inexpensive test kits are available to test for free chlorine. Excessive amounts of sodium thiosulfate may lower the pH of water.

Roccal®.—One of a group of quaternary ammonium germicidal compounds, Roccal® is an excellent disinfectant for nets when used at the manufacturer's suggested dosage for general disinfection usage and is more effective in alkaline water. Amounts carried over into fish tanks are not toxic.

Iodophors.—Sold under the trade names of Betadine, Wescodyne, and others, the disinfecting element is iodine, which is an oxidizing agent. Iodophors are used for net or surface disinfection. Follow the manufacturer's directions.

Formalin.—Formalin contains 37% by weight of formaldehyde, which is the active ingredient. Four ml (about 1 tsp) per gal. of water (about 1000 ppm) is an excellent dip for nets. Carryover into fish tanks is not enough to affect fish.

Disinfection of Ponds.—*Calcium Hydroxide.*—Calcium hydroxide, also known as slaked lime, is distributed over drained but wet ponds at the rate of about 5 lb per 100 ft^2. The material must be used fresh.

Calcium Hypochlorite.—HTH, or hi-test hypochlorite, can be distributed into ponds at the rate of 0.6 lb per 1000 ft^3 of water (10 ppm chlorine) or distributed onto wet areas of drained ponds. Refilling ponds will normally dilute and inactivate any residual chlorine; however, checking with a chlorine test kit may be advisable. HTH is noxious, and gas masks may be required in closed areas.

Calcium Oxide.—Calcium oxide is also known as quicklime. It is usually added to the bottom and sides of drained but wet ponds at the rate of about 5 lb to 100 ft^2.

Insect Control in Ponds.—Aquatic insects may effect total kills of fry or fingerlings in ponds.

Diesel Fuel.—At the rate of 2 to 4 gal. per acre, diesel fuel or kerosene will destroy aquatic beetles and other surface breathing insects. Motor oil or cottonseed oil is often added to the diesel fuel or kerosene to aid dispersion at the surface of the water.

Baytex®.—Baytex® is one of a number of organophosphates which has been used to control predatory insects. The dosage is 0.25 ppm on an indefinite basis. Larval forms of *Lernaea* and other parasitic copepods are also controlled.

Snail Control in Aquatic Systems[1].—*Physical.*—If the operation permits periodic drainage, many snails can be washed out and down the drain. Thorough drying will also kill the snails except for a probable few that will find moist spots. Heavy sprinkling of damp spots with quicklime or recently purchased slake lime will kill many snails. If possible, dry the ponds thoroughly and disc the bottom once each year to help keep snail population to a minimum.

Chemical.—*Copper.* Copper is very toxic to snails and will not kill fish if used carefully. Copper is not registered for this use but has been registered as an herbicide.

Soft Water (Methyl Orange Alkalinity Less Than 50 ppm). Use copper sulfate at 0.1 ppm (0.1 oz per 1000 ft^3, or 0.27 lb per acre-ft). Some use 2 lb of copper carbonate per 1000 ft^2 of shallow pond bottom.

Hard Water (Methyl Orange Alkalinity Greater Than 50 ppm). Use 2 lb copper sulfate plus 1 lb copper carbonate per 1000 ft^2 of shallow pond bottom, or copper sulfate at no more than 2 ppm (2 oz per 1000 ft^3, or 5.4 lb per acre-ft).

Copper sulfate is also an algicide; if too much algae is killed during hot weather, its decomposition may cause oxygen depletion and the fish may die.

Treatments of External Parasites

Ichthyophthirius multifiliis.—*Physical Methods.*—See text material.

Chemical Treatments.—(A) *Stock Solution.* Malachite green and formalin mixtures: 14 g zinc-free malachite green to 1 gal. formalin.

Pond Treatment. Use at 25 ppm at 3 to 4 day intervals (1 ml per 10 gal.)

Raceways. Use for 6 hr daily with aeration followed by flushing. Dose—1 ml stock per 10 gal.

Home Aquaria. Can be used as an indefinite treatment. Aerate during treatment but discontinue carbon filtration. Treat 3 times at 3 day intervals. Fry and skinned fish do not tolerate malachite green; also, some species of tetras may not tolerate the malachite green. On alternate

---

[1]Information provided by Dr. G.L. Hoffman.

days after each treatment, change as much water as possible. Dose—1 ml stock per 10 gal. of water.

(B) *Formalin.*

Pond or Indefinite Treatment. Use 15 to 25 ppm every other day. NOTE: 15 to 25 ppm = 0.6 to 1.0 ml formalin per 10 gal. of water or 5 to 8 gal. per acre-ft of water. Avoid use during hot weather to avoid oxygen depletion.

Bath Treatment. Use 166 to 250 ppm for 1 hr daily until deaths cease. Maintain aeration; remove fish at first signs of stress (see text). NOTE: 166 to 250 ppm = 0.66 to 1.0 ml formalin per gal. of water or 10 to 15 pt per 1000 ft^3 of water.

Home Aquaria. Normally used at 25 ppm on alternate days until the problem is over. Frequently used for tetras. On alternate days after treatments, change as much water as possible.

(C) *Malachite Green (Zinc-free in Oxalate Salt).*

Pond or Indefinite Treatment. Has been used at 0.1 ppm at 3 to 4 day intervals in ponds. NOTE: 0.25 lb of malachite green per acre-ft = 0.1 ppm.

Bath Treatments. Used at 2 ppm for 30 min. NOTE: 2 ppm = 5.4 lb of malachite green per acre-ft or 2 oz (56 g) per 1000 ft^3 of water.

Home Aquaria. Use at 0.1 ppm at 3 or 4 day intervals until fish are clear. Discontinue carbon filtration but maintain aeration. Water changes 1 day after treatment are advisable. NOTE: Make stock solution of malachite green by addition of 14 g of zinc-free malachite green oxalate to 1 gal. of water. Keep stored in a brown bottle. To get a dose of 0.1 ppm, add 1 ml to 10 gal. of water.

CAUTION: Malachite green has been shown to induce cancer in experimental animals.

(D) *Copper Sulfate.* Copper sulfate has been used as an algicide and molluscide, as well as for external protozoans. Its use in ponds during summer months should be avoided to prevent oxygen depletion through increased oxygen demand due to decomposing plants. The copper ion is very toxic to fish, and the ionization will depend on the total hardness of the water as measured by calcium carbonate.

Soft Water. If water is between 40 and 50 ppm total hardness, use less than 0.25 ppm of copper sulfate. Use one-half dose on the third day of treatment.

Moderate Hard Water. If water is between 50 and 90 ppm total hardness, use 0.5 ppm with one-half of this dose on the third day.

Hard Water. If water is between 100 and 200 ppm total hardness, use 1 ppm of copper sulfate with one-half of this dose on the third day.

NOTE: Copper sulfate is the choice method of treatment for the marine parasites *Cryptocaryon irritans* and *Oodinium* spp. The dosage is to be maintained at 0.15 to 0.18 ppm for 2 weeks.

Copper Formulation (Blasiola 1978). Add 2.23 g of copper sulfate pentahydrate and 1.5 g of citric acid to 1 liter of distilled water. From this stock solution add 1 ml per gal. to bring the copper ion level to about 0.15 mg per liter (0.15 ppm). Initially, the dose can be adjusted between 0.15 and 0.18 ppm. Thereafter, the dose should be adjusted *on a daily basis* to remain between 0.115 and 0.15 ppm. It is important to make daily tests with a copper test kit and to maintain the dosage for at least 10 days.

(E) *Potassium Permanganate.* (Used as a general treatment for ecto-parasites of fish.)

Pond Treatments. The dose depends on the organic load of the water and on the species of fish. For scaled fish use 2 to 5 ppm depending on the organic load. A biological test should be done prior to treatment of an entire pond. Use an upper dose if the organic load is high. In high organic water a pinch of the chemical will immediately turn brown; in low organic water, a pinch of the chemical will remain reddish. Overdosing will burn fish; never use more than 2 ppm on skinned fish such as channel catfish or eels.

Pond Treatment for Oxygen Depletion. Use 2 ppm and repeat.

(F) *Methylene Blue.* Not used for pond treatment. Has been shown to interfere with biological filtration in home aquaria at 5 ppm.

Treatment of Other External Protozoans of Fish

Any one of the above listed treatments will generally remove other protozoans discussed in the text. As in marine fish, freshwater *Oodinium* is best treated with copper sulfate with careful attention being given to maintaining a low dosage (0.11 to 0.15 ppm) over a 14 day period.

In experiments at the University of Georgia, we have noticed that treatment of some lots of goldfish, channel catfish, and ornamental fish infested with *Costia* and *Chilodonella* were unsuccessful when formalin-malachite green was used at 25 to 0.1 ppm, respectively. In these cases it was necessary to use bath treatments of formalin, which removed the parasites.

Treatment for Hexamita.—*Chemical Treatments.*—(A) *Metronidazole.* (1-beta (hydroxyethyl)-2-methyl-5-nitroimidazole)

This drug is sold by poultry supply companies for hexamitiasis in turkeys under the trade name Emtryl®. It is also sold under the trade name Flagyl® for human use.

In Food: Blend into food (using ground liver, cod liver oil, or egg white as an adhesive agent) at a level of 0.25%. Feed for 3 days and examine fish to assess effectiveness of treatment.

In Water: Emtryl® appears to be more soluble in water than Flagyl®. It is used at a rate of 5 ppm of active ingredient.

NOTE: Emtryl® appears to control other external protozoan parasites.

Treatment for Monogenetic Trematodes.—*Indefinite Treatments.*—(A) *Masoten.* This is an organophosphate specifically sold as a fish medicant. Other organophosphates are sold under the trade names Dylox, Dipterex, Neguvon, Chlorophos, and Trichlorfon.

Ponds. Use at 0.25 to 1.0 ppm in ponds. Treatment should be repeated 3 times at 6 to 8 day intervals.

Aquaria. Use in freshwater aquaria at 0.25 ppm; saltwater aquaria at 1.0 ppm.

NOTE: (1) In instances where organophosphates have been used on a regular basis, dosages may have to be increased. Work at the University of Georgia suggests that levels as high as 25 ppm are required to remove monogenetic trematodes from goldfish.

(2) Organophosphates are not stable in water; aqueous solutions lose their effectiveness within 24 hr.

(B) Formalin.

Indefinite Treatment. Experience at the University of Georgia suggests that 25 ppm of formalin is not effective in removing monogenetic trematodes.

Bath Treatment. Bath treatments using dosages from 125 to 250 ppm until fish are stressed are effective. Maintain aeration during bath treatment. This will also remove external protozoans.

(C) Potassium Permanganate.

Pond Treatment. 5 ppm for scaled fish.

Dip Treatment. 1% solution for 10 to 40 sec.

Treatment for Parasitic Copepods: Lernaea, Argulus.—

Masoten®.—

Ponds. Use at 0.25 ppm weekly for 4 weeks. At temperatures over 26°C (80°F), use 0.50 ppm.

Aquaria. Use at 0.25 ppm in freshwater aquaria.

Treatment for Intestinal Tapeworms.—

Drugs to Mix in Feed.

(1) Di-n-butyl tin oxide. Blend in with food at 0.3% for 3 days. Ground beef liver, cod liver oil, or egg albumin should be used to adhere drug to food.

(2) Yomesan®. Blend in with food (as above) at 0.8% for 2 days. Repeat in 2 weeks.

Treatment for Bacterial Infections.—*Individual Injection of Large Fish.* —Large fish with ulcerations can be given injections of antibiotics. Injections are given intraperitoneally with a 22 gauge needle. Dosages are calculated on the same basis as for mammals. For example, chloramphenicol is administered at the rate of 10 to 30 mg per kg. In fish (small species), antibiotics can be given with a microliter syringe or the antibiotic can be diluted in physiological saline.

Incorporation of Antibiotics with Food.—Fish chows medicated with tetracycline antibiotics are commercially available. Occasionally, because of development of antibiotic resistant strains, other antibiotics may be required.

Addition of Antibiotics to Water.—In the aquarium industry, antibiotics are frequently added to the water. Use experience suggests that many types of antibiotics are absorbed by the fish and are of therapeutic value.

Treatment of Bacterial Diseases.—*Chemical Treatments.*—*(A) Copper Sulfate.* Used for columnaris infections. Use same dosage of chemical as for external parasites.

(B) Roccal®. Use for bacterial gill disease. Use 2 ppm for soft water, 4 ppm for hard water.

Antibiotics.—The following antibiotics have been used in fish culture to control or treat bacterial infections in fish. Use of antibiotics for food fish must be done in accordance with regulations as outlined by the Food and Drug Administration and by the Environmental Protection Agency. Local and state agencies may also have regulations limiting use of chemicals or drugs.

The following list of antibiotics was excerpted from references supplied from Dr. S.F. Snieszko, Eastern Fish Disease Laboratory, Kearneysville, West Virginia. At present, only sulfamerizine and oxytetracyclines are permitted for use in food fish.

List of Antibiotics.—

Chloramphenicol. (Chloromycetin®)
  (1) Use orally with food at 50 to 75 mg per kg body weight per day for 5 to 10 days.
  (2) Use single intraperitoneal injection in soluble form of 10 to 30 mg per kg.
  (3) Add to water at 10 to 50 ppm for indefinite time as needed.

Chlortetracycline. (Aureomycin®)
(1) Use 10 to 20 ppm in water. Add to feed at a rate of 10 to 20 mg per kg of food.

Furanace (P-7138) ®. (Nifurpirinol; 6-hydroxymethyl-2-pyridine)
(1) Used as bath but may be added to food. As bath: 1 ppm for 5 to 19 min; 0.05 to 0.1 ppm may be added for indefinite period to water.
(2) Orally for treatment: 2 to 4 mg per kg fish per day for 3 to 5 days.
(3) Orally for prophylaxis: 0.4 to 0.8 mg per kg fish per day as long as needed.

Furazolidone. (Furoxone N.F. 180 N.F. 180 Hess & Clark commercial products contain furazolidone mixed with inert materials)
(1) On the basis of pure drug activity, 25 to 75 mg per kg body weight per day up to 20 days orally with food.

Kanamycin. (Antibiotic also traded as Cantrex, Kanamycin, Resistomycin)
(1) Use 50 mg per 1 kg of fish or 25 to 100 mg per kg of food. Feed for 1 week.

Nalidixic Acid. (1-Ethyl-1,4-dihydro-7-methyl-4-oxo-1,8-naphthyridine-3-carboxylic acid); NegGram®; Wintomylon®
(1) Similar in action to oxolinic acid. One tablet per 50 to 100 liters of water for treatment of 3 to 4 days duration. Infections with Gram-negative bacteria.

Nifurprazine. (HB-115); (Nitrofuran, unstable in prolonged exposure to sunlight); 1-(5-nitro-2-furyl)-2-(6-amino-3-pyridazyl)-ethylene hydrochloride); (Carofur and Aivet are water soluble formulations)
(1) As bath: For indefinite period 0.01 to 0.1 ppm.
(2) In food: 10 mg per kg of food. Feeding for 3 to 6 days at a time.

Oxolinic Acid. (1-Ethyl-1-dihydro-6,7-methyl-enedioxy-4-oxo-3-quinoline carboxylic acid)
(1) Use for control of *Aeromonas* infections. Use orally 3 mg per kg fish once daily for 5 days.
(2) As bath: 1 ppm for 24 hr for columnaris disease.

Oxytetracycline. (Terramycin®)
(1) Use 50 to 75 mg per kg body weight per day for 10 days with food. (Law requires that it must be discontinued for 21 days before fish are killed for human consumption.)

Potentiated Sulfonamide. (Sulfadimethoxine potentiated with Ormetoprim)

(1) Use for control of furunculosis and other systemic infections. Use with feed at 50 mg per kg fish per day.

Sulfamerazine.
(1) Use 200 mg per kg body weight per day with food for 14 days. (Law requires that treatment must be stopped for 21 days before fish are killed for human consumption.)

Sulfamonomethoxine. (Trade name Dimeton®; water soluble)
(1) Use with feed as is at a rate of 100 to 200 mg per kg of feed. Use as needed.

Sulfisoxazole. (Gantrisin®)
(1) Use 200 mg per kg body weight per day with food.

Treatment of External Fungi (Saprolegnia).—

On Fish Eggs.—

Malachite Green.
Dip. 1500 ppm as a 10 sec dip for eggs.
Flush. 5 ppm as a 1 hr flush for trout eggs.

Formalin. Use 2000 ppm for trout eggs for 15 min.

On Fish.—

Malachite Green.
Indefinite Treatment. 0.1 ppm in water.
Bath Treatment. 1 to 3 ppm for 1 hr.
Dip Treatment. 66 ppm for 10 to 30 sec.

Mercurochrome.
Localized Treatment. Take commercial preparation of tincture of mercurochrome and swab on external fungi using cotton tip applicator.

Copper Sulfate. Treatment may kill spores but not mycelial growths. Dose depends on water hardness; do not use in water less than 60 ppm as $CaCO_3$.
Indefinite Treatment. Use 0.5 ppm if total hardness is 60 to 90 ppm as $CaCO_3$. Use 1 ppm if total hardness is 100 to 200 ppm as $CaCO_3$. Repeat on day 3 using half of the original dose.

NOTE: MANY OF THE SPECIFIC TREATMENTS GIVEN IN THIS CHAPTER ARE NOT AUTHORIZED BY THE U.S. FOOD AND DRUG ADMINISTRATION OR THE ENVIRONMENTAL PROTECTION AGENCY, WASHINGTON, D.C. BEFORE USING (PARTICULARLY ON FOOD FISH), APPROVAL SHOULD BE ACQUIRED.

SPECIAL ACKNOWLEDGMENT

The author of this chapter is indebted to Ms. Jeannine Gilbert, laboratory assistant and Doctoral Candidate in the Department of Medical Microbiology, for providing the scanning electron micrographs, and to Mr. Daniel Beisel, Medical Illustrator, for providing excellent interpretative pen and ink drawings.

REFERENCES

BLASIOLA, G.C., JR. 1978. Coral reef disease, *O. ocellatum.* Mar. Aquarist 7, 50-58.

HOFFMAN, G.L. and O'GRODNICK, J.J. 1977. Control of whirling disease (*Myxosoma cerebralis*); effects of drying, and disinfection with hydrated lime or chlorine. J. Fish Biol. *10,* 175-179.

MEYER, F.P. 1970. Seasonal fluctuations in the incidence of disease on fish forms. *In* A Symposium on Diseases of Fishes and Shellfishes. S.F. Snieszko (Editor). American Fisheries Society, Washington, D.C.

SHOTTS, E.B., JR. and BULLOCK, G.L. 1975. Bacterial diseases of fishes: diagnostic procedures for Gram-negative pathogens. J. Fish. Res. Board Can. *32,* 1243-1247.

SMITH, C.E. and PIPER, R.G. 1975. Lesions associated with chronic exposure to ammonia. *In* The Pathology of Fishes. W.E. Ribelin and G. Migaki (Editors). University of Wisconsin Press, Madison.

WEDEMEYER, G.A., MEYER, F.P., and SMITH, L. 1976. Chemical factors in fish diseases. *In* Diseases of Fish. Environmental Stress and Fish Diseases. S.F. Snieszko and H.R. Axelrod (Editors). T.F.H. Publications, Neptune City, N.J.

WOOTEN, R. 1978. Smolts in danger from *Costia.* Fish Farmer *1,* 27.

RECOMMENDED READING

AMLACHER, E. 1970. Textbook of Fish Diseases. T.F.H. Publications, Jersey City, N.J.

ANDERSON, D.P. 1974. Immunology. *In* Diseases of Fishes, Book 4. S.F. Snieszko and H.R. Axelrod (Editors). T.F.H. Publications, Neptune City, N.J.

BULLOCK, G.L. 1971. Identification of fish pathogenic bacteria. *In* Diseases of Fishes, Book 2B. S.F. Snieszko and H.R. Axelrod (Editors). T.F.H. Publications, Jersey City, N.J.

BULLOCK, G.L., CONROY, D.A., and SNIESZKO, S.F. 1971. Bacterial diseases of fishes. *In* Diseases of Fishes, Book 2A. S.F. Snieszko and H.R. Axelrod (Editors). T.F.H. Publications, Jersey City, N.J.

DULIN, M.P. 1976. Diseases of Marine Aquarium Fishes. T.F.H. Publications, Neptune City, N.J.

ELKAN, E. and REICHENBACH-KLINKE, H. 1974. Color Atlas of the Diseases of Fishes, Amphibians and Reptiles. T.F.H. Publications, Neptune City, N.J.

HOFFMAN, G.L. and MEYER, F.P. 1974. Parasites of Freshwater Fishes. T.F.H. Publications, Neptune City, N.J.

HUCKSTEDT, G. 1973. Water Chemistry for Advanced Aquarists. T.F.H. Publications, Neptune City, N.J.

KABATA, Z. 1970. Crustacea as enemies of fishes. In Diseases of Fishes, Book 1. S.F. Snieszko and H.R. Axelrod (Editors). T.F.H. Publications, Jersey City, N.J.

REICHENBACH-KLINKE, H.H. 1973. Fish pathology. T.F.H. Publications, Neptune City, N.J.

REICHENBACH-KLINKE, H.H. 1977. All About Marine Aquarium Fish Diseases. T.F.H. Publications, Neptune City, N.J.

RIBELIN, W.E. and MIGAKI, G. 1975. The Pathology of Fishes. University of Wisconsin Press, Madison.

ROBERTS, R.J. 1978. Fish Pathology. Bailliere Tindall, London.

ROBERTS, R.J. and SHEPHERD, C.J. 1974. Handbook of Trout and Salmon Diseases. White Friars Press, London and Tonbridge.

SARIG, S. 1971. The prevention and treatment of diseases of warmwater fishes under subtropical conditions, with special emphasis on intensive fish farming. In Diseases of Fishes, Book 3. S.F. Snieszko and H.R. Axelrod (Editors). T.F.H. Publications, Jersey City, N.J.

SCHUBERT, G. 1974. Cure and Recognize Aquarium Fish Diseases. T.F.H. Publications, Neptune City, N.J.

SINDERMANN, C.J. 1966. Diseases of Marine Fishes. Academic Press, London.

SNIESZKO, S.F. 1970. A Symposium on Diseases of Fishes and Shellfishes. American Fisheries Society, Washington, D.C.

SPOTTE, S.H. 1970. Fish and Invertebrate Culture: Water Management in Closed Systems. John Wiley & Sons, New York.

SPOTTE, S.H. 1973. Marine Aquarium Keeping. John Wiley & Sons, New York.

U.S. DEP. INTERIOR. 1973. Second Report to the Fish Farmers: The Status of Warmwater Fish Farming and Progress in Fish Farming Research. F.P. Meyer, K.E. Sneed, and P.T. Eschmeyer (Editors). Bureau of Sport Fisheries and Wildlife, Washington, D.C.

WEDEMEYER, G.A., MEYER, F.P., and SMITH, L. 1976. Environmental stress and fish diseases. In Diseases of Fishes, Book 5. S.F. Snieszko and H.R. Axelrod (Editors). T.F.H. Publications, Neptune City, N.J.

8

Processing and Marketing

E.W. McCoy and M.L. Hopkins

Marketing involves those activities that occur between production and consumption. Marketing can be direct as in a fish-out operation where the marketing functions are essentially performed by the consumer; or it can be complex, where the product is changed in form and undergoes many buying and selling transactions before reaching the ultimate consumer. When an aquacultural product is marketed through normal channels, and especially if the product enters interstate commerce, relatively good records are maintained on the quantity sold. Only estimates are available on the quantity sold through direct marketing channels.

For three freshwater aquacultural products, catfish, trout, and bait fish, the degree of competition from wild fish sources varies. Cultured catfish are estimated to make up approximately 50% of the quantity sold. Cultured trout represent not only all of the trout marketed, but essentially all of the trout captured from the "wild." For bait fish, the proportion of cultured and captured is indeterminate. In each case, the wild and the cultured fish are marketed through the same channels and become indistinguishable at the customer purchase level.

EXISTING PROCESSING AND MARKETING SYSTEMS

Catfish

Catfish marketing has developed on a localized basis with the supply dependent on wild fish captured from rivers and lakes. In 1969, wild catfish comprised approximately two-thirds of the total amount sold (Natl. Mar. Fish. Inst. 1976A). Wild catfish production has stabilized at about 6400 MT (14 million lb) per year while cultured and imported catfish have increased. In 1978, both cultured and imported catfish exceeded wild catfish harvest. Future increases in production likely will come from these sources.

Catfish imports come primarily from Brazil and represent capture from the wild. There is some evidence that the Brazilian yield of wild fish for export cannot be sustained on a long term basis. The secondary import source of ocean catfish represents several species and is an entirely separate product. A product is a perfect substitute only if the consumer cannot distinguish a difference in use.

The market structure for catfish is shown in Fig. 8.1. It is estimated that approximately one-half of the total production from commercial catfish ponds is sold to processors. The remaining one-half is sold to live haulers. The proportion of catfish produced in commercial ponds is not known. At least an equal amount is marketed through fish-out operations.

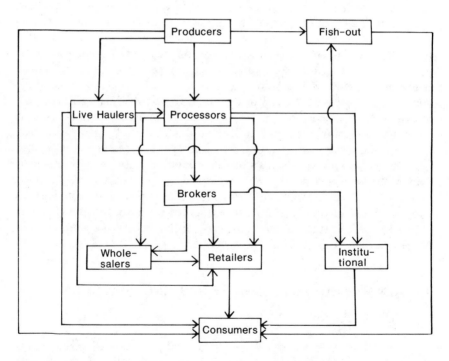

FIG. 8.1. MARKET STRUCTURE FOR CULTURED U.S. CATFISH

Commercial catfish operations are those where the fish are harvested by seining. The traditional method has been to drain some of the water from the pond and harvest the fish from a catch basin. This method entails a loss of water and requires facilities that can take the entire pond production in a very short period. Currently the partial harvest system is becoming more prevalent. The pond is seined without reducing the depth of water. The seine mesh size or mesh size of the live car allows the smaller fish to escape. The partial harvests improve cash flow, reduce pumping costs, reduce feed costs, and allow the producer to avoid the severe price seasonality of the traditional systems.

Fish-out is a recreational activity. Producers range from farmers with a sign and an honor box to operators who provide camping, picnic tables, and concession stands. Location of the pond with relation to population centers is a primary factor in profitability of fish-out operations (McCoy and Ruzic 1973). In evaluation, double counting the volume of fish sold through fish-out operations must be avoided. A substantial portion of the fish sold by commercial producers to live haulers is ultimately sold through fish-out. Some fish-out operators have their own live haul trucks, but most purchase from individuals who provide live haul service. Live hauling of fish in many instances is a cash only business. The producer receives cash on the pond bank and the live hauler receives cash when the fish are delivered. No data are available to indicate the precise scope of the business. During the spring and summer, live hauler competition bids up the price of fish. Essentially, all of the Arkansas production and much of the production from northern Mississippi are marketed to live haulers. These are two areas where catfish production is concentrated.

Combination producers exist; however, the number has declined during the last few years. Combination producers open a pond for fish-out in the spring, then drain and harvest the remaining fish for processing when fish-out demand declines. A variation is to trap or otherwise capture larger quantities of fish to supply local markets. Many producers sell to local markets after larger markets fail to develop for the entire quantity of fish. In southeastern Alabama and southwestern Georgia, failure of processing plants left many producers without a stable market for sale of fish. Innovative producers have established direct sales to restaurants, fish markets, and consumers. The volume moved through these channels is unknown. In a three county area of southeast Alabama, 100 to 200 ha (several hundred acres) of water were stocked with catfish in 1977 (Jensen 1977). None of the harvest from this hectarage (acreage) was marketed through processors.

Marketing of catfish has developed in three major directions, processing, fish-out, and direct sales. Only the growth of the processing sector has been documented. Within the processing sector, three types of ownership and two types of purchase arrangements have been attempted. In the early years of catfish production, cooperatives, including the giant Gold Kist; corporations, including Ralston Purina and Con Agra; and individuals were engaged in processing. During the late 1960s and early 1970s, the Organization for Economic Opportunity (OEO) financed several minority enterprises in catfish production. All of the government financed processing plants were organized as cooperatives. In 1978, no cooperatives remained in processing; only Con Agra remains of the original corporations. The remaining plants are operated by producer-processors. All plants are vertically integrated with substantial production hectarage (acreage) to ensure a supply during the period of competition with live haulers.

Many of the early processors envisioned catfish production as a second broiler industry. The feed company would supply all inputs except the pond and labor. Production technology in catfish was not equal to that of broilers, however, and both contractors and farmers were disappointed. No formal contract arrangements presently exist in the catfish industry except for agreements to supply fish to the processor at a specific date.

The processors have served as the market penetrators for catfish. More specifically, Con Agra and Farm Fresh have emerged as market developers. The remaining processors and the importers essentially follow into market areas developed by these two firms. The history of these two firms is illustrative of the entire processing industry.

Con Agra purchased the original catfish processing plant of Glover, Stevens, and True in Greensboro, Alabama. The plant had been developed when sales through Glover's market exceeded the capacity of the dressing facilities in the back room. Glover then developed a modern processing plant within 100 m (110 yd) of the original plant. In fact, the two plants share the same access road. Due to fluctuations in government agricultural policy and interest of the soil conservation service, catfish production also began to develop in the Mississippi delta area. Many producers in the delta area had experience with buffalo fish or shiners, and the transition to catfish was relatively easy. As the economics of catfish production and harvesting swung toward the delta type ponds, both Con Agra and Farm Fresh acquired processing plants in the area. Con Agra purchased the plant and production hectarage (acreage) of an individual while Farm Fresh acquired a bankrupt OEO plant. Neither could sustain processing on a year-round basis at either plant. Con Agra closed the Greensboro plant and committed resources to expanding production in the delta. Farm Fresh maintained its Greensboro Plant and expanded production ownership in both areas.

Without year-round production, the processors were faced with high processing costs due to inefficiencies. Only by marketing the fish in an ice pack could effective differentiation from the cheaper imported fish be maintained. Contracting for delivery during summer months proved ineffective. The processors were excluded from major markets by the inability to deliver a portion-controlled product on a regular basis. The firms aggressively expanded sales of the available product and developed new processed forms—fillets, nuggets, smoked, shanks, steaks, and other products. The firms were competing not only with other forms of fish, but with each other with varying forms of catfish. Other processing firms handled only ice pack or frozen catfish and could and did undercut prices after markets were developed. The competition drove the larger firms to develop accounts that only their assured volume could supply. The increased size and volume of the processing plant opened a gap at the lower end. The small restaurants, fish houses, or grocery stores that were the initial buyers now were unable to buy because the fish was too costly. Brokers and peddlers were introduced into catfish marketing. In 1978, the major processors were attempting to make direct sales in volume to single outlets or volume distribution to individual brokers in specific sales areas. The year 1978 was the first in which any processing plant operated at full capacity for the entire season. Assurance of supply will now allow the processors to fill institutional markets previously closed.

FIG. 8.2. HAULING RAINBOW TROUT TO A PROCESSOR

Rainbow Trout

Rainbow trout feed relatively high on the food chain. In addition, trout are native to cold water, growing most efficiently when water temperature is below 18.3°C (65°F). Waters of such low temperature are relatively nonproductive in the natural state. A natural coldwater lake or

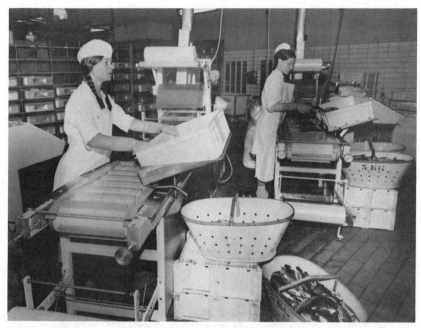

FIG. 8.3. PROCESSING RAINBOW TROUT

FIG. 8.4. SEPARATING RAINBOW TROUT BY WEIGHT WITH AUTOMATIC SEPARATOR

FIG. 8.5. PACKAGED RAINBOW TROUT READY FOR SHIPMENT

stream without fishing pressure probably could not support over 56 kg per ha or 1.5 km (50 lb per acre or 1 mi.) of lake or stream. With harvesting for market, natural populations of trout were soon dissipated. All rainbow trout presently marketed come from cultured stock. Even the trout taken from lakes and streams are cultured in private or government hatcheries and released into the wild.

Location theory suggests that *Ceteris paribus* facilities should be located near the center of the market area. However, the unique Idaho water resource has centered the major production area at the extreme edge of the United States major markets, and the location has influenced the form of the product and the type of marketing strategy.

Major markets in the United States are east of the Mississippi River and especially in the Boston-New York-Philadelphia-Washington, D.C. corridor. Any market expansion must eventually penetrate this market area. The west between the Mississippi and the Pacific Coast is sparsely populated and marketing costs per unit response are very high.

Marketing Idaho trout in eastern markets entails high transportation costs and requires a frozen product. Feed ingredients must be transported into the production area. All of these factors combine to create a high priced item at the consumer level. Unlike catfish, the rainbow trout does not have to compete with fish from the wild or with imports. Due to the exacting water quality and temperature requirements, expansion of the industry is also limited in terms of supply sources. Under these specialized conditions, rainbow trout has been identified as a luxury item that commands a luxury price.

In a market test in Alabama in 1978 (McCoy and Jensen 1978), pond raised rainbow trout were sold in the round at the pond bank for $3.30 per kg ($1.50 per lb). Channel catfish were available and sold for $1.10 per kg ($0.50 per lb). The majority of the buyers had previously eaten rainbow trout at some other location. Few southerners purchased the fish since they were not familiar with rainbow trout and were not accustomed to paying $3.30 per kg ($1.50 per lb) for pond raised freshwater fish. Many of the customers did purchase pompano at $3.30 per kg ($1.50 per lb). These buyers were aware of pompano and regarded it as a luxury product. Luxury product designation among fish thus is in the mind of the consumer.

Valid data on marketing are not available since no agency has responsibility for inland freshwater commercial fish production. Comparative production figures for trout, catfish, and salmon are listed in Table 8.1. During the 15 year period, trout production increased fivefold. A substantial portion of the increase was due to changes in production technology, i.e., greater yield from the same surface area. No major marketing changes were made to move the increased production.

TABLE 8.1. COMMERCIAL PRODUCTION OF TROUT, CATFISH, AND SALMON FOR 1960, 1970, AND 1975

| | Commercial kg (lb) | | | | | |
| | 1960 | | 1970 | | 1975 | |
Species	kg	lb	kg	lb	kg	lb
Trout	3182	7000	12,182	26,800	45,500	20,682
Catfish	145	320	18,182	40,000	62,000	28,182
Salmon	—	—	—	—	340	155

Trout, like catfish, are marketed fresh on ice, frozen, and in the round. Live sales through fish-out ponds predominate in the east. The mountainous region of Virginia, North and South Carolina, Tennessee, and Georgia have numerous fish-out operations devoted primarily to tourists.

The marketing distribution system has been described by Brown (1977). Except for the emphasis on frozen sales, the channels of distribution for trout and catfish are very similar. Trout marketing has reached the position toward which catfish is moving. Most of the sales are made through fish brokers and direct producer sales are very limited in terms of the overall quantity marketed.

Yellow Perch

The production of cultured yellow perch is still in its experimental stage since the technology is not fully developed yet. This new aquaculture research has drawn serious consideration because of the declining catch from the lakes due to increasing pollution and lack of conservation controls along with the continuous market demand. The Lake Erie, Lake Michigan and Green Bay commercial fisheries are the sources of yellow perch in this country.

Yellow perch is famous in the Midwest, especially in Wisconsin, where the bulk of the market exists. The popularity of the fish in the midwest area has created a tremendous increase in demand beyond the supply coming from the lakes. To meet the demand, imported yellow perch from Canada and Holland, including walleye pike, have also been marketed. Almost all yellow perch that are marketed commercially in the country are caught from the wild except those imported from Holland. In 1975 around 16,000 kg (35,000 lb) of cultured yellow perch fillets were imported from Holland (U.S. Bur. Census [various years]).

After research on the culture of yellow perch was initiated in 1973, a few fish farmers began experimenting with raising yellow perch. The quantity sold has been very minimal, around 250 kg (500 lb) in the last two years. The fish were marketed as fresh fillets for $9.35 per kg ($4.25 per lb) to grocery stores and friends of the fish farmers. The retail price at the grocery stores is $10.34 per kg ($4.70 per lb). The price for wild yellow perch ranges between $0.66 and $3.30 per kg ($0.30 and $1.50 per lb) in the round, depending on the season (Soderberg 1978).

Tilapia

Tilapia which are marketed in the United States are either cultured or are captured from wild stocks in Florida. The tilapia culture industry is very small with only 11 known operations producing tilapia for sales (Hopkins 1978). Four of the 11 producers culture tilapia primarily while the remainder produce other fish species as their main crop with tilapia as secondary crop. The capture fisheries harvest tilapia which have escaped into Florida waters and have established very large populations in some of the extremely fertile lakes. The main constraint on the industry is that tilapia are warmwater fish and will not survive at temperatures much below 12°C (55°F) for any significant length of time. The culture operations use geothermally heated water, heated effluents, or indoor facilities to keep the fish alive through the winter.

The cultured fish are sold to 5 markets: the food fish market; the aquatic weed control market; the experimental fish market; the bait fish market; and the catfish forage market. Except for the food fish market, all of the markets deal with live fish. Four of the operations serve only one of the markets while the rest of the operations sell to more than one market. However, each operator sells primarily through only one market.

Five operations produce tilapia primarily for the food fish market. Two other operations also sell in this market.

Dress-out percentages vary from around 65% for completely dressed fish to 80 to 85% for gutted only. On an unprocessed basis, the prices vary from $1.36 to $2.07 per kg ($0.62 to $0.94 per lb). This does not consider the expenses incurred in processing. Retail prices for cultured tilapia were not precisely measured but were estimated to be 30 to 60% higher than the price received by the producers.

The industry is very young with the longest period of marketing tilapia as food fish being only 3 to 4 years, with the average being less than 2 years, and the shortest only 6 months. In addition, the firms have been increasing their operations. This complicates estimating yearly sales. The best available estimate, based on a 50 week year and year round availability, is approximately $200,000 per year at the producers' level for 90,900 kg (200,000 lb) of fish.

The promotion of the fish has been minimal except for labels and recipes on the packages and some advertising by retailers. The selection of an appropriate name for the fish has been a source of many problems. The name tilapia is unfamiliar to consumers so the fish has also been called Colorado perch, fillets of bream, Colorado panfish, St. Peter's fish, panfish, and Florida perch. The name African perch has been used in Auburn University marketing experiments (Crawford et al. 1978). Difficulties with the government, which prohibited the use of the name Colorado perch for tilapia, are stimulating more efforts in determining the best name to use for product identification.

Six operators supply tilapia for aquatic weed control. Three operations used this market as their main outlet. The primary species are T. zilli and T. aurea. T. zilli is used primarily for macrophyte control while T. aurea is being used to control algae.

In 1977 orders for T. zilli amounted to over 1.5 million fish. The market wants small breeding adults 7.5 to 10.0 cm (3 to 4 in.) long for delivery in the spring. Current prices are $0.025 per 2.5 cm (1 in.) including delivery. Difficulties in spawning the fish in the late autumn and early winter, in order to have the desired size fish available in spring, have caused the supply to be less than the demand.

The bait fish market is currently supplied by three operations. T. aurea is used by commercial fishermen because it does not die quickly during the hot summer months when goldfish and small carp, the traditional bait, do. The desired size is 1.0 to 2.5 cm (½ to 1 in.) long. Some research is being conducted on using tilapia for chum by the tuna fisheries in Hawaii but results are not available.

The last market for cultured tilapia is the experimental market. This is a catch-all for all remaining markets. Fish are sold here for research purposes by universities and foundations and for stocking by individuals who are trying to grow tilapia. Because these individuals have not gone into commercial production, they are considered to be experimenting with the fish. Four operations sell to this market. Prices vary according to the species and size. The price range is $0.10 to $10 apiece but the common species are sold at the same rate as fish used for weed control.

Chinese Carp

There are three Chinese carps in the United States, silver carp, *Hypophthalmichthys molitrix*, grass carp, *Ctenopharyngodon idella*, and bighead carp, *Aristichthys nobilis*. Only the grass carp is being commercially produced and marketed although there has been experimental marketing of the other two species.

The grass carp is widely called white amur in reference to the native habitat of the fish and to minimize the negative connotations associated with carp. Because of fears that the fish may damage the environment, 35 states have effectively banned the fish (Sports Fish. Inst. 1977) as of March 1977. The states where the fish have not been banned as of March 1977 are Alabama, Alaska, Arkansas, Delaware, Hawaii, Kentucky, Indiana, Iowa, Massachusetts, Mississippi, New Jersey, North Dakota, Rhode Island, and South Dakota. The number of states banning and allowing grass carp is continually changing and causes large uncertainty in the market.

There are either 4 or 5 commercial grass carp producers in the United States. The two largest operations are in Arkansas and the remainder are in Alabama.

The distribution system is relatively diffuse with both individuals and dealers purchasing directly from the producers. In Alabama, there are 3 main dealers in addition to the producers. In Kansas there are 4 to 5 dealers.

The Arkansas producers sell primarily 20 to 32.5 cm (8 to 11 in.) fish. Grass carp of this size are not subject to excessive bass predation. The price is $2.00 to $2.25 per fish for small quantities and $1.50 per fish by truckload. With the recommended stocking rate of about 30 fish per ha (12 fish per acre) only dealers buy in truckload quantities. One of the producers also sells to at least 4 foreign countries. Prices of grass carp in Alabama vary from $0.50 per fish for 5.0 to 7.5 cm (2 to 4 in.) fish to $1.50 to $2.00 for 22 to 25 cm (9 to 10 in.) fish. The dealers sell at $3.00 to $3.50 per fish in Alabama and $3.00 to $5.00 per fish in Kansas.

The two main producers have been selling grass carp since 1973. The Alabama producers have been in operation since 1973 and 1976 respectively. The dealers have been selling the fish for 2 to 5 years. The production and marketing of grass carp are only secondary activities of all persons interviewed. Most of the operations are engaged in producing other fish such as minnows and catfish.

Salmon

As indicated in Table 8.1, salmon were not included among cultured fish marketed in 1960 or 1970. By 1975 salmon marketing reached approximately the same level as catfish in 1960. Cultured salmon represent an insignificant portion of the total quantity marketed. During 1975 over 90.45 million kg (201 million lb) of salmon were landed from commercial capture fisheries (Natl. Mar. Fish. Serv. 1976A). Most of the capture fisheries salmon are marketed in the canned form. Over 29 million kg (65 million lb) (product weight) were canned in 1975.

Salmon are cultured in cages in embayments and marketed when they reach about 0.45 kg (1 lb) in size. The production techniques are very similar to those used to culture other marine species. Salmon are also raised by smoltification. In effect the fish are imprinted with a specific location, then released to graze in the ocean.

The culture of salmon has taken two divergent production directions. Unlike the catfish or trout industry, the infant salmon culture industry was started by major commercial concerns.

The fish raised from cage culture are marketed in a fresh or frozen form as a competitor to rainbow trout. Most of the fish are sold in inland markets. Extensive market testing was conducted by the firms in the industry to determine consumer preference for size and form of the fish. Currently the producers have not recovered the research and development costs of market development.

Salmon are positioned as a prestige item in the market. Cost of production, processing, and marketing is relatively high and price per unit must be correspondingly high to yield sufficient returns to the factors of production.

Fish captured on return from sea grazing are marketed in a manner similar to salmon captured in commercial harvests. The fish are larger than those grown in cage culture. A portion are sold fresh or smoked while others are processed by canning.

Salmon, in 1978, represent a striking contrast to rainbow trout and catfish. The amount of harvest is relatively insignificant compared to the other two species. The amount of capital available to develop the industry is large compared to the other industries. The market development and penetration exceed that of the catfish industry and equal the trout effort. The availability of production sites and other nonmarket factors appear to be the major impediments to growth and development of the industry.

FIG. 8.6. BAIT FISH RETAIL OUTLET

FIG. 8.7. TROUT FEE FISH-OUT POND

Bait Fish

Bait fish marketing in the United States involves challenges and risks. Bait fish are only sold live. The size should be within the range acceptable to fishermen. Neither the producers nor the dealers have a perfect knowledge of the quantity of fish that will survive from the time

FIG. 8.8. "SUCCESS" AT A TROUT FEE FISH-OUT POND

FIG. 8.9. A TROPHY BASS AND CATFISH CAUGHT AT PATRICK'S FEE
FISH-OUT OPERATION, TIFTON, GA.

they leave the pond until they reach the ultimate consumers. Proper
handling of bait fish has been developed to minimize losses and to

maintain live bait, yet even to the experienced producers, there is no guarantee that losses can be counteracted.

The bait fish industry is closely tied to sports fishery. The industry developed because of the tremendous increase in the number of fishermen and fishing areas. Before the existence of cultured bait fish, the fishermen could either seine bait fish from the natural lakes and streams or buy them from bait shops whose supplies come from the wild.

In the early 1950s, the culture of bait fish received much attention because the demand far exceeded the supply coming from the wild. Moreover, the availability of bait from the wild decreased during the peak fishing season and was seasonal. The techniques for bait fish culture were developed to allow year round production. The common practice among bait producers is to produce the small size bait that is highly demanded and is not caught in the wild.

The bait fish sold throughout the country came both from the wild and from pond production. The wild bait fish are tuffy or fathead minnow harvested mostly from Minnesota and South Dakota. This bait is transported to Florida and the Tennessee Valley area (Ratliff 1978). The data on the volume of catch from the wild are unknown. The cultured bait fish are mostly golden shiner and some fathead minnow. The large and flourishing producers of golden shiner are south central states such as Arkansas, Missouri, Mississippi, and Louisiana. The largest minnow producer located in Lonoke, Arkansas, has shipped around 11.25 million kg (2.5 million lb) of live bait per year to distribution points across the southeastern and southwestern parts of the country. The production of cultured bait fish has increased 10 times in 15 years. In 1960, production was only 1.1 million kg (2.4 million lb) while in 1975, total production was 10.8 million kg (24 million lb). Between 1975 and 1977, production remained constant (Mitchell 1978).

Most of the cultured bait fish are sold within the country although there is an unspecified quantity that is exported. The largest users of bait fish are fishermen fishing for black crappie. Black crappie is noted to bite only on live bait. Bass fishermen use live bait whenever available, but the most commonly used bait for bass is artificial (Davies 1978). In lakes and reservoirs where crappie fishing is abundant, the demand for live bait fish exists. The demand is seasonal and uncertain. In most places, over 60% of the bait fish consumed in a year is sold within a 90 day period during the spring (Hudson 1968B). The uncertainty of the demand is caused by unpredictable factors such as weather conditions, especially the weekend weather, and how the fish are biting.

The marketing system for bait fish has been developed by the bait producers in such a way that the system is dominated by them. Bait producers are fish farmers in business with an extensive and expensive investment in marketing facilities and customer development. Fish farmers are those that raise bait fish but depend on the producers to harvest and market their crops (Hudson 1968B). Jobbers or distributors are those that buy bait in truckloads and distribute it to wholesalers and retailers (Ratliff 1978). The market structure for cultured bait fish is shown in Fig. 8.12.

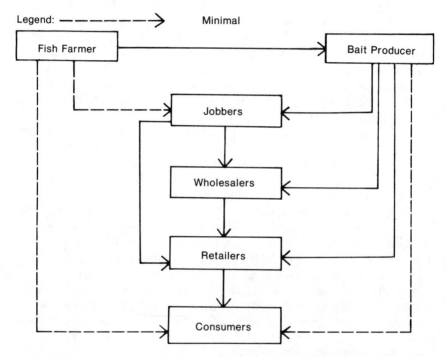

FIG. 8.10. SCHEMATIC DIAGRAM OF MARKET STRUCTURE FOR BAIT FISH IN THE UNITED STATES

The bait producers establish their own distributors throughout the areas where live bait fishing is located. In areas where abundant fishermen and fishing areas exist, more distributors are established. The largest minnow producer had 7 distributors in Texas, which is the largest bait fish market in the country.

The distributors or jobbers handle the bulk of the bait fish produced by the producers. They in turn distribute the bait to wholesalers and retailers. The payment arrangement between producers and distributors or jobbers is mostly on a cash basis. This arrangement protects the producers from not being paid because of heavy losses due to the buyers' negligence. Losses due to death of the bait fish during transport are shouldered by the producers. The common practice is to replace the dead fish on the next delivery.

The bait fish retailers, unlike other retail stores, are scattered around strategic places where fishermen are most likely to come by. Live bait is sold at service stations near the lakes where fishermen are likely to stop to fill up their boats with gas and at bait shops near lakes and reservoirs.

Consumers of bait fish in general are more concerned with the quantity received and the lowest cost possible than they are with the quality. Their rationale is that they still come up ahead even if they lose some at a low price. This attitude of the consumers enables the wild bait catch to compete with cultured bait.

Competition from the wild catch to some degree is hurting the cultured bait business because the prices for wild bait are lower than for cultured bait. But, in areas other than price, the cultured bait business has grown tremendously over the years because the supply is reliable, the bait are hardy, and the size and color satisfy the desires of the fishermen. It had also been predicted that catch from the wild would decline in the near future which in effect would lessen this competition. Apparently, the situation still continues. The anticipation of a decrease in wild catch, which would be caused by the decline in stock, did not happen because the problem was not with the stock but with the harvesting method used (Gordon 1968).

In the bait fish business, price is associated with size. Different sizes of

FIG. 8.11. MACHINE AUTOMATICALLY SIZING BAIT FISH

bait have different prices. The smallest size bait within the saleable size range has the highest price per kg or lb, while big size bait is priced lower (Table 8.2). The difference in prices between sizes might be attributed to the absence of competition from the wild in the small size bait and the degree of risk due to mortality in handling small delicate fish. Moreover, the closer the size of the cultured bait to the size of the wild bait, the lower the price because the cultured bait are sold at the same price as the wild bait. The price of the bait fish at the wholesale level is expressed in dollars per kg or lb of fish while it is dollar per dozen (12) at the retail level.

TABLE 8.2. SIZE RANGE OF LIVE BAIT FISH ACCORDING TO USE AND WHOLESALE PRICE, 1978

Size[1] (in.)	Trade Name	Use	Price/lb[2]	Price/kg
1–1.5	Small crappie	Crappie	$2.25	$4.95
1.5–2	Large crappie	Crappie	1.85	4.07
2–3	Small bass	Bass, perch	1.35	2.97
3–4.5	Large bass	Bass, pike	1.15	2.53
6	Troutline	Catfish, carp, pike and muskellunge	1.00	2.20

[1]2.5 cm = 1 in.
[2]Source: Anderson Minnow Farms (undated).

Fishermen establish the size range for live bait according to their use (Table 8.2). Each size range is distinct from another so that they do not overlap and so that the prices are different. Fishermen are very particular about the size of bait they use. When the size does not fall within the range desired, they refuse to buy them. Such problems happened during an initial marketing study on bull minnow, *Fundulus grandis*, at Gulf Shores, Alabama. Based on the survey, 5.0 to 7.5 cm (2 to 2½ in.) live bait were desired for flounder. When the desired bait were put on the market, the fishermen complained that the bait were too small. It was observed that the technique of measurement used by the fishermen was different from that of the biologists. The tail of the fish is not included in the measurement by fishermen based on a practical reason that the tail is transparent and cannot be seen under water. Size ranges among states are different. In the north, the medium size minnow would be the small size in the southeast. Size ranges of bait fish are commonly expressed in number of kg (lb) per 1000 fish.

Within the market structure, the flow of supply from bait producers to ultimate consumers is associated with a price margin of around 200%. The wholesale level takes a margin between 80 and 100% while at the retail level the margin is at 100% and above. The large price margin between each level is attributed to the seasonal and uncertain nature of demand, high transportation and production costs, and the perishability of the product (Hudson 1968B). This pattern of price margin exists only where competition is keen. In areas where competition is less keen while demand is great, lesser price margins are applied without hurting the business.

When the bait fish business started, most production was geared only for local consumption. As the business developed, the market was getting farther from the source. The increase in production along with competition in the local area encouraged the producers to transport the product to areas with less competition. It was then cheaper to transport bait to these areas because the price was high, large volumes of bait were sold, and the cost of transportation was lessened.

Over the years, as even more people entered the business, in addition to the increased volume of catch from the wild, transportation became an integral part of bait fish marketing. Most of the markets are scattered throughout the country so that the only way to get the product to consumers is by transporting it. Unlike other products, bait is not storable or processable. It is only marketed live; therefore, it is highly perishable. Transporting bait to distant areas involves timing, extra care in handling, and requires specialized transport facilities to minimize mortality. All established large-scale producers maintain their own transportation facilities to ensure delivery when needed.

The energy crisis experienced in the United States has greatly affected the bait fish business. Due to the increase in transportation cost, some large-scale producers are not expanding in either hectarage (acreage) or production. At the same time, they are cutting down on the delivery service and some are diversifying their operation.

The main problem faced by bait producers is the low price received for their premium quality bait. Competition from the wild catch has always been pointed out as the primary cause for the lowering of price, but there are other factors that will affect price that do not concern the competitors. Ultimate consumers also influence the price. Some of the determinable factors that affect price are the demand for bait fish in the local area, prevailing wage rates, frequency of fishing, general economic condition of the area, availability of substitutes, and the cost involved in fishing (Hudson 1968B). These factors concern the consumers who are not part of the marketing schemes practiced by bait sellers.

Tropicals

See tropical fish section written by Ross Socolof in Chapter 5.

Goldfish

Goldfish sales by producers are usually in 2.5 cm (1 in.) classes. Classes run from under 2 in., 2 to 3 in., etc., up to the 10 to 12 in. class. Different prices prevail for each class.

Some farm sales are direct from farm producer to retailer. In these cases, the retailer may have 4 to 5 stores and be close enough to the farm that picking up 5 to 10 boxes of fish is economical. The fish are packed in polybags with oxygen added and the bags are packaged into cardboard boxes for shipment. This procedure is followed whether sales are direct to retailers or to wholesalers.

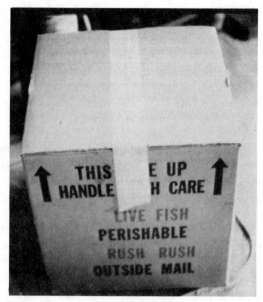

FIG. 8.12. PACKAGED GOLDFISH READY FOR SHIPMENT

Most farm sales go initially to wholesalers. Nearby wholesalers or wholesalers with large orders may buy at the farm. Most sales are by air freight. In the past some shipments have been made by United Parcel Service. After arriving at the wholesalers, the fish are stored in vats having running water and a dependable oxygen supply. Sometimes in case of a fungus problem the fish may be given a saline treatment. The wholesaler tries to move the fish in less than a week in order to make room for his next purchase. In a few cases, fish may remain at the wholesaler for 2 weeks. The wholesaler then delivers the fish to retail shops in poly-bags and boxes. Individual routes may go out 300 km (200 mi.).

The wholesale markup is usually about 100%. The retail shop may then mark up the wholesale price by 100 to 200%. Mortality from farm to wholesale is usually less than 1%, which is replaced free of charge on the next delivery; the wholesaler to retailer mortality is less then 1%; average mortality in retail shops is usually less then 5%.

Some states regulate the sales of goldfish for bait or for release in private or public waters. The producer needs to contact state officials before attempting to sell goldfish for bait.

MARKETING SUGGESTIONS FOR NEW ENTERPRISERS

Catfish

There are a limited number of catfish processors as marketing outlets. Processors also pay the lowest price per kg (lb) and will purchase the entire production from a pond, thus simplifying problems with harvesting.

A producer with a relatively small surface area of water 8 ha or less (20 acres or less) should first evaluate the local supply and demand situation. In an area of few catfish producers, primary outlets for sales would be local restaurants, supermarkets, and individuals. The restuarants and supermarkets would require dressed fish, and individual sales could be increased if dressing service were available. The producer should check with the local health department to determine the requirements for selling dressed fish.

Fish-out should also be considered as an alternative in areas of high population density. Recreational sales require additional facilities at the pond but returns per kg (lb) of fish caught are higher than with other forms of sales. Capital costs of equipment for complete pond harvest are high and should be avoided by small producers unless processor or live hauler sales are the only viable alternatives.

Rainbow Trout

Except for producers in Idaho, most rainbow trout farmers should attempt to market fish through fish-out operations. Surplus production can be marketed in local areas. Rainbow trout are much more sensitive to oxygen and temperature changes and require skilled handling for live transport. The fish are aggressive feeders and well suited for recreational fishing. Since the fish are a gourmet item, limited quantities are sold through small supermarkets or local small town restaurants. Producers should establish the local demand for the fish and attempt to supply sufficient quantities on a regular basis. With large volume operations, producers will be in competition with frozen trout. Fresh trout generally can favorably compete; however, as weight increases, the frozen form is preferred by buyers due to storage and handling.

Tilapia

The market for tilapia is not well established outside of Florida. The fish resembles several sport fish caught in the south and the taste is comparable. Producers in areas where tilapia are not common may have to market fish on consignment. Market penetration can be attempted by point of sales advertising and use of local media. It is particularly useful to gain local exposure by feature articles in local newspapers or on television. The unique nature of the fish as a mouth brooder, plankton eater, fast grower, and of African origin may lead to news articles.

No national markets or processor exists for tilapia as food fish. Thus each potential producer or group of producers must carefully assess the market situation before committing resources to tilapia production.

Chinese Carp

The principal market for Chinese carp as food fish is with specific ethnic groups. A potential market for fingerlings exists for use in catfish production ponds and sewerage lagoons. Some states have banned production of grass carp due to fear of ecological damage. The silver, bighead, and grass carps have excellent flavor and texture, but the boniness of the silver and bighead create problems in marketing the fish to American consumers.

The fish are excellent grazers and appear to enhance the production of channel catfish. Producers might consider stocking grass carp in ponds with excessive weed growth. Silver or bighead carp would be most appropriate in ponds with heavy growths of algae. The fish can be produced at minimal cost and would be excellent for home consumption with limited sales of excess fish in local areas.

Salmon

Small scale salmon producers, as with all small scale producers, should first establish market sales in the local area. Continuity of supply is crucial in marketing to stores and restaurants. If the supply is small the producer should establish on premise sales—advertising the availability of the fish during the limited time period.

Large scale producers have followed the traditional fish broker route. Fish brokers have contracts with major markets, including food and restaurant chains. The broker operates on commission from sale of fish. Most brokers would not be interested in handling small volumes; however, a group of producers might pool fish for sales through a broker.

Bait Fish

The changes in energy costs have greatly altered production and marketing possibilities for bait fish. Fishing with live bait occurs throughout the United States; however, the major production area for bait has been in Arkansas. Increased transportation costs and a rapidly declining water table in the production area have greatly increased costs so local producers can now fill local market needs at a competitive price.

The demand for bait is seasonal and varies by region of the country. The prospective producer should visit fish camps and other bait sales areas to establish the scope of the market. Direct sales to fishermen can be made if the production facility is convenient to the fishing area or on a well traveled route to fishing areas.

Many live bait dealers have discontinued sales due to lack of supply over the past two years (1977–78), but these dealers may be interested in purchasing bait fish if assured of an adequate supply during high demand periods. As with any enterprise, the prospective entrant should very carefully evaluate marketing before committing resources to production.

Tropicals

See section on tropical fish written by Ross Socolof in Chapter 5 for marketing suggestions.

Goldfish

Producers contemplating entry into the goldfish market usually have few problems with entry. Nearly any wholesaler or retailer will order a trial shipment. If supply is dependable, quality high, and prices reasonable and competitive, a market is available.

Lists of wholesalers can be found in annual directories published by: (1) Pets-Supplies-Marketing, 1 East First Street, Duluth, Minnesota 55802; or (2) Pet Age, 2561 North Clark Street, Chicago, Illinois 60614.

Several other directories are published by other firms. Retail names can be obtained from the yellow pages of telephone directories.

GENERAL MARKET RECOMMENDATIONS

Any individual contemplating fish culture should first determine primary goals. If the venture is planned for profit, then market appraisal is necessary. Freshwater fish represent an insignificant portion of total United States fish sales; only rainbow trout are marketed nationwide. There is no assurance that fish produced can be sold at a price sufficient to cover costs, so only in areas where major processors are located can a producer be assured of a sale of fish. Even in those areas the time of sale cannot be guaranteed. Producers with limited supplies should always consider assuming some of the marketing functions to increase the revenue from fish production. By processing and direct sale of fish the producer should be able to realize about 100% over the live weight sales price.

REFERENCES

ANDERSON, M.R. 1978. Personal correspondence. Yazoo City, Miss.

ANDERSON, R. 1978. Personal correspondence. Anderson Minnow Farms, Lonoke, Ark.

ANDERSON MINNOW FARMS. (Undated). Promotional brochure from Anderson Minnow Farms, Lonoke, Ark.

ANON. 1975. Aquaculture: raising perch for the midwest market. Univ. Wis. Sea Grant Coll. Program, Madison, Advis. Rep. *13*.

BAILEY, W.M. 1975. Commercial fishery industry survey July 1, 1974 to June 30, 1975. Fed. Aid Commer. Fish. Proj. *2-243 R*, Segment 1.

BROWN, E.E. 1977. World Fish Farming: Cultivation and Economics. AVI Publishing Co., Westport, Conn.

BURTLE, G. 1978. Personal correspondence. Auburn Univ., Auburn, Ala.

CRAWFORD, K.W., DUNSETH, D.R., ENGLE, C.R., HOPKINS, M.L, McCOY, E.W., and SMITHERMAN, R.O. 1978. Marketing tilapia and chinese carps. Symp. Culture Exotic Fish., Fish Culture Section, Am. Fish. Soc., Atlanta, Jan. 4, 1978, 241-257.

DAVIES, W. 1978. Personal correspondence. Auburn Univ., Auburn, Ala.

EDDY, B. 1978. Personal correspondence. Quitman Minnow Farm, Quitman, Ga.

GORDON, W.G. 1968. The bait minnow industry of the Great Lakes. U.S. Dep. Interior Fish Wildl. Serv., Bur. Commer. Fish. Leaflet *608*.

HARRY, G.V. 1968. Handling and transporting golden shiner minnows. Proc. Commer. Bait Fish. Conf. Texas A & M Univ., College Station, Tex., March 19–20, 37-42.

HILL, L. 1978. Personal correspondence. Lonoke, Ark.

HOPKINS, K.D. 1978. The marketing of tilapia and chinese carps in the United States. Auburn Univ. Dep. Agric. Econ., Auburn, Ala. (unpublished)

HUDSON, S. 1968A. Guidelines for marketing. Proc. Commer. Bait Fish Conf., Texas A & M Univ., College Station, Tex., March 19–20, 43-45.

HUDSON, S. 1968B. Personal correspondence. Cheney, Kan.

JENSEN, J.W. 1977. Feasibility of commercial aquaculture in the Piedmont Region of Alabama. Unpublished M.S. Thesis. Auburn Univ., Auburn, Ala.

McCOY, E.W. and BOUTWELL, J.L. 1977. Preparation of financial budget for fish production, catfish production in areas with level land and adequate ground water. Auburn Univ. Agric. Exp. Stn., Auburn, Ala. Circ. 233.

McCOY, E.W. and CRAWFORD, K.W. 1975. Catfish are not the only fish in the pond. Highlights Agric. Res. 22 (2) 10.

McCOY, E.W. and HOPKINS, M.L. 1977. Establishing a market for an exotic fish species. Highlights Agric. Res. 24 (1) 4.

McCOY, E.W. and JENSEN, J.W. 1978. Market test of rainbow trout grown in still water ponds. Annu. Rep. Dep. Fish. Allied Aquacultures, Auburn Univ., Auburn, Ala. 13.

McCOY, E.W. and RUZIC, J.E. 1973. Alabama's recreational catfish ponds. Ala. Agric. Exp. Stn. Bull. 451.

MITCHELL, A. 1978. Personal correspondence. Fish Farming Exp. Stn., Stuttgart, Ark.

NATL. MAR. FISH. SERV. 1976A. A Review of Catfish Processing. National Marine Fisheries Service, Little Rock, Ark.

NATL. MAR. FISH. SERV. 1976B. Fisheries of the United States, 1975. National Marine Fisheries Service, NOAA, U.S. Dep. Commer. Current Fish. Stat. 6900.

NATL. RES. COUNC. 1978. Aquaculture in the United States: Constraints and Opportunities. National Academy of Sciences, Washington, D.C.

RATLIFF, J.B. 1978. Personal correspondence. Ozark Fisheries, Guntersville, Ala.

SODERBERG, R.W. 1978. Personal correspondence. Univ. Wis. Sea Grant Coll. Program, Madison.

SPORTS FISH. INST. 1977. Grass carp outlawed in 35 states. Sports Fish. Inst. Bull. 282.

U.S. BUR. CENSUS. (Various Years). Imports: general consumption, Schedule A. U.S. Dep. Commerce, Washington, D.C.

Appendix

TABLE A.1. CONVERSION FACTORS

Metric and American Systems

10 millimeters (mm)	=	1 centimeter (cm)
1 centimeter (cm)	=	0.3937 inches (in.)
1 meter (m)	=	39.37 inches (in.)
1 kilometer (km)	=	0.62137 mile (mi.)
1 square meter (m^2)	=	10.76 square feet (ft^2)
10,000 square meters (m^2)	=	1 hectare (ha)
1 hectare (ha)	=	2.471 acres
1 acre	=	4047 square meters (m^2)
1 liter	=	1.0567 liquid quarts (liq qt)
1 quart (qt)	=	0.25 gallon (gal.)
28.4 grams (g)	=	1 ounce (oz)
16 ounces (oz)	=	1 pound (lb)
454 grams (g)	=	1 pound (lb)
1 kilogram (kg)	=	2.20 pounds (lb)
1 metric ton (MT)	=	2204.6 pounds (lb)
1 short ton (ST)	=	2000 pounds (lb)
1 short ton (ST)	=	0.907 metric tons (MT)

Index

A